Advances in

ECOLOGICAL RESEARCH

VOLUME 34

CLASSIC PAPERS

Advances in ECOLOGICAL RESEARCH

Compiled by
H. CASWELL

Biology Department,
Woods Hole Oceanographic Institution, USA

VOLUME 34

2004

ELSEVIER
ACADEMIC
PRESS

Amsterdam Boston Heidelberg London New York Oxford
Paris San Diego San Francisco Singapore Sydney Tokyo

ELSEVIER B.V.
Sara Burgerhartstraat 25
P.O. Box 211, 1000 AE
Amsterdam, The Netherlands

ELSEVIER Inc.
525 B Street, Suite 1900
San Diego, CA 92101-4495
USA

ELSEVIER Ltd
The Boulevard, Langford Lane
Kidlington, Oxford OX5 1GB
UK

ELSEVIER Ltd
84 Theobalds Road
London WC1X 8RR
UK

First edition 2004

ISBN: 0-12-013934-0

♾ The paper used in this publication meets the requirements of ANSI/NISO Z39.48-1992 (Permanence of Paper).
Printed in The UK

Contributors to Volume 34

N.H. ANDERSON, *Department of Entomology, Oregon State University, Corvallis, OR 97331-5752*

N.G. AUMEN, *Everglades National Park, c/o Arthur R. Marshall Loxahatcheee National Wildlife Refuge, 10216 Lee Road, Boynton Beach, Florida 33437-9741*

S.P. CLINE, *US Environmental Protection Agency, National Health and Environmental Effects Research Laboratory, Western Ecology Division, 200 SW 35th St., Corvallis, Oregon 97333*

K. CROMACK, JR., *Department of Forest Science, Oregon State University, 321 Richardson Hall, Corvallis, Oregon 97331-5752*

K.W. CUMMINS, *Appalachian Environmental Laboratory, University of Maryland, Frostburg, Maryland 21532*

D. EAMUS, *Institute for Water and Environmental Resource Management, University of Technology, Sydney, Broadway, NSW 2007, Australia*

J.F. FRANKLIN, *U.S. Department of Agriculture, Forest Service, Forestry Sciences Laboratory, Corvallis, Oregon 97331*

S.V. GREGORY, *Department of Fisheries and Wildlife, Oregon State University, 104 Nash Hall, Corvallis, Oregon 97331-5752*

M.E. HARMON, *Department of Forest Science, Oregon State University, 321 Richardson Hall, Corvallis, Oregon 97331-5752*

P.G. JARVIS, *Department of Forestry and Natural Resources, University of Edinburgh, The King's Buildings, Mayfield Road, Edinburgh EH9 3JU, United Kingdom*

H. LAMBERS, *Department of Plant Ecology and Evolutionary Biology, PO Box 800.84, NL-3508 TB, Utrecht, The Netherlands*

J.D. LATTIN, *Systematic Entomology Laboratory, Department of Entomology, Oregon State University, 2046 Cordley Hall, Corvallis, Oregon 97331-5752*

G.W. LIENKAEMPER, *USGS Forest and Rangeland Ecosystem Science Center, 3200 SW Jefferson Way, Corvallis, Oregon 97331*

H. POORTER, *Department of Plant Ecology and Evolutionary Biology, P.O. Box 800.84, NL-3508 Utrecht, The Netherlands*

I.C. PRENTICE, *Department of Earth Sciences, University of Bristol, Wills Memorial Building, Bristol, BS8 1RJ, United Kingdom*

J.R. SEDELL, *US Department of Agriculture, Forest Service, Forestry Sciences Laboratory, Corvallis, Oregon 97331*

P. SOLLINS, *Department of Forest Science, Oregon State University, 321 Richardson Hall, Corvallis, Oregon 97331-5752*

F.J. SWANSON, *USDA Forest Service, Pacific Northwest Research Station, 3200 Jefferson Way, Corvallis, Oregon 97331*

C.J.F. TER BRAAK, *Plant Research International, Box 16, 6700 AA Wageningen, The Netherlands*

Preface

Collected in this volume are four classic papers from *Advances in Ecological Research*, chosen by the publisher based on citation counts and continued importance. Despite all the risks of using citation counts to measure impact, there is no doubt that these papers have had, and continue to have, significant impact in their respective fields.

There is no reason to expect that a set of classic papers will exhibit a theme, but these do. This volume is a bit like a walk through the woods. Come along. But be careful not to stumble over that dead branch. Dead branch? That's a piece of coarse woody debris (CWD). The paper by Harmon *et al.* (1986) takes this branch and all the other pieces of CWD in the forest and explores their importance to ecosystems. They review the roles of CWD as a creator of habitat (what do you suppose is living *under* that log?), as a contributor to energy and nutrient dynamics, and as a determinant of geomorphic processes in streams and rivers. They do not, alas, resolve the long-standing question of what happens if CWD falls in the forest with no one there to hear it.

As we continue our walk, you might notice the composition of the forest changing around us. As Ter Braak and Prentice (1988) point out, communities do that as you move along any kind of a gradient, as species respond to changes in their environment. Their paper reviews techniques for gradient analysis, a family of statistical methods that characterize and, hopefully, explain these changes. In direct gradient analysis, species abundances are related to measured environmental variables. In indirect gradient analyses, samples are arranged along axes that reflect the variation in composition within those samples, in hopes of interpreting the axes in terms of the environment. Both kinds of methods are complicated by the non-linear response of species to gradients, but have become a powerful tool in community analysis.

Some of the changes taking place around us in the forest probably reflect differences in the potential growth rate of the trees and other plants. Plants adapted to high-nutrient, high-productivity locations often have higher growth rates than do plants from low-nutrient environments. The differences persist even when the slow-growing plants are transplanted to more comfortable situations. Lambers and Poorter (1992) review the causes of these differences and the possible ecological advantages conferred by each syndrome. "Growth" is, of course, a deceptively simple word for what plants do. In their review, Lambers and Poorter identify a series of growth rate components (leaf area ratio, net assimilation rate, specific leaf area, leaf weight ratio, etc.) and explore the factors affecting each and how they influence biomass allocation, root growth, chemical composition, photosynthesis, and respiration. The plants around us in the forest are doing this all the time; they just do not know all that goes into it.

Plants obtain energy through photosynthesis, for which CO_2 is one of the substrates. If we were taking our walk in the mid-19th century, the concentration of CO_2 in the air around us would be been only about 75% of what it is now, and it is rising each year. It seems obvious, as Eamus and Jarvis (1989) begin by pointing out, that this increase in CO_2 would translate into an increase in plant growth and yield. But there are many processes involved, operating on scales from the single leaf (where stomata are controlled and photosynthesis takes place) to the individual tree to, eventually, the ecosystem. They review these processes and attempt to link them across scales, a type of investigation that has only become more important since the publication of this article.

Each of these classic papers looks at the forest, or the trees that make it up (or, to be fair to the authors, at plant communities in general) from a different perspective. Together they are an example of the lasting impact that good synthetic review papers can have in ecology.

Hal Caswell

Contents

This volume has been composed of chapters in previous volumes of this series.

The Direct Effects of Increase in the Global Atmospheric CO_2 Concentration on Natural and Commercial Temperate Trees and Forests

Originally Published in Volume 19 (this series), pp 1–55, 1989

D. EAMUS and P.G. JARVIS

Ecology of Coarse Woody Debris in Temperate Ecosystems

Originally Published in Volume 15 (this series), pp 133–302, 1986

M.E. HARMON, J.F. FRANKLIN, F.J. SWANSON, P. SOLLINS, S.V. GREGORY, J.D. LATTIN, N.H. ANDERSON, S.P. CLINE, N.G. AUMEN, J.R. SEDELL, G.W. LIENKAEMPER, K. CROMACK, JR. and K.W. CUMMINS

A Theory of Gradient Analysis

Originally Published in Volume 18 (this series), pp 271–317, 1988

CAJO J.F. TER BRAAK and I. COLIN PRENTICE

Inherent Variation in Growth Rate Between Higher Plants: A Search for Physiological Causes and Ecological Consequences

Originally Published in Volume 23 (this series), pp 187–261, 1992

HANS LAMBERS and HENDRIK POORTER

Originally Published in Volume 19 (this series), pp 1–55, 1989

The Direct Effects of Increase in the Global Atmospheric CO_2 Concentration on Natural and Commercial Temperate Trees and Forests

D. EAMUS AND P.G. JARVIS

ADVANCES IN ECOLOGICAL RESEARCH VOL. 34
0065-2504/04 $35.00 DOI 10.1016/S0065-2504(03)34001-2

I. INTRODUCTION

A. The Problem

There is an extensive literature documenting the evidence for an increase in global mean CO_2 concentrations from the mid 18th century through to the present day. The evidence is based upon ice core data (Neftel *et al.*, 1985; Pearman *et al.*, 1986; Fifield, 1988); inferences from tree ring data (Lamarche *et al.*, 1984; Hamburg and Cogbill, 1988; Hari and Arovaara, 1988); climate modelling based on fossil fuel consumption and the CO_2 airborne fraction (see Gifford, 1982; Keepin *et al.*, 1986); and most recently measurements of global atmospheric CO_2 concentrations at different sites around the world (see Crane, 1985). Much of this evidence has been presented and evaluated elsewhere (e.g. Gifford, 1982; Ausubel and Nordhaus, 1983; Gates, 1983; Bolin, 1986; Keepin *et al.*, 1986; Wigley *et al.*, 1986). There is a consensus that prior to the industrial revolution in northern Europe about 130 years ago, the global atmospheric CO_2 concentration was about 270 to 280 $\mu mol\ mol^{-1}$. Today (1989) the concentration is *ca* 350 $\mu mol\ mol^{-1}$ and is increasing at *ca* 1.2 $\mu mol\ mol^{-1}$ per year (Conway *et al.*, 1988). Throughout this review, it will be presumed that atmospheric CO_2 concentrations will double to approximately 700 $\mu mol\ mol^{-1}$ by the mid to late 21st century and we shall not discuss here the evidence for past or future increases in the atmospheric CO_2 concentration.

Other reviews of this topic include Kramer (1981), Strain and Bazzaz (1983), Sionit and Kramer (1986), Kramer and Sionit (1987).

B. Aims of this Review

Because CO_2 is the substrate for photosynthesis, one would expect that elevated CO_2 concentrations must result in enhanced rates of photosynthesis which in turn must lead to enhanced plant growth and yield. This simplistic view does not take into account the complex interactions between CO_2 concentration, photosynthesis and other environmental variables and the more complex relationships between carbon assimilation and plant respiration, growth and yield. The major aim of this review is to assess what we know of these relationships in trees and to predict the consequences of an increase in CO_2 upon temperate zone forests.

Information concerning the reaction of trees and forests to increase in the atmospheric CO_2 concentration is particularly important because forests cover about one-third of the land area of the world and carry on about two-thirds of the global photosynthesis (Kramer, 1981). The total amount of carbon stored in terrestrial ecosystems has diminished over recent centuries as a result of anthropogenic actions, especially forestry clearance (Bolin, 1977; Woodwell *et al.*, 1978; Houghton *et al.*, 1983). Recently, about 2.5×10^6 m^3 of wood have been consumed annually (FAO, 1982). FAO predicts an increasing global rate of forest clearance for the remainder of this century. If this occurs, the present area of forest may be further reduced by as much as 20% by the year 2000. On a global scale, this further reduction in area of forest will both exacerbate the rise of CO_2 in the atmosphere through oxidation of wood and wood products, and reduce the sink strength for CO_2 (Houghton *et al.*, 1987; Marland, 1988; Jarvis, 1989). By contrast, the area of temperate and boreal forest has been increasing in Europe, North America and some parts of Asia; in the UK, for example, the area of afforested land is projected to continue to increase at a rate of about 30 000 ha per year. However, the increase in area of temperate and boreal forest is too small to compensate for the amount of CO_2 released into the global atmosphere by clearance of tropical forests (Houghton *et al.*, 1987).

Although the database is small, there have been a sufficient number of studies of the effects of CO_2 on northern temperate forest and woodland species to demonstrate that the primary effects are on photosynthesis and stomatal action, although direct effects on other processes such as leaf initiation and microbial action may yet be convincingly demonstrated. Because of the complexity of forest ecosystems, there may be many consequences of long-term changes in the rates of carbon gain and water loss by trees and stands. We identify four main reasons for being concerned about the rise in CO_2 and its effect on trees and forests:

(*a*) the enhancement of biological knowledge about the functioning of tree species of major ecological and economic importance,
(*b*) the impact on the productivity and value of the economic product,
(*c*) the impact on the ecology and environment of woods and forests, and
(*d*) the downstream, socio-economic consequences.

C. Spatial and Temporal Scales

The effects of a major environmental variable on plants and ecological systems can be examined at spatial scales ranging from sub-cellular through to the biome or geographical region and with timescales ranging from seconds or parts of seconds for rapid biophysical processes, to centuries for evolutionary changes affecting land-use systems on the larger, spatial scales. The consequences of a doubling in the global, atmospheric CO_2 concentrattion may be seen over these ranges of both spatial and temporal scales (Figure 1). For example, an increase in CO_2 may affect the primary photosynthetic carboxylation at the spatial scale of the chloroplast over the timescale of seconds or minutes, whereas assimilation of CO_2 and transpiration of water by leaves of trees is influenced by CO_2 at a timescale of hours, or of weeks if allowance is made for acclimation of the plants to a high CO_2 environment. The growth of seedlings and individual trees is influenced by an increase in CO_2 concentrations over periods of weeks or years, although particular growth processes such as carbon allocation or nutrient uptake may have shorter timescales. The functioning of stands may be affected at timescales of the order of hours, but the main impact of the effects of a rise in CO_2 will be seen in changing properties of stands with timescales of a year or more. Effects on forest ecosystem processes are likely to be seen over decades and effects on regional land-use over centuries (Strain, 1985).

This review examines the likely effects of the projected increase in global atmosphere CO_2 concentration on these spatial and temporal scales. However, the information we have to draw upon is severely restricted to certain spatial and temporal scales, determined largely by the convenience and feasibility of experimental methods with the current levels of technology and resources.

D. Physiological Action of CO_2

The main plant physiological roles of CO_2 are as a substrate and activator for photosynthetic carbon assimilation, with concomitant effects upon photorespiration, and as an environmental variable determining stomatal aperture. Processes of dark fixation of CO_2 predominantly in roots may also be concentration-dependent, but the CO_2 concentration in soil environments dominated by respiration is far in excess of the range of atmospheric concentrations under consideration. All species of major ecological and economic importance in northern temperate forests, both overstorey and understorey, have photosynthesis of the typical C_3 pattern, and, therefore we may expect the main effects of rising CO_2 concentrations to be similar to effects seen on other C_3 species. However, the amount of data for temperate and

	Temporal scale				
Spatial scale	Seconds/Minutes	Hours/Days	Weeks/Months	Years/Decades	Centuries
Cell	enzyme activation; fluorescence; carboxylation; transport/ partitioning	enzyme kinetics; organelle structure	acclimation of cellular processes		
Leaf/Shoot		assimilation; transpiration; stomatal action	acclimation of assimilation; senescence		
Seedling/Tree			growth and carbon allocation; nutrient uptake; root/shoot dynamics	crown properties; branching	
Plantation/ Woodland		canopy properties of photosynthesis, transpiration, light interception	nutrient uptake; WUE; productivity	canopy structure; competition; harvest index	
Forest/ Ecosystem				yield; rotation length; WUE	natural and artificial selection; land use; species composition

Figure 1 The range of temporal and spatial scales available in the study of the effects of elevated CO_2 upon trees. WUE = water use efficiency.

boreal forest species is small, especially in comparison with comparable data for agricultural crops (cf. Warrick *et al.* (1986) with Shugart *et al.* (1986)).

It has been inferred from several experiments on the growth response of herbaceous plants to elevated CO_2 concentrations that CO_2 has a direct stimulating effect on leaf growth, not driven through increased availability of substrate (e.g. Morison and Gifford, 1983) and the same inference has been drawn for tree seedlings (Tolley and Strain, 1984a). However, these inferences are unsubstantiated at present, although possible mechanisms have been proposed (Sionit *et al.*, 1981; Sasck *et al.*, 1985).

The allocation of carbon within plants is clearly influenced by the supply of CO_2 and, as will be discussed later, this appears to result from change in the balance between rates of uptake of CO_2 and inorganic nutrients, especially nitrogen.

There are suggestions, too, that increase in atmospheric CO_2 concentration will lead to changes in mycorrhizal development and in root exudations, thus affecting rhizosphere activities (Strain and Bazzaz, 1983). Such effects remain to be clearly demonstrated and are most probably the result of greater availability of carbohydrates. There is, however, a stronger supposition, also still awaiting clear demonstration, that changes in the composition (especially the C:N ratio) of leaf and other litter, will lead to significant changes in rates of mineralisation and hence recycling of nutrients through the soil (Luxmore, 1981).

E. The Data Base

1. The leaf and shoot scale

There has now been an appreciable number of short-term experiments with a timescale of hours or days on single leaves, shoots or young trees of temperate forest origin in assimilation chambers, where the response of CO_2 uptake and stomatal action has been characterised in relation to a range of ambient CO_2 concentrations, and many more on agricultural crops and other herbaceous species (see Strain and Cure, 1986). However, the majority of these experiments has been on unacclimated plants that have been grown in air containing the current, atmospheric CO_2 concentration. There have been only a few experiments on woody plants that have been grown for a substantial period in higher CO_2 concentrations, so that the plants were at least partially acclimated anatomically and physiologically to an appropriate CO_2 environment. Much of this work and the growth studies considered below, has been carried out in North America and only a few studies of physiological processes have been made at the leaf scale on species important in other forests.

2. The seedling, sapling and individual tree scale

Prior to 1983, the motivation for experiments in which young trees were grown at elevated CO_2 concentrations was derived from horticultural practice and the aims of the experiments generally were to produce improved planting stock. For this purpose, young trees were often grown in unusual conditions with the CO_2 increased by a factor of $\times 4$ or more (Hårdh, 1967; Funsch *et al.*, 1970; Yeatman, 1970; Krizek *et al.*, 1971; Siren and Alden, 1972; Tinus, 1972; Laiche, 1978; Canham and McCavish, 1981; Lin and Molnar, 1982; Mortensen, 1983). These experiments covered a range of broadleaved and coniferous species, mostly ornamentals of horticultural interest or forest trees of importance n North America. The evidence from these experiments indicates that growth in height, leaf area, stem diameter and dry weight of temperate trees is generally increased by large increases in the ambient CO_2 concentration but growth was extremely variable and in some species no growth response was elicited. Because of the unusual environmental conditions and the high concentrations of ambient CO_2 used, the results are of marginal relevance to the situation we are addressing.

It is only since the realisation of an appropriate scenario for the projected rise in the atmospheric CO_2 concentration that relevant experiments have been done. Since 1983 some nine species of conifers and 15 species of broadleaves have been subjected to elevated ambient CO_2 in the range $\times 2$ to $\times 2.5$ and the results published in each case in 15 papers. Regarding each species in each paper as a separate experiment, three experiments were done in glasshouses, five in open-topped chambers and 23 in controlled environment growth cabinets, chambers or rooms. In 32 cases plants were grown in pots or boxes and in only one case were plants grown directly into soil, and this was, necessarily, in an open-topped chamber. Two experiments lasted for three years, another for two and a half years and another for one year. The remainder of the experiments lasted 32–287 days with a median of around 120 days. Nutrients were added in 22 experiments but there was a series of nine in which no nutrients were added. In one experiment nitrogen supply was a variable and in another phosphorus supply. In five experiments water supply was a variable and light was a variable in three others. In one experiment, competition was studied; in all others it was excluded.

Most of these experiments were done with native North American species, of little or no economic or ecological importance in Europe or elsewhere. Only three were with species grown extensively in Europe, for example (Higginbotham *et al.*, 1985 with *Pinus contorta*; Mortensen and Sandvik, 1987 with *Picea abies*; and Hollinger, 1987 with *Pseudotsuga menziesii*) whilst only two others were with Australasian species. Few of the experiments were with native species from elsewhere or with other commercial forest species grown widely outside North America, such as *Picea sitchensis*. These experiments are, thus, limited in their direct relevance to temperate and boreal forests worldwide, although substantial information about the effects of CO_2 on young trees can be adduced from them.

3. *The plantation and woodland scale*

There have been no observations of *stand* physiology on any timescale in relation to ambient CO_2 concentration. The nearest approach to this is a single study of the assimilation and transpiration of an individual tree of *Eucalyptus*, growing on a lysimeter and enclosed in a tall, open-topped, plastic chamber in south-east Australia (Wong and Dunin, 1987). Essentially, measurement was made of the CO_2 and water vapour exchange of an unacclimated tree in a giant cuvette. This experiment was, therefore, a physiological study of an individual tree and is not of direct relevance either to stand functioning or, because of the short timescale of the experiment, to the rise in atmospheric CO_2. It did, however, demonstrate that a single tree behaves much as one would anticipate from cuvette studies on individual leaves.

Although open-air exposure to CO_2 is now being practised in agricultural crops in several places, it is hardly surprising that uncontained stands of trees, or areas of woodland, have not been exposed to high CO_2 concentrations in experiments. The sheer scale of the problem renders this a phenomenally expensive approach: Allen *et al.* (1985), have estimated the annual cost of CO_2 alone for this to exceed \$US4 $\times 10^6$ in a 40 m tall stand of tropical forest! In the absence of any directly determined data and with the likelihood that none will be forthcoming, recourse must be made at this and larger scales to modelling the functioning of the system with respect to the atmospheric CO_2 concentration. We would emphasise, however, that inability to address the problem directly at the scale of concern imposes a severe constraint on our capability to make adequate assessment of the consequences of the rise in atmospheric CO_2 concentration for stand-scale processes.

F. Problems with the Data

It is evident that there is a serious dilemma with respect to the acquisition of data regarding the response of trees and forest to increased levels of CO_2. For technical reasons experiments have been confined to small numbers of comparatively small and young trees *enclosed* in relatively small volumes of space. There are significant problems with this approach and these problems are necessarily exacerbated with respect to trees and forest, because of the large size of mature trees and the areal extent of woodlands and forest.

Juvenile trees are known to be more sensitive to environmental change (Higginbotham *et al.*, 1985), and consequently the response elicited may be larger in juvenile trees than in mature trees. Conversely, the long-term totally acclimated response to elevated CO_2 may be larger than anticipated, *if* there are accumulative effects on tree morphology and ecosystem processes. In either case, the present experiments may introduce bias in the estimation of the predicted response for mature stands of trees unless such possibilities are taken into account.

The majority of experiments so far have been carried out in controlled environment chambers or rooms, and these have the advantage that the foliage of the plants is generally well-coupled to the atmosphere prescribed by the operator. In comparison, in glasshouses and open-topped chambers, ventilation can be very much poorer and there may be substantial feedback between the exchanges of mass and energy at the plant surfaces and the local, leaf environment. Consequently in glasshouses and open-topped chambers the plants are usually very poorly coupled to the atmosphere and substantial practical problems may result, particularly with respect to overheating (e.g. Surano *et al.*, 1986). More importantly, however, the poor coupling to the atmosphere may result in a substantial shift in the importance of the driving variables for gas exchange by the plants and, hence, lead to fluxes of water vapour and carbon dioxide that are substantially different from the fluxes that would be expected in response to the weather from well-coupled vegetation such as trees and forests in the field (Jarvis and McNaughton, 1986).

Controlled environment chambers and rooms do, however, have other problems, not least that the plants must be grown in pots and that the volume of space available is usually rather small. This puts a major constraint on the length of the experiment and this is very evident in the relatively short periods over which young trees have been exposed to elevated CO_2 concentrations. Whilst large responses to CO_2 have been elicited in many experiments during the initial few weeks of exposure, the responses have diminished with time and after more than a year have virtually disappeared. It is not yet clear whether this is an artefact of the restricted rooting conditions or a physiological response of the plants. Possible interaction between the response of young trees to an increase in CO_2 and the availability of nutrients is an important question (Kramer, 1981), but it is unlikely that it can be elucidated by growing plants with their roots constrained in a fixed volume of soil.

Even with plants growing in open-topped chambers, the opportunity of rooting the plants into the soil directly has only been taken in one experiment, and then only as a secondary aspect (Surano *et al.*, 1986).

G. Acclimating, Acclimated and Adapted Trees

Most of the early studies of the effects of CO_2 on trees used non-acclimated plants and leaves. Trees were grown at the current global CO_2 concentration (*ca* 325 to 340 $\mu mol\ mol^{-1}$) and leaf or shoot responses to elevated CO_2 were measured in the short-term over periods of minutes or hours. Such a protocol is no longer acceptable, since we do not know whether the response observed after a few hours will be maintained by tissue grown in elevated CO_2 concentrations over a period of years. There is evidence that the responses of photosynthesis, transpiration and growth to increase in the ambient CO_2 concentration change with time. However, since much of the information available has been obtained on non-acclimated trees, the data will be briefly reviewed.

The process of acclimation is multi-faceted and may occur over a timescale of hours to generations, depending upon the process studied. Photosynthesis and stomatal conductance *may* acclimate over a period of hours or days; changes in stomatal density in response to elevated CO_2 concentrations have been recorded over periods of weeks (Woodward and Bazzaz, 1988) and months (Oberbauer *et al.*, 1985), but also, it has been claimed, over generations (Woodward, 1987). Whether the latter is the result of acclimation or of natural selection and genetic adaptation is open to question. In trees and forests both complete acclimation and genetic adaptations are likely to require centuries.

The use of the term ‘acclimated’ may perhaps be better replaced by ‘acclimating’. This is particularly pertinent to many studies where trees have been grown for months or years at the current global CO_2 concentration and then transferred to an elevated ambient CO_2 atmosphere. Such a sudden, single-step perturbation may induce oscillations in many plant processes and it is likely that subsequent measurements are made on *acclimating* trees. In some studies (Tolley and Strain, 1984a; Higginbotham *et al.*, 1985; Oberbauer *et al.*, 1985) seeds have been sown directly into an elevated ambient CO_2 concentration and the resulting plants may resemble more closely the acclimated trees of the next century.

Two further points concerning the experimental data must be made. Trees growing today with a lifespan of several decades are subject to a *gradually* increasing CO_2 concentration. Long-term physiological responses to a slow increase in CO_2 concentration (*ca* 1.2 $\mu mol\ mol^{-1}\ a^{-1}$ presently) are likely to differ somewhat from the responses generated by large, immediate, single-step increase in CO_2 concentration, even though plants are accustomed to wide short-term local, diurnal and seasonal variations (usually $< 100\ \mu mol\ mol^{-1}$), depending on location and latitude (Jarvis, 1989). This problem of acclimation was highlighted recently for crop species, for which it was shown that the weighted average, short-term (non-acclimated) CO_2 assimilation rate was increased by 52% for a doubling of the ambient CO_2 concentration, whilst the acclimated assimilation rate was enhanced by only 29% (Cure and Acock, 1986). This result was ascribed to the lack of *sustained*, active sinks for photosynthate in the acclimated plants.

It is evident that the data available at present are severely limited in their applicability by the conditions in which the experiments have been done. Not only must the results be interpreted with great caution, but also considerable care must be taken in using information obtained in the experiments to parameterize models used for predicting the consequences of increased CO_2 levels at larger, spatial and temporal scales.

The following sections review the results obtained so far about the influence of an increase in CO_2 concentration on seedlings, trees and forests over the range of spatial scales shown on the left-hand side of Figure 1 and at the range of temporal scales along the top of Figure 1. We shall, necessarily, concentrate upon the entries in the top, left-hand area of Figure 1 for which substantial information is available, even though this information is of limited relevance to the likely responses of stands of trees over one or more generations.

II. THE CELLULAR SCALE

Measurements at the cellular scale include the measurement of fluorescence induction phenomena, enzyme activity, carbon partitioning within cellular pools and cellular transport properties. The latter two processes, although directly relevant to the understanding of growth and differentiation with respect to the influence of environmental perturbations (such as changes in CO_2 concentration), have not received any attention to date, and we can only highlight this deficiency. It will be shown later that source–sink relationships (and hence, carbon partitioning and cellular transport) directly influence photosynthesis and assimilate allocation and this gap in our knowledge is critical to our poor understanding and central to our relative inability to extrapolate with confidence from leaf and shoot scale studies to a stand or regional scale.

Enzyme activities have been better studied, generally as part of wider programmes investigating leaf/shoot responses. The major emphasis by far has fallen on the primary carboxylating enzyme. In addition to acing as a substrate for ribulose bisphosphate carboxylase-oxygenase (rubisco), CO_2 combines with this enzyme as the first step in a two-step chemical activation process. The inactive enzyme can bind with a molecule of CO_2 in a slow, reversible reaction. When this enzyme–CO_2 complex subsequently reacts with Mg^{2+} in a rapid, reversible reaction, the active enzyme–CO_2–Mg^{2+} complex is formed. Carboxylation requires a second CO_2 molecule to bind at a third site. The concentration of CO_2 therefore, influences enzyme activity *via* its influence upon the activation state as well as through its role as a substrate. Changes in activation state alter V_{max} and K_m (Edwards and Walker, 1983). Growth at elevated CO_2 concentration directly influences this activation process (Sage and Sharkey, 1987; Sage *et al.*, 1988). What little is known about trees in these connections will be taken up in the following sections that treat acclimation and adaptation at the leaf scale.

III. THE LEAF AND SHOOT SCALE

A. Stomatal Conductance

1. Non-acclimated tree studies

With some exceptions, stomatal conductance decreases in response to increases in CO_2 concentration. The mechanism by which CO_2 affects stomatal aperture and conductance (g_s) remains speculative (see Eamus, 1986a) and will not be dealt with in this review. However, it is clear that stomata respond to the intercellular space CO_2 concentration (C_i) rather than to the ambient CO_2 concentration (C_a) (Mott, 1988). Responses of g_s to CO_2 concentration follow one of the curves shown in Figure 2. Broadly, we may classify trees into those with stomata that are quite sensitive to CO_2 (− −) and those, which have little sensitivity (+, 0, −).

Table 1 summarizes some of the available data for tree species using non-acclimated shoots. There are two points to note from Table 1. Firstly, there is a paucity of data, and secondly, interactions between other environmental variables and CO_2 concentration on g_s have been poorly studied. The latter, in particular, requires investigation, since in one study (Beadle *et al.*, 1979), the direction of response of g_s was reversed by imposing water stress. Modification of the response of g_s to elevated CO_2 concentrations by the environment is likely, and without this information, extrapolation from single shoot studies in optimal conditions in cuvettes in the laboratory to the field remains impossible.

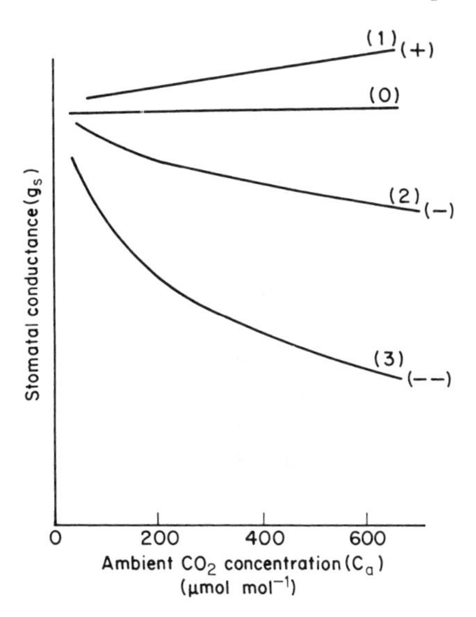

Figure 2 Diagram to show the major responses of stomatal conductance (g_s) to change in the ambient CO_2 concentration (C_a).

Table 1 Response of stomatal conductance to changing CO_2 concentration in a range of unacclimated tree species. Refer to Figure 2 for curve form. C_a: ambient CO_2 concentration. C_i: mean intercellular space CO_2 concentration

Species	Response curve	Max C_a or C_i (μmol mol^{-1})		Source
Picea sitchensis	−	600	C_a	Ludlow and Jarvis (1971)
Picea sitchensis				
unstressed	−	600	C_a	Beadle *et al.* (1979)
water stressed	+	600	C_a	
Picea rubra	−	600	C_i	D. Eamus (unpub)
Pinus taeda	+	400	C_i	Teskey *et al.* (1986)
Fraxinus pennsylvanica	+	400	C_i	Davis *et al.* (1987)
Populus deltoides	−	1000	C_a	Regehr *et al.* (1975)
Eucalyptus pauciflora	−	300	C_i	Wong *et al.* (1978)
Malus pumila	−	400	C_i	Warrit *et al.* (1980)

2. Acclimating tree studies

Table 2 summarizes recent data concerning the response of g_s to elevated CO_2 levels in acclimating trees. Several points can be noted from this table. The length of time allowed for acclimation is short; in all but two cases, it is

Table 2 Response of stomatal conductance (g_a) (mmol m^{-2} s^{-1}) to elevated CO_2 concentrations (μmol mol^{-1}) in a range of acclimating tree species

Species	g_a at current C_a	g_a at elevated C_a	Elevated C_a	Period of study	Source
Pinus radiata					
well watered	20	20	660	22 weeks	Conroy *et al.* (1986b)
stressed	20	20	660		
nutrient stressed	22	12	660		
Pinus taeda					
high light	41–74	41–74	675, 1000	56 days[a]	Tolley and Strain (1985)
low light	82–131	41–82			
Pinus taeda	48	30	500	15 months	Fetcher *et al.* (1988)
Liquidambar styraciflua					
high light	74–131	41–90	675, 1000	56 days[a]	Tolley and Strain (1985)
low light	49–115	20–61			
Liquidambar styraciflua	700	350	500	15 months	Fetcher *et al.* (1988)
Ochroma lagopus	299	127	675	60 days[a]	Oberbauer *et al.* (1986)
Pentaclethra macroloba	156	45	675	123 days	
Pinus radiata	105	68	640	120 days	Hollinger (1987)
Nothofagus fusca	78	66	640	120 days	
Pseudotsuga menziesii	160	166	640	120 days	
Pinus ponderosa					
07.00 h	47	63	*ca* 500	1 year	Surano *et al.* (1986)
11.00 h (time of day)	98	55	500	1 year	
13.00 h	47	63	500	1 year	
18.00 h	63	29	500	1 year	

[a] started from seed.

less than one growing season. It is unknown whether acclimation of g_s to elevated CO_2 levels occurs over a longer time period (Morison, 1985). Oberbauer *et al.* (1985) observed changes in stomatal density over 123 days for *Pentaclethra macroloba*, but no change over 60 days for *Ochroma lagopus*. Decreases in stomatal density for eight temperate, arboreal species in the past 200 years have been found (Woodward, 1987). Long-term studies of acclimation of g_s and stomatal index [(100 × no. of stomatal pores)/(no. of stomatal pores + total number of epidermal cells)] to elevated CO_2 concentration are required.

The general response of g_s to elevated CO_2 levels is a decrease of 10–60% (Oberbauer *et al.*, 1985; Tolley and Strain, 1985; Hollinger, 1987). However, notable exceptions to this are apparent for well-watered and water-stressed *Pinus radiata* (Conroy *et al.*, 1986a); *Pinus taeda* grown under high light levels (Tolley and Strain, 1985) and *Pseudotsuga menziesii* (Hollinger, 1987). The response of g_s to elevated CO_2 concentrations varied according to the time of day in *Pinus ponderosa* (Surano *et al.*, 1986), possibly the result of variations in needle temperature. It is not possible to state whether the apparent variation in response amongst species is a true reflection of inter-species differences, differences in tree age, duration of the experiment, CO_2 concentrations used, degree of acclimation or other additional variables.

Stomatal sensitivity to CO_2 increased in *P. radiata* grown at elevated CO_2 concentrations (Hollinger, 1987) but decreased in *Betula pendula* and *Picea sitchensis* (P. G. Jarvis, A. P. Sandford and A. Brenner, unpublished). Stomatal sensitivity to CO_2 also varies with photon flux density: in response to an increase, stomatal sensitivity to CO_2 may increase (Morison and Gifford, 1983) or decrease (Beadle *et al.*, 1979). Abscisic acid both reduces g_s and increases stomatal sensitivity to CO_2 in crop plants but whether this is true for trees in unknown. In view of the long-term influence of drought upon subsequent stomatal behaviour in crop plants (Eamus, 1986b) and trees (Schulte *et al.*, 1987), this area requires further study in trees.

As the difference in vapour pressure between leaf and ambient air (D_l) increases, g_s decreases (e.g. Watts and Neilson, 1978; Johnson and Ferrell, 1983; Roberts, 1983) and stomatal sensitivity to D_l varies between species and with plant water status (Ludlow and Jarvis, 1971; Johnson and Ferrell, 1983; Schulte *et al.*, 1987). Growth at elevated CO_2 significantly influences stomatal response to D_l. The relative closure induced by a stepwise increase in D_l was decreased by growth at elevated CO_2 for *P. radiata* and *P. menziesii*, but not for *Nothofagus fusca*, so that the stomata of trees grown in an elevated CO_2 concentration were less sensitive to dry air (Hollinger, 1987). Temperature can also modify the response of g_s to D_l and possibly also to CO_2 (Johnson and Ferrell, 1983).

It is clear from the above that a detailed study of the interactions amongst temperature, CO_2 concentration, D_l photon flux density and plant water status is required to determine more realistically the influence of elevated CO_2 upon g_s. Stomatal sensitivity to CO_2 may well be reduced under high photon flux densities in plants of high relative water content (that is, not water stressed). These are the conditions frequently used in experiments in which stomatal sensitivity to CO_2 is assessed and this may explain the observation that in many species of trees stomata show a low sensitivity to CO_2 concentration. In crop plants a role for indole-3-acetic acid and calcium in the control of stomata has been postulated. We are not aware of any studies investigating these regulating factors with respect to stomatal response to elevated CO_2 concentrations in trees.

B. Photosynthesis

1. Non-acclimated tree studies

The relationship between CO_2 concentration and rate of photosynthesis has the form shown in Figure 3. The exact form of the relationship varies somewhat amongst species and with other environmental variables but is essentially the same for all C_3 species (von Caemmerer and Farquhar, 1981, 1984). Table 3 summarizes some of the available data for non-acclimated trees. From this table, it can be seen that doubling the ambient CO_2

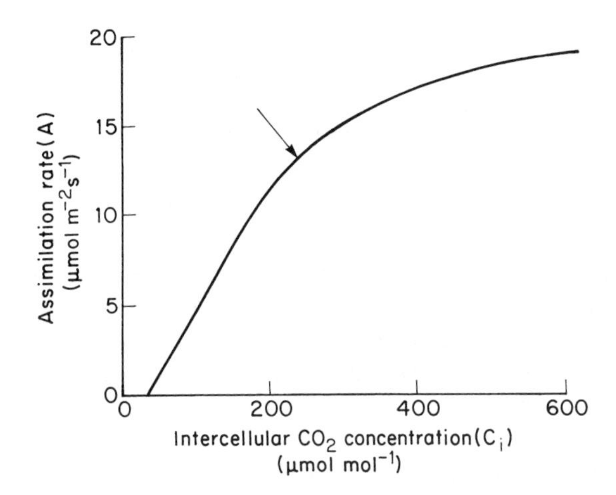

Figure 3 The relationship between rate of assimilation of CO_2 in photosynthesis (A) and mean intercellular space CO_2 concentration (C_i) in *Picea sitchensis*. (Adapted from Jarvis and Sandford, 1986.)

Table 3 Influence of a doubling of CO_2 concentration on assimilation rate (A) (μmol m^{-2} s^{-1}) determined for a range of non-acclimated tree species. All figures are approximate having been read from graphs

Species	A at present C_a	A at doubled C_a	Percentage increase	Source
Picea engelmanii				
warm	3.5	4.8	37	De Lucia (1987)
chilled	0.5	0.6	20	
Picea rubra	10.0	12.0	20	D. Eamus (unpubl)
Picea sitchensis				
unstressed	9.1	11.4	25	Beadle *et al.* (1981)
water stressed	1.1	3.4	209	
Picea sitchensis	12.6	20.2	60	Ludlow and Jarvis (1971)
Pinus sylvestris				
warm	6.0	7.5	25	Strand and Öquist (1985)
chilled	3.0	5.0	66	
Pinus taeda	6.0	8.0	33	Teskey *et al.* (1986)
Populus deltoides	15.8	18.9	20	Regehr *et al.* (1975)
Pseudotsuga menziesti				
high light	1.6	2.6	60.5	Brix (1968)
low light	0.2	0.3	50	
Quercus suber	2.0	4.0	100	Tenhunen *et al.* (1984)
Malus pumila				
unstressed	24.0	45.0	87	Jones and Fanjul (1983)
water stressed	4.0	10.0	150	
Malus pumila				
extension shoot	18.2	27.3	50	Watson *et al.* (1978)
spurs + fruit	9.1	15.9	75	
spurs−fruit	8.2	15.9	94	

concentration invariably increases photosynthetic rate. However, the percentage increase in carbon dioxide assimilation (A) varies widely from 20% to 300%. Furthermore, although the absolute increase in assimilation at elevated CO_2 concentration decreases with the imposition of additional stress variables, the percentage increase is frequently larger, so that an increase in CO_2 concentration may ameliorate the effects of stresses of various kinds (see below).

2. Acclimating tree studies

Table 4 summarizes recent data on the influence of elevated CO_2 concentrations upon CO_2 assimilation in acclimating trees. Although there

Table 4 Influence of elevated CO_2 concentrations (μmol mol^{-1}) upon assimilation rate (A) for a range of tree species acclimating for various periods to elevated CO_2

Species	% change in A	Elevated C_a	Duration of experiment	Source
Liquidambar styraciflua				
well watered	ns	675, 1000	56 days[a]	Tolley and Strain (1984a)
water stressed	ns			
Liquidambar styraciflua				
high light	17–100	675, 1000	56 days[a]	Tolley and Strain (1986)
low light	20–100			
Liquidamabar styraciflua (high light)				
1–4 weeks	41, 44	675, 1000	112 days	Tolley and Strain (1984b)
4–8 weeks	33, 43			
8–12 weeks	22, 10			
12–16 weeks	−15, −23			
Liquidambar styraciflua	*ca* −10	500	15 month	Fetcher *et al.* (1988)
Nothofagus fusca	+45	620	120 days[c]	Hollinger (1987)
Ochroma lagopus	−49	675	60 days[a]	Oberbauer *et al.* (1985)
Pentaclethra macroloba	−, ns	675		
Carya ovata	*ca* 80	700	90 days[d]	Williams *et al.* (1986)
Liriodendron tulipifera	*ca* 80	700		
Acer saccharinum	*ca* 80	700		
Fraxinus lanceolata	*ca* 30	700		
Platanus occidentalis	0	700		
Quercus rubra	*ca* 80	700		
Populus euramericana	−29	660	15 days[e]	Gaudillère and Mousseau (1989)
Picea abies				
24 h day	67	1650	75 days	Mortensen (1983)
10 h day	68	1650	75 days	

Pinus contorta				
saturating light	96	1000	*ca* 150 days[a]	Higginbotham *et al.* (1985)
	43	2000		
saturating CO_2	−5	1000		
	−55	2000		
Pinus radiata	55	620	120 days[a]	Hollinger (1987)
Pinus radiata				
adequate P	200	660	22 weeks[b]	Conroy *et al.* (1986b)
low P	−35	660		
Pinus taeda				
well watered	207	675	56 days[a]	Tolley and Strain (1984a)
water stressed	83	675		
well watered	−, ns	1000		
water stressed	−53	1000		
Pinus taeda				
high light	0–38	675, 1000	56 days[a]	Tolley and Strain (1986)
low light	10–38			
Pinus taeda (high light)				
1–4 weeks	ns, ns	675, 1000	84 days[a]	Tolley and Strain (1984b)
4–8 weeks	ns, 47			
8–12 weeks	207, ns			
12–16 weeks	−45, −34			
Pinus taeda	12.4	500	15 months	Fetcher *et al.* (1988)
Pseudotsuga menziesii	32	620	120 days[c]	Hollinger (1987)

[a] started from seed.
[b] started with 8-week-old plants.
[c] started with 10-month-old plants.
[d] started with 1-year-old plants.
[e] started with cuttings.

are some exceptions, in the majority of these experiments a doubling of the CO_2 concentration resulted in enhance rates of assimilation, which sometimes more than doubled. The large variability (e.g. Tolley and Strain, 1984a) may reflect true inter-species differences, differences in the CO_2 concentration used, duration of the experiment, stress induced by other variables, attack by pathogens, mites and insects, or other aspects of the cultural conditions.

The increase in assimilation rate to leaves grown in elevated CO_2 is the result of both increased substrate concentration for rubisco and increased activation of the rubisco, since CO_2 activates this enzyme. In addition the quantum yield (ϕ) is increased and photorespiration is reduced (Figure 4) with increasing CO_2 concentration, as a result of enhanced competition by CO_2 relative to O_2 for active sites on rubisco (Pearcy and Björkman, 1983; Ehleringer and Björkman, 1981). Elevated CO_2 can also cause a reduction in the compensation photon flux density and this may possibly also delay leaf senescence and lead to a larger population of more effective shade leaves in the tree crown.

Tolley and Strain (1984a,b) showed for *Liqidambar styraciflua* and *P. taeda* that the rate of net assimilation declined as the duration of the experiment increased, particularly at high photon flux densities. From analysis of the relation between A and C_i, it could be shown that the observed reduction in photosynthesis was not the result of stomatal closure,

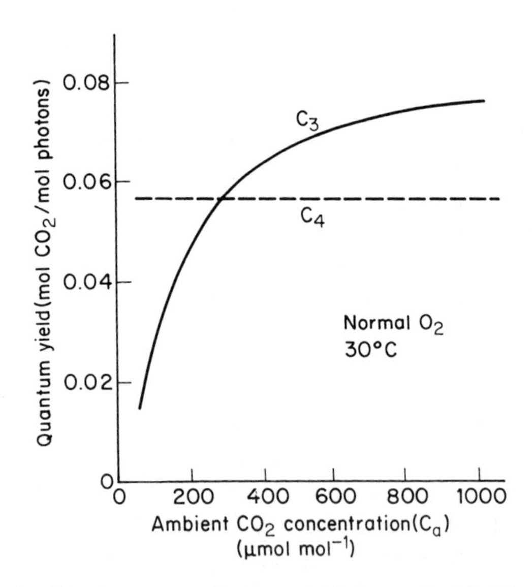

Figure 4 Relationship between quantum yield (ϕ, mole of CO_2 assimilated per mole of photons absorbed) and ambient CO_2 concentration (C_a) in C_3 plants. (Adapted from Pearcy and Björkman, 1983.)

since C_i increased. After 16 weeks, *L. styraciflua* grown at a CO_2 concentration of 1000 μmol mol^{-1} had a rate of CO_2 assimilation 23% *less* than the control trees. For low-light grown trees, a peak in photosynthetic rate was observed, followed by a decline in the control values. A similar reduction has been observed in crop plants (Raper and Peedin, 1978; Sionit *et al.*, 1981; Wulff and Strain, 1981; DeLucia *et al.*, 1985; Ehret and Jollife, 1985) as well as in some woody species (Oberbauer *et al.*, 1985; Williams *et al.*, 1986).

Several possible causes of this reduction in photosynthetic capacity have been proposed. Starch accumulation in the leaves of plants grown in elevated CO_2 concentration has been noted (DeLucia *et al.*, 1985; Ehret and Jollife, 1985). An inverse relationship between leaf starch levels and phtotosynthesis has been observed in annual crop species (Mauney *et al.*, 1979) where a decrease in quantum yield has been associated with starch accumulation. Such a decline indicates inhibition of the production and/or the consumption of NADPH and ATP (DeLucia *et al.*, 1985) which may in extreme cases be the result of chloroplast disruption by starch accumulation (Wulff and Strain, 1981).

Where starch accumulation is observed, this clearly indicates that photosynthate production exceeds demand. There is accumulating evidence that the maintenance of high rates of photosynthesis in response to elevated CO_2 concentrations requires the existence of *sustained*, active carbon sinks (Koch *et al.*, 1986; Downton *et al.*, 1987). In the absence of sustained sinks, assimilation declines as a result of a series of feedback processes. This point is particularly pertinent to experiments with pot-grown seedlings, in which continuing growth of both root and shoot is constrained by a fixed (generally small) volume of soil and amount of nutrients.

Carboxylation efficiency has been observed to decline following growth at elevated CO_2 in a number of species (see Berry and Downton, 1982). A decline in photosynthetic capacity of tree species as a result of growth at high CO_2 has been noted on several occasions by comparing the rates of photosynthesis of trees grown at elevated CO_2 with the rates of control trees, both *measured* at the control concentration of CO_2, or through comparison of the entire A/C_i response function (Oberbauer *et al.*, 1985; Williams *et al.*, 1986; Hollinger, 1987; Fetcher *et al.*, 1988; P. G. Jarvis, A. P. Sandford and A. Brenner, unpublished).

Such reductions in efficiency may be the result of a decrease in the amount or activity of rubisco as has been observed in crop plants (Pearcy and Björkman, 1983; Porter and Grodzinsky, 1984; Peet *et al.*, 1985; Sage and Sharkey, 1987). A reduction in the amount or activity of this enzyme to match the rate of production of ATP and NADPH set by the rate of electron transport through the photosystems (Pearcy and Björkman, 1983), would release nitrogen to downstream enzyme systems concerned with the utilization of assimilate. Field and Mooney (1986) have argued that plants maximize efficiency of resource use by the reallocation of resources, principally nitrogen, to maintain a balance between all the components of the photosynthetic system. Thus, changes in the growth CO_2 concentration influence the orthophosphate limitation of photosynthesis and can cause changes in the amount and activity of rubisco, *via* the reallocation of nitrogen. Conroy *et al.* (1986b) have shown that electron flow was not enhanced by elevated CO_2 concentrations in *P. radiata*. However, in a study of fruiting *Citrus* trees, Koch *et al.* (1986) found that the activity of rubisco was increased in one species and not affected in another, and postulated that it was the rate of turnover of the enzyme that was influenced by CO_2 concentrations. Sage (1989), however, has proposed that the activation state of rubisco is an effective indicator of the acclimation potential of a species to elevated CO_2. This proposal deserves further attention.

Additional hypotheses have been proposed to explain the reduction in assimilation observed in crop plants after extended periods of growth at elevated CO_2. These include increase in the amount or activity of carbonic anhydrase (Chang, 1975); sequestration of phosphate as sugar phosphates (Pearcy and Björkman, 1983); and inability of the Calvin cycle to regenerate ribulose bisphosphate or orthophosphate (von Caemmerer and Farquhar, 1981, 1984). Indeed, in a study of five crop species, Sage (1989) showed that a period of growth at elevated CO_2 increased the extent to which assimilation was stimulated by a decrease in O_2 concentration or an increase in CO_2 concentration. This indicates that the limitation imposed upon assimilation rate by the capacity to regenerate orthophosphate (Pi) was decreased by growth in elevated CO_2. This result could be attributed to one or all of (*a*) an absolute increase in Pi regeneration capacity, resulting from an increase in the enzymes synthesizing sucrose and starch, (*b*) an increase in Pi regeneration relative to RuBP regeneration capacity, or (*c*) a decline in rubisco and RuBP-regeneration capacity. This last possibility may be a stress response. Little is known about the mechanisms by which photosynthesis of herbaceous plants respond to long-term CO_2 enhancement (Sage, 1989) and even less is known about these mechanisms within trees. It seems clear, however, that elevated CO_2 will influence photosynthesis through the nitrogen and phosphorus economies both of the entire tree and the tree stand (Evans, 1988).

3. Compensation for stress

Elevated CO_2 has been shown to ameliorate the effects of water stress (Tolley and Strain, 1984b; 1985; Conroy *et al.*, 1986), such that the percentage increase in biomass accumulation following a period of drought is larger in elevated CO_2 than in ambient CO_2. This amelioration has several possible causes. The onset of drought is often delayed, because plants grown in elevated CO_2 have a larger root mass and are able to acquire water from a larger soil volume. CO_2-induced stomatal closure may also delay the onset of drought, although this may well be offset by increase in leaf area. A delay in the onset of drought is associated with a delay in drought-induced stomatal closure, so that photosynthesis is maintained. The rate of assimilation after the period of drought is generally higher than in control plants and also recovers faster (Tolley and Strain, 1984; Sionit *et al.*, 1985; Hollinger, 1987). A possible effect of elevated CO_2 upon osmoregulatory processes and maintenance of turgor remains untested, although it has been noted that the solute potential of trees grown in elevated CO_2 is more negative than in controls (Tolley and Strain, 1985).

Electron flow, subsequent to PS II, was affected in drought-stressed *P. radiata* grown at a CO_2 concentration of 330 μmol mol^{-1}, but was not affected in plants grown at 660 μmol mol^{-1} (Conroy *et al.*, 1986b), indicating one mechanism whereby elevated CO_2 may ameliorate the effect of drought stress.

The ameliorating effect of elevated CO_2 is apparently species-specific, since under identical growth conditions, elevated CO_2 moderated the effects of a drying soil on *L. styraciflua*, but not on *P. taeda* (Tolley and Strain, 1985).

In C_4 weeds the thermal stability of phospho-enol-pyruvate carboxylase has been shown to increase in response to high CO_2 concentrations (Simon *et al.*, 1984) and the temperture optimum for photosynthesis in C_3 *Larrea divaricata* also increased in elevated CO_2 concentrations (Pearcy and Björkman, 1983). Similarly, the thermal tolerance of *P. ponderosa* increased after growth at a CO_2 concentration of 500 μmol mol^{-1} but decreased after growth at 650 μmol mol^{-1} (Surano *et al.*, 1986). It is unclear whether this thermal effect in *P. ponderosa* resulted directly from the higher CO_2 concentrations or whether it was influenced by the thermal environment within the open-topped chambers, since needle abscission and chlorosis developed during the experiment. This point requires clarification since it is likely that global mean temperatures will also increase as CO_2 levels increase, and this increase may be larger in temperate regions than nearer the equator.

Nutrition can also influence the photosynthetic response to CO_2. In well-watered *P. radiata* adequately supplied with P, CO_2 assimilation was enhanced by 221% by doubling the CO_2 concentration (Conroy *et al.*, 1986a, 1986b). With insufficient P, however, photosynthesis was decreased by approximately 35% (Conroy *et al.*, 1986b). Significantly, acclimation of the trees to low P supply occurred after 21 weeks growth in a CO_2 concentration of 330 μmol mol^{-1} but not in plants grown in 660 μmol mol^{-1}

CO_2. The decrease in photosynthesis at elevated CO_2 concentrations in P-deficient trees was attributed to structural changes in the chloroplast thylakoid membranes and a concomitant decrease in the ability of the photon-harvesting proteins to trap and transfer and transfer energy (Conroy *et al.*, 1986b).

C. Photorespiration

CO_2 and O_2 compete for active sites on rubisco. When CO_2 is the substrate, carbon enters the photosynthetic carbon reduction cycle (the PCR or Calvin cycle): when O_2 is the substrate, phosphoglycolate and 3-phosphoglycerate are formed and carbon enters the photorespiratory carbon oxidation cycle (the PCO cycle), in which O_2 is consumed and CO_2 released (hence photorespiration). The release of CO_2 represents a significant loss of previously fixed carbon but this can be inhibited either by reducing the O_2 concentration or by raising the CO_2 concentration, since it is the ratio of O_2:CO_2 concentrations that determines the relative rates of the PCR and PCO cycles (Figure 5). A reduction in photorespiration also results in an enhanced availability of NADPH and ATP (which are consumed in photorespiration) and this may have significant, positive feedback effects on photosynthesis.

Whilst a significant reduction in photorespiration as a result of increasing the ambient CO_2 concentration has scarcely been demonstrated in trees, there is every reason to believe that the observed increases in net CO_2 assimilation, resulting from elevated CO_2 concentration, are partially attributable to reduction in photorespiration. Mortensen (1983) has shown

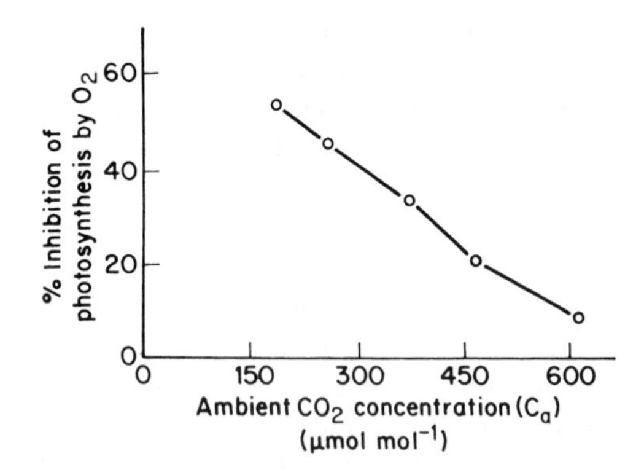

Figure 5 Influence of ambient CO_2 concentration (C_a) on inhibition of CO_2 assimilation by oxygen in *Solanum tuberosum*. (Adapted from Ku *et al.*, 1977.)

that elevated CO_2 concentrations caused a significant reduction in photorespiration of *Picea abies*.

IV. THE SEEDLING, SAPLING AND INDIVIDUAL TREE SCALE

A. Growth over Periods of Weeks and Months

In almost all experiments done to date, growth of both broadleaves and conifers was increased by an approximate doubling of the ambient CO_2 concentration from between 20 to 120% with a median of *ca* 40%. An exception was *Pseudotsuga menziesii* which showed no increase in growth in response to CO_2, in contrast to *Pinus radiata*, in a 120-day experiment (Hollinger, 1987). Some of this variability is the result of experiments of different length and some is related to stress associated with the experimental treatments, but much of it is not readily related to species and is largely inexplicable and can only be attributed to differences in experimental procedure and environments. Koch *et al.* (1986) suggested that effects of increased CO_2 on growth are likely to be large and persistent when there is a large sink capacity for carbon in the plant. In the longer experiments reviewed here, lack of sink capacity may well have attenuated the response and caused much of the variability (see Tables 5 and 6).

In most of the experiments, relative growth rates (1/W.dW/dt) were less than 5% per day and in many of the experiments, less than 2% per day. These rates are very low in comparison with reported rates of up to 10% per day for seedlings of conifers and 20% per day for seedlings of broadleaves (Jarvis and Jarvis, 1964; Ingestad, 1982; Ingestad and Kahr, 1985). In experiments in which nutrients were not supplied, the relative growth rates were lower still, less than 1% per day. The relative growth rate of tree seedlings does decline rapidly with age (Rutter, 1957) and this is particularly evident in the experiments by Brown and Higginbotham (1986) on interactions between CO_2 enrichment and nitrogen supply with *Populus tremuloides* and *Picea glauca*. None the less, the suspicion remains that very low relative growth rates indicate that the experiments were, in many cases, being done in far from optimum growth conditions. This may well, of course, reflect conditions in the real world, as in the experiments in which nutrients were not added, but the reasons for low relative growth rates in the experiments are often not clear.

Table 5 Available data for the influence of elevated CO_2 concentrations on the growth of conifers

Species	C_a	Growth location	Duration	Nutrients added	Other variables	Growth response	Author
Picea abies	× 3	GR; P	118 days	×	light	+	Mortensen and Sandivk (1987)
Picea glauca	× 2	GR; P	100 days		nitrogen	+	Brown and Higginbotham (1987)
Pinus contorta	× 3, × 5	GR; P	150 days	×		+	Higginbotham *et al.* (1985)
Pinus echinata	× 2	GC; P	168 days			+	O' Neil *et al.* (1987)
Pinus echinata	× 2	GC; P	287 days			+	Norby *et al.* (1987)
Pinus ponderosa	< × 2	OTC; soil; P	2.5 years	×		+	Surano *et al.* (1986)
Pinus radiata	× 2	GR; P	120 days	×		+	Hollinger *et al.* (1987)
Pinus radiata	× 2	GC; P	254 days	×	water; phosphorus	+	Conroy *et al.* (1986a)
Pinus taeda	< × $2\frac{1}{2}$	OTC; P	*ca* 200 days	×			Thomas and Harvey (1983)
Pinus taeda	× $2\frac{1}{2}$	GC; P	84 days	×	light	+	Tolley and Strain (1984a)
Pinus taeda	× $2\frac{1}{2}$	GC; P	84 days	×	water	+, −	Tolley and Strain (1984b)
Pinus taeda	× 2	GH; P	3 years	×		+	Sionet *et al.* (1985); Telewski and Strain (1987); Fetcher *et al.* (1988)
Pinus taeda	× $2\frac{1}{2}$	OTC; P	90 days	×		+	Rogers *et al.* (1983)
Pinus virginiana	< × 3	OTC; P	122 days			+	Luxmore *et al.* (1986)
Pseudotsuga menziesii	× 2	GR; P	120 days	×		+	Hollinger (1987)

[a] OTC—open-topped chamber; GC—growth chamber; GR—growth room; GH—glasshouse; P—pot.

Table 6 Available data for the influence of elevated CO_2 concentrations on the growth of broadleaves

Species	C_a	Growth location	Duration	Nutrients added	Other variables	Growth response	Author
Populus tremuloides	×2	GR; P	100 days	×		+	Brown and Higginbotham (1987)
Nothofagus fusca	×2	GR; P	120 days	×		+	Hollinger (1987)
Carya ovata					light and competition	No effect on community biomass	Williams *et al.* (1986)
Liriodendron tulipifera		GC;	90 days				Williams *et al.* (1986)
Quercus rubra		GC;	90 days				Williams *et al.* (1986)
Platanus occidentalis		GC;	90 days				Williams *et al.* (1986)
Acer saccharinum		GC;	90 days				Williams *et al.* (1986)
Fraxinus lanceolata		GC;	90 days				Williams *et al.* (1986)
Quercus alba	×2	GC; P	280 days			+	Norby *et al.* (1986a)
Quercus alba	×2	GC; P	210 days			+	O'Neil *et al.* (1987)
Quercus alba	×2	GC; P	280 days			+	Norby *et al.* (1986b)
Liriodendron tulipifera	×2	GC; Box	32 days			+	O'Neil *et al.* (1987)
Liquidambar styraciflua	$< \times 2\frac{1}{2}$	OTC; P	90 days	×		+	Rogers *et al.* (1983)
Liquidambar styraciflua	$< 2\frac{1}{2}$	OTC; P	*ca* 200 days	×			Thomas and Harvey (1983)
Liquidambar styraciflua	$< 2\frac{1}{2}$	OTC; P	*ca* 200 days	×	light	+	Tolley and Strain (1984a)
Liquidambar styraciflua	$< 2\frac{1}{2}$	GC; P	113 days	×	water	+	Tolley and Strain (1984b)
Liquidambar styraciflua	×2	GH; P	3 years	×		+	Sionit *et al.* (1985); Telewski and Strain (1987); Fetcher *et al.* (1988)
Citrus sinensis	×2	GC; P	305 days	×			Downton *et al.* (1987)
Robinia pseudoacacia	×2	GC; P	150 days			+	Norby (1987)
Alnus glutinosa	×2	GC; P	90 days				Norby (1987)

OTC—open-topped chamber; GC—growth chamber; GR—growth room; GH—glasshouse; P—pot.

B. Growth Partitioning

In a number of the experiments, and increase in CO_2 concentration resulted in an increase in leaf number, leaf area and leaf weight per plant (e.g. Tolley and Strain, 1984a; Sionit *et al.*, 1985; Conroy *et al.*, 1986a; Koch *et al.*, 1986; Brown and Higginbotham, 1986), leaf thickness (Rogers *et al.*, 1983) and leaf weight per unit area (e.g. Conroy *et al.*, 1986a; Brown and Higginbotham, 1986). Increases in leaf thickness and in weight per unit area may be associated with both an increase in cell size (e.g. Conroy *et al.*, 1986a) and an increase in the number of layers of cells in the mesophyll (Thomas and Harvey, 1983). It is, however, still an open question as to whether these effects result from direct action of CO_2 on leaf initiation, as suggested by Tolley and Strain (1984a), for example, or whether they result from an enhanced supply of substrate. Whilst a direct effect of CO_2 upon root initiation and growth has been observed (J.F. Farrar, pers. comm.), such an effect upon leaf primordial activity remains to be shown.

In most of the experiments too, an increase in CO_2 concentration led to an increase in the weight of both coarse and fine roots and this was especially large in *Pinus contorta* (Higginbotham *et al.*, 1985). Fruit production was increased by 70% in Valencia oranges (Downton *et al.*, 1987) and there were also concomitant increases in soluble solids. Stem diameter, weight or volume also generally increased (Tolley and Strain, 1984a; Conroy *et al.*, 1986a; Surano *et al.*, 1986; Mortensen and Sandvik, 1987) as well as height. In the longer experiments, it was noticeable that ring width was larger but there was no change in the average density of the wood (Telewski and Strain, 1987; Kienast and Luxmore, 1988). However, such observations on the juvenile wood of very young trees provide little guide to eventual changes in mature wood.

In addition to the changes in the amounts of leaves and roots, changes in response to an increase in CO_2 concentration were also observed in the dynamics of the leaf population and have been inferred in the dynamics of the population of fine roots. In experiments with *Quercus alba* the leaf area duration increased as a result of delayed leaf abscission (Norby *et al.*, 1986a,b), whereas in the relatively long-term experiment with *Pinus ponderosa* (Surano *et al.*, 1986) the sapling exposed to the highest CO_2 concentration had by the end of the study lost most two-year-old and many one-year-old needles. This last result may be an experimental artefact caused by very poor coupling between the needles and the atmosphere in the open-topped chambers and the excessively high needle temperatures that resulted.

The allocation of carbon within the plant is usually expressed as the root:shoot ratio, although this is not explicit in terms of whether there is, for example, an increase in the weight of roots or a decrease in the weight of leaves. In most of the experiments reviewed here, when nutrients have been supplied in adequate amounts sufficiently often, increase in CO_2 concentration resulted in a decrease or in no change in the root:shoot ratio (Tolley and Strain, 1984a; Sionit *et al.*, 1985; Koch *et al.*, 1986; Conroy *et al.*, 1986a; Brown and Higginbotham, 1986; Hollinger, 1987; Mortensen and Sandvik, 1987). One marked exception to this was noted with *P. contorta* in which root weight increased by $\times 15$ in response to a $\times 2$ increase in CO_2 concentration, although regularly fertilized (Higginbotham *et al.*, 1985). A similar result was obtained with *Castanea sativa* in which root weight not only increased substantially but shoot weight also decreased in response to a doubling of ambient CO_2 concentration, although apparently well-fertilized (Mousseau and Enoch, 1989).

In contrast, in the experiments in which no nutrients were added and trees were growing on a fixed capital of nutrients in low fertility soil, an increase in the atmospheric CO_2 concentration led to a substantial increase in the root:shoot ratio (Norby *et al.*, 1986a,b; Luxmore *et al.*, 1986; Norby *et al.*, 1987; O'Neill *et al.*, 1987a). This was shown in some experiments to be made up particularly of an increase in the amount of fine roots (Luxmore *et al.*, 1986; Norby *et al.*, 1986).

These results are in general accordance with the hypothesis that allocation of carbohydrates within the plant depends on the balance between the rate of supply of carbon to the leaves and the rate of supply of nutrients, particularly nitrogen, to the roots (Reynolds and Thornley, 1982; McMurtrie and Wolf, 1983; Makela, 1986; Ågren and Ingestad, 1987). Essentially, the response of a tree to a low rate of supply of nitrogen is to increase the amount of roots, whereas the response to a low rate of supply of carbon is to increase the area of leaves. When the rate of supply of carbon is increased, as in these experiments, the proportion of leaf mass declines somewhat and the proportion of root mass increases, since relatively, the rate of supply of nitrogen to the roots is insufficient in relation to the supply of carbon.

A general conclusion from these measurements of growth is that increase in CO_2 concentration primarily leads to plants getting larger more quickly and that the majority of the changes observed are normal ontogenetic changes associated with growth and development.

C. Compensation for Stress

Increase in CO_2 concentration can compensate for stress-induced reduction in growth. Compensation in plants grown in low photon flux densities and double the current atmospheric CO_2 concentration was complete for *Liquidambar styraciflua* but only partial for *Pinus taeda* (Tolley and Strain, 1984a), but a tripling in CO_2 concentration led to over-compensation for *Picea abies* (Mortensen and Sandvik, 1987).

Full compensation for water stress by a doubling of the CO_2 concentration was also shown in *L. styraciflua* and partial compensation in *P. taeda* by Tolley and Strain (1984b), although Sionit *et al.* (1985) demonstrated only partial compensation in older seedlings of *L. styraciflua* growing in glasshouse sections. Compensation for water stress was also found with *P. radiata* (Conroy *et al.*, 1986b).

D. Water Use Efficiency

An increase in atmospheric CO_2 concentration is likely to result in a decreased stomatal conductance, and this will reduce the rate of transpiration per unit leaf area. Conversely, rates of photosynthesis are generally increased in elevated CO_2 environments. Consequently, an increase in water use efficiency (WUE), defined as (mol CO_2 assimilated)/(mol H_2O transpired), would be expected as a result of growing plants in elevated ambient CO_2 concentrations. Only a few reports on trees actually quantify WUE (e.g. Oberbauer *et al.*, 1985; Norby *et al.*, 1986; Hollinger, 1987), but an increase in WUE can be inferred from many reports. Where measured, water use efficiency has been shown to increase in response to the increase in CO_2 concentration, largely because water use did not increase in proportion to the increase in plant size (Rogers *et al.*, 1983; Norby *et al.*, 1986a). Figure 6 shows the positive linear relationship between CO_2 concentration and WUE for *L. styraciflua* observed by Rogers *et al.* (1983).

A possible benefit from increased WUE as a result of stomatal closure is a reduced rate of water consumption per unit leaf area, and hence a decrease in the likelihood of drought developing. However, since elevated CO_2 concentrations frequently result in an increase in total leaf area, this increase in WUE may be offset by increase in leaf area per tree. Increased WUE may also result in increased growth during drought in comparison to droughted control trees, since drought is delayed and less severe in elevated CO_2 environments (see Acclimating Tree Studies).

E. Nutrient Stress—a Limiting Factor?

It has been suggested that the nutrient stress commonly experienced by trees growing in woods and forests could completely negate any benefits to

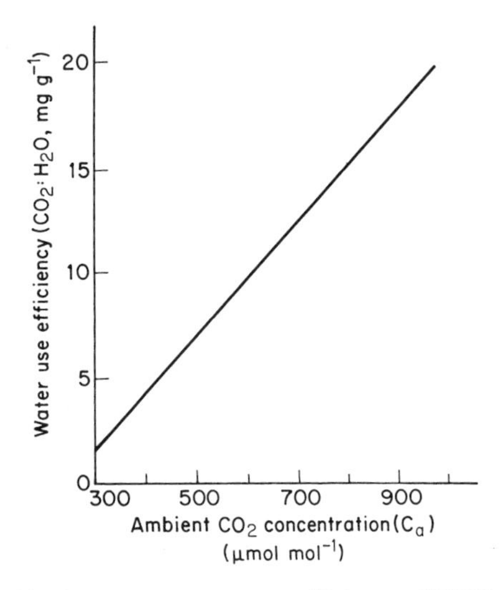

Figure 6 Relationship between water use efficiency (WUE) and ambient CO_2 concentration (C_a) for *Liquidambar styraciflua*. (Adapted from Rogers *et al.*, 1983 and Tolley and Strain, 1985.)

growth that might result from the increase in CO_2 concentration. In the majority of experiments referred to so far, attempts have been made to eliminate nutrient stress from consideration by supplying nutrients regularly to the plant-pots. In most cases this seems to have been successful and nutrient stress has been avoided, although in some cases the growth responses do suggest that some nutrient stress did occur. In two experiments the limiting factor hypothesis has been tested by investigating the response to increase in CO_2 concentration by seedlings provided with a range of nitrogen concentrations in sand culture with *Populus tremuloides* and *Picea glauca* (Brown and Higginbotham, 1986) and additions of different amounts of phosphorus to a phosphorus-deficient soil with *Pinus radiata* (Conroy *et al.*, 1986a,b). Unfortunately, in both experiments the *amount* of nutrients added was controlled and no attempt was made to follow Ingestad principles and to control the *rate of supply* of the nutrients in relation to the endogenous capacity for growth (Ingestad, 1982; Ingestad and Lund, 1986).

In both these experiments growth increased in response to an increase in CO_2 concentration at each amount of added nutrient. Plant weight, leaf weight and plant height were increased particularly at the higher additions of nutrients and root weight at the lower additions. In both experiments the increase in growth was proportionately larger in the treatments with the larger additions of nutrients, but substantial growth increases (13% with P and 31% with N) were observed at the lowest additions of nutrients.

The response of *P. tremuloides* was, however, only temporary and did not persist after 50 days.

In a series of experiments carried out at Oak Ridge by R. J. Luxmore and his colleagues, seedlings of *Q. alba* (Norby *et al.*, 1986a,b; O'Neill *et al.*, 1987b), *Liriodendron tulipifera* (O'Neill *et al.*, 1987a). *Pinus virginiana* (Luxmore *et al.*, 1986) and *Pinus echinata* (O'Neill *et al.*, 1987b) have been grown on soils of low fertility with a fixed, small capital of nutrients and their growth and uptake of a range of nutrients investigated in relation to root and rhizosphere properties. The four species investigated at Oak Ridge showed large (37–99%) increases in growth in response to a doubling of the ambient CO_2 concentration: the corresponding increases in relative growth rates ranged from 23 to 40%. These increases were largely the result of an increase in root weight, which was consistently larger than the increase in total dry weight. Analysis of the plant nutrient contents showed four patterns of nutrient uptake as follows:

(*a*) a larger increase in uptake of the nutrient than in the assimilation of carbon, leading to an increase in the nutrient concentration in the plant;
(*b*) an increase in uptake of the nutrient proportional to the increase in carbon assimilation, leading to no change in the nutrient concentration in the plant;
(*c*) an increase in uptake of the nutrient less than the increase in carbon assimilation, leading to a decline nutrient concentration in the plant; and
(*d*) no increase in uptake of the nutrient, so that the nutrient was progressively diluted as carbon was assimilated, leading to a large, progressive decline in the concentration of the nutrient in the plant.

The two broadleaves behave similarly, showing no increase in uptake of N and S at the high CO_2 concentration (pattern *d*), but with all other nutrients (P, K, Ca, Mg, Al, Fe, Zn) following pattern (*b*) or (*c*). In contrast, in *P. virgniana* N and Ca showed no change in concentration (i.e. pattern *b*) and P, Mg and K declined in concentration (pattern *c*, *c* and *d*, respectively) whilst the trace elements increased in concentration in the plant (pattern *a*). Patterns (*c*) and (*d*) must inevitably lead to severe deficiency as the nutrients in the plant are progressively diluted by increasing amounts of carbon and this cannot continue indefinitely. To survive, the plant must either extract previously unavailable quantities of the nutrient or exploit new volumes of soil which is not, of course, possible, in pot experiments. It is also questionable with respect to patterns (*a*) and (*b*), how the plants continue to take up certain nutrients in excess of or in proportion to the carbon assimilated, from a declining resource as in a pot experiment. Possibly the capital is not fixed and the plants extract progressively more nutrients from the soil, even though it is of low fertility. The dilution of nutrients through growth in (*c*) and (*d*) leads to apparent increase in growth per unit of nutrient (i.e. nutrient use efficiency), which appears to increase in response to an increase in ambient CO_2 concentration for nutrients such as nitrogen. This apparent increase in efficiency is, however, no more than an indication of the limited ability of the plants to extract nutrients from a limited source.

F. Extraction of Nutrients

Mechanisms by which plants extract nutrients from a limted source were investigated in the same series of papers. In a number of experiments a greater proliferation of fine roots was observed in response to an increase of CO_2 concentration and this was, in some cases, associated with the increase in the amount of nutrients taken up (Norby *et al.*, 1986a; Luxmore *et al.*, 1986; O'Neill *et al.*, 1987a).

In *Q. alba* and *P. echinata*, this increase in fine roots was associated also with increase in the rate of establishment and density of mycorrhizal symbiosis (O'Neill *et al.*, 1987a; Norby *et al.*, 1987). Increased root exudation of carbohydrates may explain this, as there was a temporary increase in exudation of carbon-containing compounds from roots of *P. echinata* (Norby *et al.*, 1987). The observed increase in carbon allocation to roots and concomitant root exudation may have stimulated rhizosphere microbial and symbiotic activity and, hence, nutrient availability, although conclusive proof of this point was not provided by Norby *et al.* (1987) and no response of the microbial populations in the rhizosphere of *L. tulipifera* was found by O'Neill *et al.* (1987a). Luxmore *et al.* (1986) suggested that increased acidification of the rhizosphere of *P. virginiana* as a result of increase in the atmospheric CO_2 concentration might have been responsible for increase in the uptake of certain nutrients such as Zn, as a result of proton exchange.

None of these studies demonstrated in a convincing way how nutrients are taken up from a limited source of low availability in response to an increase in atmospheric CO_2 concentration, so that the internal plant concentration of nutrient does not change. A possibility, not so far considered, is that root surface enzymes, such as phosphatases, may be effective in solubilising bound nutrients, enabling them to be extracted more effectively by the expanded network of fine roots and mycorrhizas.

An enhanced role of soil microbial activity in the release of nutrients through mineralisation and decomposition is a possibility, since there is some evidence that the composition of leaves abscissed from growing trees is affected by CO_2 fertilization. Leaves abscissed from seedlings of *Q. alba* grown in increased CO_2 concentration were found to contain larger amounts of soluble sugars and tannin and a smaller amount of lignin. However, the differences were small and unlikely to increase significantly the rates of litter decomposition (Norby *et al.*, 1986b).

G. Nitrogen Fixation

Nitrogen-fixing trees are important for providing a stable long-term source of nitrogen in many silvicultural systems. Norby (1987) investigated the influence of elevated CO_2 on nodulation and nitrogenase activity of three nitrogen-fixing trees growing in an infertile forest soil and found that total nitrogenase activity per plant increased because of a larger root system, whereas specific activity was not influenced. Symbiotic nitrogen fixation is closely linked to photosynthetic capacity, and Norby concluded that the observed increase in nitrogenase activity per unit leaf area resulted from the increase in photosynthesis stimulated by CO_2 enrichment. Since the contribution of nitrogen fixation to the total nitrogen budget of the tree increases with age (Akkermans and van Dijk, 1975), the small increase in nodulation observed in young trees may have significantly larger effects in mature trees.

Two nitrogen-fixing species, *Robinia pseudoacacia* and *Alnus glutinosa*, also showed significant increases in growth, nodule weight and in nitrogenase activity in reponse to a doubling of atmospheric CO_2 concentration (Norby, 1987), perhaps because nitrogen fixation was carbohydrate-limited for other reasons.

H. Seedling Regeneration

Compensation for light stress, water stress and nutrient stress by seedlings, as observed in some of the experiments, is likely to enhance the ability of seedlings to establish and grow in a competitive environment and, hence, to promote seedling regeneration. It is evident from the foregoing, that seedlings of different species respond differently to an increase in atmospheric CO_2 concentration and it is possible, therefore, that the balance amongst regenerating species may change. This was investigated in an experiment in which three species from an upland habitat and three from a lowland habitat were grown together in competition on a fixed nutrient capital with increase in CO_2 concentration in low and high quantum flux densities (Williams *et al.*, 1986). Increase in CO_2 concentration did not increase the overall biomass of either population, but the relative weight of each species changed in a complex way depending on the atmospheric CO_2

concentration, photon flux density and the community. In this experiment, too, N and P became progressively diluted within the plants as a result of growth, so that nutrient acquisition and physiological response to nutrients may well have been responsible for the development of dominance by one or other species. In the upland community, *Quercus rubra* became dominant over *Carya ovata* and *L. tulipifera*, whereas in the lowland community *Fraxinus lanceolata* dominated *Platanus occidentalis* and *Acer saccharinum*.

Differences amongst provenances with respect to increase in CO_2 concentration are also likely. Surano *et al.* (1986), in their $2\frac{1}{2}$-year experiment showed significant differences in the growth response of provenances of *P. ponderosa* from the Sierra Nevada and the Rocky Mountains to CO_2 but this experiment was complicated by leaf chlorosis and abscission, perhaps as a result of excessive temperatures in the open-topped chambers.

Changes in old-field secondary succession have been predicted from observations of *L. styraciflua* and *P. taeda* grown under elevated CO_2 with imposed drought (Tolley and Strain, 1984a). First-year survival of seedlings of *L. styraciflua* in drier sites, an increased tolerance to drought of *L. styraciflua* compared to *P. taeda* and the overall greater, CO_2-enhanced seedling growth of *L. styraciflua* may favour the establishment of *L. styraciflua* in drier habitats presently dominated by *P. taeda* (Tolley and Strain, 1985).

I. Growth at Timescales of Years and Decades

Only two experiments have been reported lasting for more than two years (Sionit *et al.*, 1985, and Fetcher *et al.*, 1988; Surano *et al.*, 1986) and it is difficult, if not dangerous, to extrapolate from these experiments to longer time scales. For the two species, *P. taeda* and *L. styraciflua*, differences in growth in response to increase in CO_2 concentration were established early on and persisted but did not increase with time. It seems very likely that this was a result of the constrained growing conditions, even though nutrients were added, and would, perhaps, not occur with unrestricted rooting.

It is readily shown that the quite small differences in relative growth rate observed in many of the experiments discussed, should lead to very large differences in plant size after quite short periods, if they persist. For example a 20% increase in relative growth rate from 0.8 to 1.0% per day should, after ten years, lead to trees that are over $\times$ 1000 larger. There is, however, at present no evidence as to whether the increases in seedling growth rates will persist and large differences in tree size will eventuate.

There is some indication in the experiments on seedlings that crown structure may change as a result of increase in CO_2 concentration. For example, shoot length increased in *P. abies* (Mortensen and Sandvik, 1987) and plants of *L. styraciflua* developed increased branching (Sionit *et al.*, 1985). However, the extent to which these differences might persist is unknown. A change in shoot structure and degree of branching is likely to lead to a change in the distribution of leaf area density within the tree crown and it has been shown with models that this can have a major influence on the total amount of crown photosynthesis in stands of *P. sitchensis* (Russell *et al.*, 1988; Wang, 1988).

In some cases, stem growth was increased and this could, conceivably, lead to enhanced harvest index later in life. In *P. radiata* seedlings Conroy *et al.* (1986a) found an increase in stemwood density but in the longer experiment with *P. taeda* (Sionit *et al.*, 1985) average stemwood density did not change, although there was an increase in latewood density (Telewski and Strain, 1987). Whether increase in atmospheric CO_2 concentration will lead to changes in harvest index and wood quality is, at the present time, entirely speculative.

V. THE PLANTATION AND WOODLAND SCALE

For practical and technical reasons, direct measurement of the effects of increase in CO_2 concentration on processes at this scale are extremely difficult, if not impossible, and at present there are no directly determined data and little likelihood of any in the immediate future. Purely on account of the large volume occupied by a tree, the scale of the problem in exposing large enough areas of forest to increased atmospheric CO_2 concentration is very much greater than with agricultural crops and it seems, unlikely that exposure without enclosure will be possible.

Whilst the maintenance of increased CO_2 concentration around individual trees is technically feasible for extended periods, given adequate resources, the results would not bear directly on the functioning of the plantation or woodland system as a whole. Thus, at the present time, we have the problem of estimating the likely effects of an increase in CO_2 concentration on acclimating stands from measurements made on unacclimated stands at the present day CO_2 concentrations or from measurements made on seedlings and young trees partially acclimated to increased CO_2 concentrations in artificial surroundings. The use of models provides the only way forward in this situation.

A. At the Timescales of Hours and Days

Measurements of the exchanges of energy, momentum, water vapour, carbon dioxide and various pollutants by a range of plantation and woodland canopies in current atmospheric conditions have been made (Jarvis *et al.*, 1976; Jarvis, 1986; Verma *et al.*, 1986; Baldocchi *et al.*, 1987) and these data can be used to test models of the influence of environmental, structural and physiological variables on stand carbon dioxide balance and water use efficiency. A suitable model for this purpose exists (Grace *et al.*, 1987; Wang, 1988). This model, called MAESTRO, is straightforward to run for an existing canopy structure, stand structure and environment and can, in principle, provide answers to the question: what would be the consequence for assimilation, transpiration and water use efficiency of a stand of a doubling of the ambient CO_2 concentration? It is difficult, however, to know what aspects of acclimation to the increased CO_2 concentration should be taken into account. It is clear that acclimation of photosynthetic and stomatal parameters at the leaf scale should be considered, but it is far from clear as to what other aspects of tree or stand structure, canopy leaf area density distribution, soil properties, decomposition processes etc. need to be included. Figure 7 shows calculated canopy CO_2 assimilation for a stand of *Picea sitchensis* at the current ambient CO_2 concentration, using appropriate leaf parameters in an assimilation sub-model based on that of von Caemmerer and Farquhar (1981), in comparison with estimates in an atmosphere of double CO_2 concentration, again using appropriate parameters. The parameters were determined on seedlings grown in controlled environment rooms at current and twice the current ambient CO_2 concentrations. Canopy assimilation is increased substantially by a doubling of the atmospheric CO_2 concentration, but not as much as would be expected if parameters for unacclimated plants were used in the predictions at the doubled CO_2 concentration. Such predictions cannot be verified directly but depend for their acceptance on adequate verification of the model against measurements of canopy processes in the current CO_2 environment. MAESTRO, for example, has been shown to predict accurately the radiation environment beneath the canopy of stands of *Pinus radiata* and *Picea sitchensis* (Wang and Jarvis, 1989), and the predicted response of CO_2 assimilation to photon flux density with current parameters in the current CO_2 environment agrees reasonably well with measurements of CO_2 influx by a similar stand of *P. sitchensis* at similar, small water vapour saturation deficits (Jarvis and Sandford, 1986).

B. At the Timescales of Weeks and Months

In 1977 Monteith introduced a simple model of the growth of agricultural and horticultural tree crops as a function of intercepted radiation and the efficiency of utilization of that intercepted radiation. This model

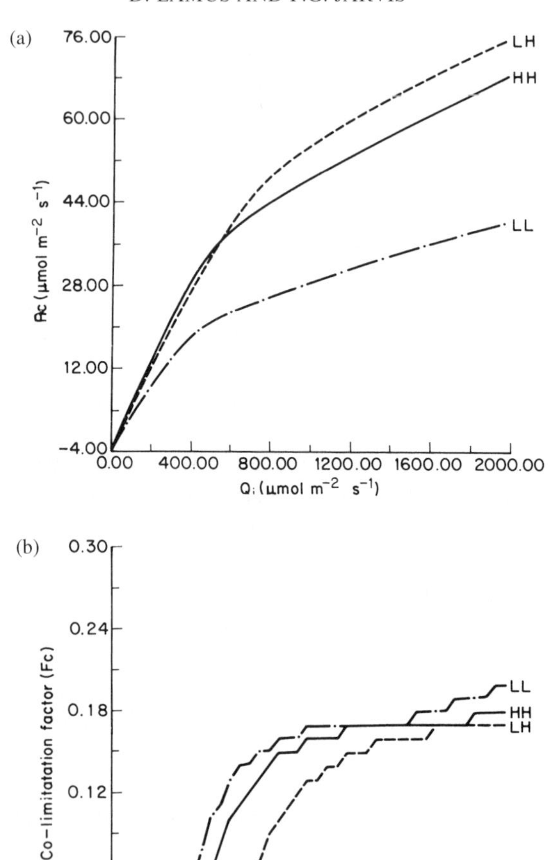

Figure 7 (a) Response of canopy CO_2 assimilation (A_c) to the incident photon flux density (Q_i) calculated using MAESTRO for a pole stage stand of *Picea sitchensis* in central Scotland with appropriate parameters for the current ambient CO_2 concentration (345 µmol mol^{-1}) (LL) and double the current ambient CO_2 concentration (HH). The line LH shows what is obtained if current parameters are used in combination with a doubling in the atmospheric CO_2 concentration. The beam fraction of the incident quantum flux density was assumed to be 0.5, the leaf area index 9.0 and the zenith angle of the sun 45°. Other assumptions and a list of parameters are given by Wang (1988) (From Jarvis, 1989). (b) The calculated increase in co-limitation by RUPB-carboxylase in relation to photon flux density for the conditions in (a). Limitation by electron transport is absolute ($A_c = 0$) up until an incident photon flux density of *ca* 400 µmol m^{-2} s^{-1} at which limitation by the carboxylase begins. A change in ambient CO_2 concentration has little or no effect on co-limitation.

has been extended to plantations and woodlands (Jarvis and Leverenz, 1983) and has been shown to be applicable to stands of conifers and broadleaves (Linder, 1985, 1987), to biomass plantations of willow and poplar (Cannell *et al.*, 1987, 1988) and to commercial pole-stage plantations of *P. sitchensis* in Scotland (Wang, 1988). The model predicts that the rate of increase of dry matter is linearly proportional to the amount of photosynthetically active radiation (PAR) absorbed by the crop over the same period (Russell *et al.*, 1988).

For *P. sitchensis* in Tummel Forest in Central Scotland the average value of the dry matter: radiation absorption coefficient (efficiency of utilization of radiation) is 0.41 g of dry matter above ground per mole of photosynthetically active quanta and the relationship is linear over the range of radiation experienced. Thus, we may suppose that if an increase in CO_2 concentration leads to an increase in the efficiency with which PAR is utilized without compensating increases in respiration or turnover of fine roots, leaves and other plant parts, there will be an equally large increase in growth rate, i.e. a 10% increase in the net efficiency of PAR utilisation as a result of the increase in CO_2 concentration, will lead to a 10% increase in growth rate. If, in addition, the effect of an increase in CO_2 concentration is to increase the leaf area that the canopy can sustain (because an increase in CO_2 concentration may compensate for shading of the lower leaves in the canopy and reduce leaf senescence), so that more PAR is absorbed per hectare, then there will be a further increase in growth. This is exemplified in Figure 8 where Point A defines the present growth rate in relation to absorbed radiation and Point Z indicates the likely growth rate as a result of increase in both the efficiency of utilization of radiation (Point Y) and the amount of radiation absorbed (Point X). That the growth of forest may increase in relation to the increase in CO_2 concentration, as a result of increases in both these processes, seems likely but is speculative.

To take this approach further, the two major processes discussed here require to be dissected into constituent, partial processes and the effects of an increase in CO_2 concentration on those processes investigated in detail, so that the likely changes in both efficiency of utilization and light absorption can be made quantitative. At the stand scale, we must necessarily be concerned about the consequences of an increase in CO_2 concentration on mineralization processes in the soil and on nutrient cycling through the system, since these processes may ultimately limit the extent to which photosynthetic efficiency can be increased or leaf populations maintained. Similarly, an increase in stand leaf area will lead to increases in both interception loss and transpiration loss of water (Jarvis and McNaughton, 1986), so that the availability of water may act to limit the projected increase in growth resulting from the increase in CO_2 concentration. This point

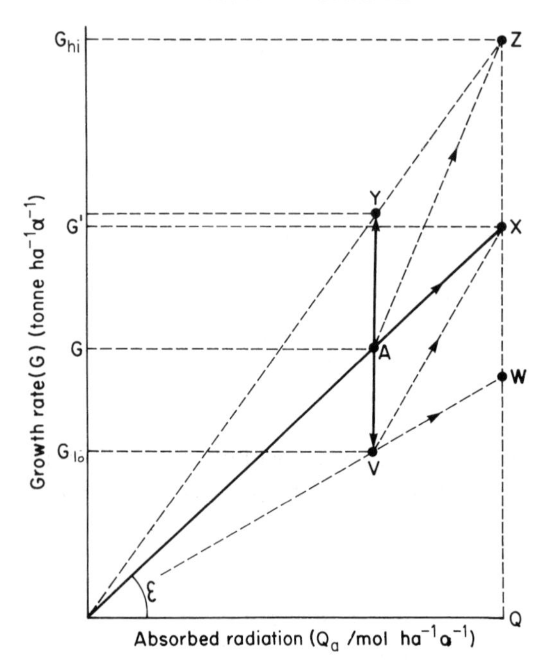

Figure 8 Hypothetical relationship between stand growth rate (G) and absorbed photon flux density (Q_a). For a thining and fertilizer experiment in a pole-stage stand of *Picea sitchensis* in central Scotland, $\varepsilon = 0.3$ to 0.5 with an overall mean of 0.41 g of above-ground dry matter per mole of absorbed PAR. An increase in ambient CO_2 concentration, leading to an increase in either ε or the quanta absorbed could lead to an increase in growth rate from G to G′ (A→Y or A→X): an increase in both would lead to a further rise in G to G_{hi} (A→Y). A decline in ε, without a compensating increase in the quanta absorbed, could lead to a reduction in stand growth rate to G_{lo} (A→V). The likely magnitude of any such effects is unknown.

is further complicated, since increases in water use efficiency and total leaf area per plant may increase, with opposing effects upon total plant water use. More complex models of stand processes than are available at present are required to take these possible feedbacks into account. Models for this purpose are under development at several places at the present time.

C. At the Timescales of Years and Decades

Accelerated growth during early life of a stand may lead to earlier canopy closure and ultimately to a reduced rotation length. At the present time, for example, this can readily be achieved by more intensive fertilizer applications (Axelsson, 1985; Linder, 1987) but this is costly and generally not in accordance with the economic strategy underlying current forest management practice. It is likely that small gains in this regard will result from the increase in CO_2 concentration but these gains are likely to be very much smaller and less immediate than the gains that could be obtained from more intensive stand management. Models capable of predicting the likely reduction in rotation length as a result of an increase in CO_2 concentration, with an appropriate mechanistic content, are not available, but there are, of course adequate models to predict the economic benefits resulting from a reduction of rotation lengths of, say, 10% or five years.

It seems unlikely that the increase in CO_2 concentration will change the size-class distribution of stemwood within the stand, but this is of some economic importance and should be approached through the development of a suitable model.

It is probable that in natural woodlands an increase in the atmospheric CO_2 concentration will lead to a change in the species composition of both overstorey and understorey, but a natural woodland is a very complex ecological system and models adequate to test this are presently unavailable.

VI. THE FOREST AND ECOSYSTEM SCALE

A. At the Timescales of Years and Decades

Substantial changes are presently occurring in both the European and North American countryside, largely as a result of policy instruments of national governments and the European Community. Changes in forest management are likely to occur rapidly in the near future, as a result of the expansion of forest onto better quality, ex-agricultural soils, the introduction of new genotypes, a wider range of species and improved fertilizer regimes. In the British Isles, for example, the area of forest is expanding rapidly and is likely to continue to do so, and with the advent of progressively more second-rotation forest, the diversity of species and of stand and compartment composition is likely to increase substantially. Future growth enhancement resulting from increase in CO_2 concentration is likely to be subsumed into growth enhancement resulting from these and other changes, so that there are unlikely to be any directly consequential changes in forest management practice: the effects of an increase in CO_2 concentration will be superimposed upon, and difficult to distinguish from, the major changes in land management presently in progress. The scale of the increase in productivity over the next 25 years that may result from the increase in atmospheric CO_2 concentration is likely to be of the same magnitude as the increases in productivity that have occurred over the last 25 years, as the result of changes in fertilizer application regimes, for example. A change of this order of magnitude certainly could be expected to improve the economics of afforestation and may enhance the prospects for industrial utilization of wood in particular areas, but is unlikely to have a socio-economic impact of the same magnitude as the present expansion in area of the temperate forest estate.

There are other well-known consequences of afforestation, for streamflow and the yield of water, for amentity, wildlife and recreation. It seems likely that the effects of increase in CO_2 on these aspects of forests will also not be noticeable in relation to the large change resulting from fiscal policy and political instruments.

None the less, there is a strong probability that CO_2 fertilisation, accompanied by nitrogen fertilisation resulting from anthropogenic emissions, has been responsible in part for the substantial increases in temperate forest productivity over the past 100 years that have been observed in many places (e.g. the French Vosges region). Conversely, there is some evidence that the temperate and boreal forests are now affecting the global atmospheric CO_2 concentration (see discussion by Jarvis, 1989).

B. Timescales of a Century or More

Over much longer timescales we may expect the increase in CO_2 concentration to have a noticeable impact on the species composition and character of natural woodlands as a result of changes in the competitive balance among species, leading to ecological selection and, possibly, ultimately to natural selection, and evolution (Shugart *et al.*, 1986). What will happen in the future can, at present, only be guessed at and it is difficult to see how simplistic experiments in which three or more species are grown together in controlled environments can lead to useful predictions as to what may happen in natural woodland systems in the long term. Our understanding of natural ecosystems is still so elementary and likely to remain so for the foreseeable future, that sensible predictions about the likely effects of increase in CO_2 concentration on ecosystems are out of the question (Jarvis, 1987).

VII. SUMMARY

A. Assessing the Impact of the Rise in CO_2

A doubling of the present atmospheric CO_2 concentration is likely to occur by the end of the next century and will have many effects on trees and forests. There may be major changes in climate, especially temperature and precipitation, that will affect the growth and ecology of trees and forests: there will be substantial effects on the growth and ecology of trees and forests as a result of the direct effects of CO_2 on the physiology of trees. To a considerable extent the consequences of an increase in atmospheric CO_2 concentration on climate are speculative, whereas many of the direct effects of an increase in CO_2 concentration on the physiology and growth of plants have been established through experimentation and are quite well known and understood.

There are several reasons for showing active concern about the consequences of the rise in CO_2 concentration for trees and forests: to enhance knowledge, at a fundamental level, about the functioning of woodlands and forests; to assess the likely impact on their ecology, productivity and value; and to anticipate any significant downstream, socio-economic consequences.

The consequences of the rise in CO_2 concentration may be assessed at a range of temporal and spatial scales: from seconds to centuries, from cell to region. Biological information about the effects of CO_2 on plants is generally available at short time scales and small spatial scales, whereas ecological and socio-economic concern is largely expressed with respect to the much larger scales of the stand and the forest over tens or hundreds of years.

There are major problems with the data currently available. The environmental conditions pertaining in the majority of the experiments (e.g. soil type, temperature) differ significantly from forest conditions throughout the boreal and temperate regions. Secondly, the use of juvenile trees in growth cabinets and rooms makes extrapolation to the field situation difficult. Finally, the use to pot-grown plants with a limited soil volume, and all the problems of nutrient depletion that this entails, imposes severe constraints on the interpretations of changes in plant nutrient contents, assimilation rates and growth.

B. Photosynthesis and Stomatal Action

The primary role of CO_2 as a substrate for photosynthesis is well understood and there are no reasons to suppose that effects of CO_2 concentration on the primary carboxylation are any different in trees from other plants with the C_3 pathway. However, the role of CO_2 as an effective activator and regulator of rubisco is only currently being unravelled.

In general, photosynthesis increases substantially, but less than proportionately (up to 75%), in response to a doubling in ambient CO_2 concentration, but a lack of response, or even a reduction in rate of photosynthesis, has been observed. There is increasing evidence that maintenance of active sinks for assimilate is necessary for the effective stimulation of photosynthesis by an increase in CO_2 concentration. Constrained rooting of pot-grown trees and the lack of alternative sinks as a result of determinate growth, may be the reason why a strong, positive response has not always been seen and why the assimilation rate declines with time in many of the experiments. Closure of stomata in response to stress may also negate the increase from CO_2 fertilization.

When plants are grown for extended periods in elevated CO_2 so that they may become acclimated, the increase in rate of photosynthesis is less than in short-term experiments. Analysis of A/C_i response functions indicates that this may result from a reduction in the activity of rubisco, although it is not clear whether the amount or the activation state of the enzyme is reduced, or from Pi limitation.

Stomatal conductance generally declines with an increase in CO_2 concentration to between 0 to 70% of the control. There is considerable variation in response and in some trees, particularly conifers, stomatal conductance is rather unresponsive to changes in ambient CO_2 concentration. The lack of a mechanistic understanding of the role of CO_2 in stomatal action prevents explanation at present.

There is some evidence that a reduction in the number of stomata per unit area or per unit epidermal cell may also contribute to the reduction in stomatal conductance of plants grown at high CO_2 concentrations, but both increases and decreases have been observed in trees. It is not clear whether these changes are the result of acclimation of developmental processes, ecological selection or genetic adaptation.

There may be other presently unknown effects of CO_2 concentration on growth processes at the cellular level. In particular, the effects of CO_2 concentration on leaf growth may not be explicable solely in terms of substrate supply and there may be effects of CO_2 on leaf initiation, development and expansion in other ways.

C. Tree Growth

Growth of seedlings and young trees of both broadleaves and conifers is increased by a doubling of the ambient CO_2 concentration. The observed magnitude of the response ranges from 20 to 120% with a median of *ca* 40%. Both root and shoot growth are increased. Young trees grow faster in high CO_2 but normal ontogenetic processes occur.

In general, the experiments suffer from a number of inadequacies and this may explain the variability of response. In particular nearly all the experiments have been short-term (less than 12 months) on very young trees, that are often pot-bound and with growth restricted by lack of sinks and in a nutrient-deficient condition. Very low relative growth rates indicate far from good growth conditions.

Increases in growth in response to increase in CO_2 concentration occur in stress conditions and can ameliorate to a large extent the influence of stresses (water, nutrient, low light) on growth. Nutrient deficiency enhances biomass partitioning to the roots, whereas with free nutrient supply, an increase in CO_2 concentration leads to a relative enhancement of leaf growth.

Acclimation to nutrient depletion, evident as increased nutrient use efficiency, may occur but its extent and long-term significance remains uncertain. Enhanced fine-root mass and increased mycorrhizal infection may be significant in this respect, in addition to changes in root uptake processes.

The combination of enhanced assimilation and reduced stomatal condutance at elevated CO_2 leads to an increase in water use efficiency at the leaf and seedling scales in controlled environments, but it is not known whether this occurs at the stand scale in the field.

Because trees are perennial, a small increase in relative growth rate of 0.1% per day, at the prevailing low growth rates in many of the experiments, should lead to a large difference in individual tree size after a number of years. Whether CO_2 fertilization will have this effect is however, uncertain.

A major gap in our knowledge is related to the influence of elevated CO_2 levels on yield quality (wood density, knottiness, etc.).

D. Stand Processes and Production

The experimental approach is not practical with older trees, stands and forests. The effects of CO_2 on assimilation, water use and other processes at the stand or larger scales, can only be assessed through models. Models of stand processes in relation to a doubling of CO_2 concentration, taking into account acclimation to CO_2 at the leaf scale, indicate a substantial increase in stand CO_2-assimilation and in water use efficiency. However, the information available to parameterize such models may not be appropriate because it has been derived from seedlings that were only partially acclimated.

Experimentation on mature trees is technically feasible but logistically and financially difficult. A more feasible approach available may be to use portions of mature trees, such as branches, in large cuvettes to obtain data required for the parameterization of stand-scale models that are used to make predictions of the likely consequences of the increase in the atmospheric CO_2 concentration for stand functioning. Micro-meteorological changes within the cuvettes and changes in feedback loops between the enclosed portion and the atmosphere, must, of course, be taken into account as well as the physiological consequences of treating only part of the plant. Such models require to be carefully tested at the present CO_2 concentration and in relation to other environmental variables, since they cannot be tested in a doubled CO_2 environment in forests.

Changes in tree canopy structure may also influence canopy functioning as well as understorey species composition and growth: changes in the pattern of branching and in internode length, as a result of enhanced CO_2 concentrations, have been observed, for example. However, models that include feedbacks such as interactions between CO_2 and crown structure, growth and respiration of trees, or leaf composition and rates of litter decomposition, are not presently available. Whether changes in harvest index and wood quality will ensue at the stand scale cannot be predicted at present.

E. Forest and Region

At larger scales, a very complex network of processes must be taken into consideration and there is quite insufficient information about the effects of CO_2 on these processes to permit reasonable predictions. Future changes in forest management practice, the selection of new genotypes, extensive fertilizer treatments and an increase in the quality of land available for afforestation, are likely to lead to significant changes in forest production. The changes attributable to elevated CO_2 concentrations will be superimposed upon those caused by these other factors. It does, however, seem that the effects of the rise in CO_2 may be relatively small in relation to impending changes in land use and management practices.

VIII. RECOMMENDATIONS FOR FUTURE RESEARCH

In 1981 Kramer wrote: "We cannot make reliable predictions concerning the global effects of increasing CO_2 concentration, until we have information based on long-term measurements of plant growth from experiments in which high CO_2 concentration is combined with water and nitrogen stress on a wide range of species."

Whilst some progress has been made in this direction, as evidenced by the work referred to here, this conclusion of Kramer's is still generally valid, and it is particularly valid with respect to trees and forests.

Arising from this review, we have identified particular needs for research on trees and forests with respect to the increase in the global atmospheric CO_2 concentration in the following areas:

(*a*) The basic action of CO_2 on rubisco activity. Exploration of direct effects of CO_2 on leaf initiation and growth.
(*b*) Characterisation of the acclimation of photosynthetic processes to elevated CO_2 with associated development of leaf scale models. Investigation of the role of CO_2 in stomatal action. Comparison of the effects of CO_2 on seedling and mature foliage using bagged branches on trees and potted grafts of mature branches.
(*c*) The *long-term* effects of exposure to CO_2 on the growth and growth processes of forest species in appropriately controlled conditions; controlled environment rooms or well-ventilated open-topped chambers, with experimental control of nutrient supply rate according to Ingestad principles, control of water supply rate, and unconstrained rooting. Long-term acclimation of photosynthesis, stomatal conductance, water use efficiency, biomass partitioning, leaf and root dynamics and crown development in relation to high CO_2. Development of models of growth and partitioning to predict likely effects of high CO_2.
(*d*) Development and parameterization of models of stand processes incorporating multiple feedbacks. Rigorous testing of such models at the stand scale using eddy correlation measurements of water and CO_2 fluxes in present conditions to give confidence to predictions. Further development of more empirical stand-scale light-interception and growth models and parameterization with respect to predictions regarding elevated CO_2 effects.
(*e*) Tests of genotype/CO_2 interactions and selection of genotypes for growth in high CO_2 environment.
(*f*) Finally, we see the development of models as a matter of particularly urgent priority where forests are concerned and recommend the development of a new modelling framework to encompass a wide range of ecosystem processes for the purpose of assessing the consequences of the increase in CO_2 on forest stands at the present time.

ACKNOWLEDGEMENTS

We are grateful to Professor B. R. Strain for useful comments on the manuscript and to Dr R. A. Houghton and many other colleagues for helpful discussions and assistance with the literature. The substance of this review was first produced for the Department of the Environment (UK) in connection with the 1988 Toronto Conference. We gratefully acknowledge their financial sponsorship. The Department of the Environment is not in any way responsible for the opinions expressed here.

REFERENCES

Ågren, G.I. and Ingestad, T. (1987) Root:shoot ratio as a balance between nitrogen productivity and photosynthesis. *Plant, Cell and Environment* **10**, 579–586.

Akkermans, A.D.L. and van Dijk, C. (1975) The formation and nitrogen-fixing ability of *Alnus glutinosa* under field conditions. In: *Symbiotic Nitrogen Fixation in Plants* (Ed. by P.S. Nutman), pp. 511–520. Cambridge University Press, Cambridge.

Alden, T. (1971) Influence of CO_2, moisture and nutrients on the formation of lammas growth and prolepsis in seedlings of *Pinus sylvestris* L. *Studia Forestalia Suecica* **93**, 21.

Allen, L.M., Beladi, S.E. and Shinn, J.H. (1985) Modelling the feasibility of free-air carbon dioxide releases for vegetation response research. *17th Conference on Agriculture and Forest Meteorology and 7th Conference on Biometeorology and Aerobiology, Scottsdale, Arizona.* A&F 9.4, pp.161–164. American Meteorological Society, Boston.

Arovaara, H., Hari, P. and Kuusela, K. (1984) Possible effect of changes in atmospheric composition and acid rain on tree growth. *Communicationes Instituti Forestalis Fennicae* **122**, 15.

Ausubel, J.H. and Nordhaus, W.D. (1983) A Review of Estimates of Future CO_2 Emissions. In: *Changing Climate.* Report of the CO_2 assessment committee. National Academy Press: Washington.

Axelsson, B. (1985) Increasing Forest Productivity and Value. In: *Forest Potentials, Productivity and Value* (Ed. by R. Ballard, P. Farnum, G.A. Ritchie and J.K. Winjum), Weyerhaeuser Science Symposium 4, pp. 5–37. Weyerhaeuser Co: Centralia.

Axelsson, E. and Axelsson, B. (1986) Changes in carbon allocation patterns in spruce and pine trees following irrigation and fertilization. *Tree Physiology* **2**, 189–204.

Baldocchi, D.D., Verma, S.B. and Anderson, D.E. (1987) Canopy photosynthesis and water use efficiency in a deciduous forest. *Journal of Applied Ecology* **24**, 251–260.

Beadle, C.L., Jarvis, P.G. and Neilson, R.E. (1979) Leaf conductance as related to xylem water potential and carbon dioxide concentration in Sitka spruce. *Physiologia Plantarum* **45**, 158–166.

Beadle, C.L., Neilson, R.E., Jarvis, P.G. and Talbot, H. (1981) Photosynthesis as related to xylem water potential and CO_2 concentration in Sitka spruce. *Physiologia Plantarum* **52**, 391–400.

Berry, J.A. and Downton, W.J.S. (1982) Environmental Regulation of Photosynthesis. In: *Photosynthesis II. Development, Carbon Metabolism, and Plant Productivity* (Ed. by Govindjee), pp. 263–344. Academic Press, New York.

Bolin, B. (1977) Changes of land biota and their importance to the carbon cycle. *Science* **196**, 613–615.

Bolin, B. (1986) How much CO_2 will remain in the atmosphere? In: *The Greenhouse Effect, Climatic Change and Exosystems* (Ed. by B. Bolin, B.R.O. Doos, J. Jager and R.A. Warrick), Scope 29, pp. 93–156. Wiley, Chichester.

Brix, H. (1968) The influence of light intensity at different temperatures on rate of respiration of Douglas-fir seedlings. *Plant Physiology* **43**, 389–393.

Brown, K. and Higginbotham, K.O. (1986) Effects of carbon dioxide enrichment and nitrogen supply on growth of boreal tree seedlings. *Tree Physiology* **2**, 223–232.

Caemmerer, S. von and Farquhar, G. (1981) Some relationships between the biochemistry of photosynthesis and the gas exchange of leaves. *Planta* **153**, 376–387.

Caemmerer, S. von and Farquhar, G. (1984) Effects of partial defoliation, changes in irradiance during growth, short term water stress and growth at enhanced p(CO_2) on photosynthetic capacity of leaves of *Phaseolus vulgaris*. *Planta* **160**, 320–329.

Canham, A.E. and McCavish, W.J. (1981) Some effects of CO_2, day length and nutrition on the growth of young forest tree plants. I. In the seedling stage. *Forestry* **54**, 169–182.

Cannell, M.G.R., Milne, R., Sheppard, L.J. and Unsworth, M.H. (1987) Radiation interception and productivity of willow. *Journal of Applied Ecology* **24**, 261–278.

Cannell, M.G.R., Sheppard, L.J. and Milne, R. (1988) Light use efficiency and woody biomass production of poplar and willow. *Forestry* **61**, 125–136.

Chang, C.W. (1975) Carbon dioxide and senescence in cotton plants. *Plant Physiology* **55**, 515–519.

Conroy, J., Barlow, E.W.R. and Bevege, D.I. (1986a) Response of *Pinus radiata* seedlings to carbon dioxide enrichment at different levels of water and phosphorus: growth, morphology and anatomy. *Annals of Botany* **57**, 165–177.

Conroy, J.P., Smillie, R.M., Kuppers, M., Bevege, D.I. and Barlow, E.S. (1986b) Chlorophyll *a* fluorescence and photosynthetic growth responses of *Pinus radiata* to P deficiencies, drought stress and high CO_2. *Plant Physiology* **81**, 423–429.

Conway, T.J., Tans, P., Waterman, L.S., Thoning, K.W., Masarie, K.A. and Gammon, R.M. (1988) Atmospheric carbon dioxide measurements in the remote global troposphere, 1981–1984. *Tellus* **40B**, 81–115.

Crane, A.J. (1985) Possible effects of rising CO_2 on climate. *Plant, Cell & Environment* **8**, 371–379.

Cure, J.D. and Acock, B. (1986) Crop responses to CO_2 doubling: a literatue survey. *Agricultural and Forest Meteorology* **38**, 127–145.

Davis, J.E., Arkebauer, T.J., Norman, J.M. and Brandle, J.R. (1987) Rapid field measurement of the assimilation rate versus internal CO_2 concentration relationship in green ash (*Fraxinus pennsylvanica* Marsh): the influence of light intensity. *Tree Physiology* **3**, 387–392.

Dc Lucia, E.II. (1987) The effect of freezing nights on photosynthesis, stomatal conductance and internal CO_2 concentration in seedlings of Engelmann spruce (*Picea Engelmannii* Parry). *Plant, Cell and Environment* **10**, 333–338.

DeLucia, E.H., Sasek, T.W. and Strain, B.R. (1987) Photosynthetic inhibiton after long term exposure to elevated levels of CO_2. *Photosynthesis Research* **7**, 175–184.

Downton, W.J.S., Grant, W.J.R. and Loveys, B.R. (1987) Carbon dioxide enrichment increases yield of Valencia orange. *Australian Journal of Plant Physiology* **14**, 493–501.

Eamus, D. (1986a) Further evidence in support of an interactive model in stomatal control. *Journal of Experimental Botany* **37**, 657–665.

Eamus, D. (1986b) The response of leaf water potential and leaf diffusive resistance to abscisic acid, water stress and low temperature in *Hibiscus esculentus*. The effect of water stress and ABA pre-treatments. *Journal of Experimental Botany* **37**, 1854–1862.

Edwards, G. and Walker, D. (1983) *C_3, C_4: Mechanisms, and Cellular and Environmental Regulation, of Photosynthesis*. Blackwell Scientific Publications, Oxford.

Ehret, I.L. and Jollife, P.A. (1985) Leaf injury to bean plants grown in carbon dioxide enriched atmospheres. *Canadian Journal of Botany* **63**, 2015–2020.

Evans, J.R. (1988) Photosynthesis and nitrogen relations in leaves of C_3 plants. *Oecologia* **78**, 9–19.

FAO (1982) *World Forest Products Demand and Supply 1990 and 2000*. FAO, Rome.

Fetcher, N., Jaeger, C.H., Strain, B.R. and Sionit, N. (1988) Long-term elevation of atmospheric CO_2 concentrtion and the carbon exchange rates of saplings of *Pinus taeda* L. and *Liquidambar styraciflua* L. *Tree Physiology* **4**, 255–262.

Fifield, R. (1988) Frozen assets of the ice cores. *New Scientist* **1608**, 28–29.

Funsch, R.W., Mattson, R.H. and Mowry, G.R. (1970) CO_2-supplemented atmosphere increases growth of *Pinus strobus* seedlings. *Forest Science* **16**, 459–460.

Gates, D.M. (1983) An overview. In: *CO_2 and Plants: The Response of Plants to Rising Levels of Atmosphere Carbon Dioxide* (Ed. by E.R. Lemon) AAAS Selected Symposia, 84, pp. 7–20. Westview Press: Boulder.

Gaudillère, J.-P. and Mousseau, M. (1989) Short-term effect of CO_2 enrichment in leaf development and gas exchange of young poplars (*Populus euramericana*) cv I 214). *Acta Oecologia/Oecoogia Plantarum* **10**, 95–105.

Gifford, R.M. (1982) Global Photosynthesis in Relation to our Food and Energy Needs. In: *Photosynthesis, Vol. 2, Development, Carbon metabolism and Plant Production* (Ed. by Govindjee), pp. 459–495. Academic Press: London.

Grace, J.C., Jarvis, P.G. and Norman, J.M. (1987) Modelling the interception of solar radiant energy in intensively managed stands. *New Zealand Journal of Forestry Science* **17**, 193–209.

Hamburg, S.P. and Cogbill, C.U. (1988) Historical decline of red spruce populations and climatic warming. *Nature* **331**, 428–430.

Hårdh, J.E. (1967) Trials with carbon dioxide, light and growth substances on forest tree plants. *Acta Forestalia Fennica* **82**, 1–10.

Hari, P. and Arovaara, H. (1988) Detecting CO_2-induced enhancement in the radial increment of trees. Evidence from the Northern Timer line. *Scandinavian Journal of Forest Research* **3**, 67–74.

Higginbotham, K.O., Mayo, J.M., L'Hirondelle, S. and Krystofiak, D.K. (1985) Physiological ecology of lodgepole pine (*Pinus contorta*) in an enriched CO_2 environment. *Canadian Journal of Forest Research* **15**, 417–421.

Hollinger, D.Y. (1987) Gas exchange and dry matter allocation responses to elevation of atmospheric CO_2 concentration in seedlings of three tree species. *Tree Physiology* **3**, 193–202.

Houghton, R.A. (1987) Terrestrial metabolism and CO_2 concentrations. *BioScience* **37**, 672–678.

Houghton, R.A., Hobbie, J.E., Melillo, J.M., Moore, B., Peterson, B.J., Shaver, G.R. and Woodwell, G.M. (1983) Changes in the carbon content of terrestrial biota and soils between 1860 and 1980: a net release of CO_2 to the atmosphere. *Ecological Monographs* **53**, 235–262.

Houghton, R.A., Boone, R.D., Fruchi, J.R. *et al.* (1987) The flux of carbon from terrestrial ecosystems to the atmosphere in 1980 due to changes in land use: geographic distribution of the global flux. *Tellus* **39B**, 122–139.

Ingestad, T. (1982) Relative addition rate and external concentration: Driving variable used in plant nutrition research. *Plant, Cell and Environment* **5**, 443–453.

Ingestad, T. (1987) New concepts in soil fertility and plant nutrition as illustrated by research on forest trees and stands. *Geoderma* **40**, 237–252.

Ingestad, T. and Ågren, G.I. (1988) Nutrient uptake and allocation at steady-state nutrition. *Physiologia Plantarum* **72**, 450–459.

Ingestad, T. and Kahr, M. (1985) Nutrition and growth of coniferous seedlings at varied relative nitrogen addition rates. *Physiologia Plantarum* **65**, 109–116.

Ingestad, T. and Lund, A.-B. (1986) Theory and techniques for steady state mineral nutrition and growth of plants. *Scandinavian Journal of Forest Research* **1**, 439–453.

Jarvis, P.G. (1985a) Increasing productivity and value of temperate coniferous forest by manipulating site Water Balance. In: *Forest Potentials, Productivity and Value* (Ed. by R. Ballard, P. Farnum, G.A. Ritchie and J.K. Winjum), Weyhaeuser Science Symposium, 4, pp. 39–74. Weyerhaeuser Co: Centralia.

Jarvis, P.G. (1985b) Transpiration and assimilation of tree and agricultural crops: the 'omega factor'. In: *Attributes of Trees as Crop Plants* (Ed. by M.G.R. Cannell and J.E. Jackson), pp. 460–480. Institute of Terrestrial Ecology: Abbots Ripton.

Jarvis, P.G. (1986) Coupling of carbon and water interactions in forest stands. *Tree Physiology* **2**, 347–369.

Jarvis, P.G. (1987) Water and carbon fluxes in ecosystems. *Ecological Studies* **61**, 50–67.

Jarvis, P.G. (1989) Atmospheric carbon dioxide and forests. In: *Forests, Weather and Climate. Philosophical Transactions of the Royal Society, Series B* (Ed. by P.G. Jarvis, J.L. Monteith, M.H. Unsworth and J. Shuttleworth), (in the press).

Jarvis, P.G., James, G.B. and Landsberg, J.T. (1976) Coniferous Forest. In: *Vegetation and the Atmosphere, Vol. 2 Case Studies* (Ed. by J.L. Monteith), pp. 171–240. Academic Press, London.

Jarvis, P.G. and Jarvis, M.S. (1964) Growth rates of woody plants. *Physiologia Plantarum* **17**, 654–666.

Jarvis, P.G. and Leverenz, J.W. (1983) Productivity of Temperate, Deciduous and Evegreen Forest. In: *Encyclopedia of Plant Physiology New Series Vol 12, Physiological Plant Ecology IV* (Ed. by O.L. Lange, P.S. Nobel, C.B. Osmond and H. Ziegler), pp. 233–280. Springer-Verlag, Berlin.

Jarvis, P.G. and McNaughton, K.G. (1986) Stomatal control of transpiration. *Advances in Ecological Research* **15**, 1–29.

Jarvis, P.G. and Sandford, A.P. (1986) Temperate forests. In: *Photosynthesis in Contrasting Environments* (Ed. by N.R. Baker and S.P. Long), pp. 199–236. Elsevier, Amsterdam.

Jia, H.-J. and Ingestad, T. (1984) Nutrient requirements and stress response of *Populus simonii* and *Paulownia tomentosa*. *Physiologia Plantarum* **62**, 117–124.

Johnson, J.D. and Ferrell, W.K. (1983) Stomatal response to vapour pressure deficit and the effect of plant water stress. *Plant, Cell and Environment* **6**, 451–456.

Jones, H.G. and Fanjul, L. (1983) Effects of water stress on CO_2 in apple. In: *Effects of Stress on Photosynthesis* (Ed. by R. Marcelle, H. Clijsters and M. van Poucke), pp. 75–93. Martinus Nijhoff/Dr W. Junk, The Hague.

Jones, P.D., Wigley, T.M.L. and Wright, P.B. (1986) Global temperature variations between 1861 and 1984. *Nature* **322**, 430–434.

Jones, P.D., Wigley, T.M.L., Folland, C.K. *et al.* (1988) Evidence for global warming in the past decade. *Nature* **332**, 790.

Keepin, W., Mintzer, I. and Kristoferson, L. (1986) Emission of CO_2 into the atmosphere. The rate of release of CO_2 as a function of future energy developments. In: *The Greenhouse Effect, Climatic Change and Ecoystems* (Ed. by B. Bolin, B.R. Doos, J. Jager *et al.*), Scope 29, pp. 35–92. Wiley, Chichester.

Kienast, F. and Luxmore, R.J. (1988) Tree-ring analysis and conifer growth responses to increased atmospheric CO_2-levels. *Oecologia* **76**, 487–495.

Koch, K.E., Jones, P.H., Avigne, W.T. and Allen, L.H. (1986) Growth, dry matter partitioning, and diurnal activities of RuBP carboxylase in citrus seedlings maintained at two levels of CO_2. *Physiologia Plantarum* **67**, 477–484.

Kramer, P.J. (1981) Carbon dioxide concentration, photosynthesis, and dry matter production. *BioScience* **31**, 29–33.

Kramer, P.J. (1983) *Water Relations of Plants.* Academic Press, New York.

Kramer, P.J. and Sionit, N. (1987) Effects of increasing CO_2 concentration on the physiology and growth of forest trees. In: *The Greenhouse Effect, Climate change and U.S. Forests* (Ed. by W.E. Shands and J.S. Hoffman). The Conservation Foundation, Washington DC.

Krizek, D.T., Zimmerman, R.H., Klueter, H.H. and Bailey, W.A. (1971) Growth of crab apple seedlings in controlled environments. Effect of CO_2 levels, and time and duration of CO_2 treatment. *Journal of American Society for Horticultural Science* **96**, 285–288.

Ku, S.B., Edwards, G.E. and Tanner, C.B. (1977) Effects of light, carbon dioxide and temperature on photosynthesis, oxygen inhibition of photosynthesis and transpirtion in *Solanum tuberosum. Plant Physiology* **59**, 868–872.

Laiche, A.J. (1978) Effects of refrigeration, CO_2 and photoperiod on the initial and subsequent growth of rooted cuttings of *Ilex cornuta* Lindl. et Paxt. cv. Burfordii. *Plant Propagation* **24**, 8–10.

Lamarche, V.C., Graybill, D.A., Fritts, H.C. and Rose, M.R. (1984) Increasing atmospheric carbon dioxide: tree ring evidence for growth enhancement in natural vegetation. *Science* **225**, 1019–1021.

Lin, W.C. and Molnar, J.M. (1982) Supplementary lighting and CO_2 enrichment for accelerated growth of selected woody ornamental seedlings and rooted cuttings. *Canadian Journal of Plant Science* **62**, 703–707.

Linder, S. (1985) Potential and Actual Production in Australian Forest Stands. In: *Research for Forest Management* (Ed. by J.J. Landsberg and W. Parsons), pp. 22–35. CSIRO, Melbourne.

Linder, S. (1987) Responses to water and nutrients in coniferous ecosystems. *Ecological Studies* **61**, 180–202.

Linder, S., McMurtrie, R.E. and Landsberg, J.J. (1986) Growth of eucalyptus: a mathematical model applied to *Eucalyptus globulus*. In: *Crop Physiology of Forest Trees* (Ed. by P.M.A. Tigerstedt, P. Puttonen and V. Koshi), pp. 107–127. Helsinki Univesity Press, Helsinki.

Ludlow, M.M. and Jarvis, P.G. (1971) Photosynthesis in Sitka spruce (*Picea sitchensis* (Bong) Carr.) I. General characteristics. *Journal of Applied Ecology* **8**, 925–953.

Luxmore, R.J. (1981) CO_2 and phytomass. *BioScience* **31**, 626.

Luxmore, R.J., O'Neil, E.G., Ellis, J.M. and Rogers, H.H. (1986) Nutrient uptake and growth responses of Virginia pine to elevated atmospheric carbon dioxide. *Journal of Environmental Quality* **15**, 244–251.

Luxmore, R.J., Tharp, M.L. and West, D.C. (1989) Simulating the physiological basis of tree ring responses to environmental changes. In: *Forest Growth: Process Modelling of Responses to Environmental Stress*. Timber Press, Alabama.

Makela, A. (1986) Partitioning coefficients in plant models with turn-over. *Annals of Botany* **57**, 291–297.

Marland, G. (1988) *The prospect of solving the CO_2 problem through global reforestation*. US Department of Energy TRO39, p. 66. US Department of Commerce: Springfield, Virginia.

Mauney, J.R., Guinn, G., Fry, K.E. and Hesketh, J.D. (1979) Correlation of photosynthetic carbon dioxide uptake and carbohydrate accumulation in cotton, soyabean, sunflower and sorghum. *Photosynthetica* **13**, 260–266.

McMurtrie, R.E. and Wolf, L. (1983) Above and below-ground growth of forest stands: a carbon budget model. *Annals of Botany* **52**, 437–448.

Monteith, J.L. (1977) Climate and efficiency of crop production in Britain. *Philosophical Transactions of the Royal Society of London, Series B* **281**, 277–294.

Morison, J.I.L. (1985) Sensitivity of stomata and water use efficiency to high CO_2. *Plant, Cell and Environment* **8**, 467–474.

Morison, J.I.L. and Gifford, R.M. (1983) Stomatal sensitivity to CO_2 and humidity. *Plant Physiology* **71**, 789–796.

Mortensen, L.M. (1983) Growth response of some greenhouse plants to environment. VIII. Effect of CO_2 on photosynthesis and growth of Norway spruce. *Meldinger fra Norges Landbrukshogskole* **62**(10), 1–13.

Mortensen, L.M. and Sandvik, M. (1987) Effects of CO_2 enrichment at varying photon flux density on the growth of *Picea abies* (L.) Karst. seedlings. *Scandinavian Journal of Forest Research* **2**, 325–334.

Mott, K.A. (1988) Do stomata respond to CO_2 concentrations other than intercellular? *Plant Physiology* **86**, 200–203.

Mousseau, M. and Enoch, H.Z. (1989) Effect of doubling atmospheric CO_2 concentration on growth, dry matter distribution and CO_2 exchange of two-year-old sweet chestnut seedlings (*Castanea sativa* Mill.). *Plant, Cell and Environment* (in the press).

Neftel, A., Moor, E., Oeschger, H. and Stauffer, B. (1985) Evidence from polar ice cores for the increase in atmospheric CO_2 in the past 2 centuries. *Nature* **315**, 45–47.

Norby, R.J. (1987) Nodulation and nitrogenase activity in nitrogen-fixing woody plants stimulated by CO_2 enrichment of the atmosphere. *Physiologia Plantarum* **71**, 77–82.

Norby, R.J., O'Neill, E.G. and Luxmore, R.J. (1986a) Effects of atmospheric CO_2 enrichment on the growth and mineral nutrition of *Quercus alba* seedlings in nutrient poor soil. *Plant Physiology* **82**, 83–89.

Norby, R.J., Pastor, J. and Melillo, J. (1986b) Carbon–nitrogen interactions in CO_2-enriched white oak: physiological and long-term perspectives. *Tree Physiology* **2**, 233–241.

Norby, R.J., O'Neill, E.G., Hood, W.G. and Luxmore, R.J. (1987) Carbon allocation, root exudation and mycorrhizal colonization of *Pinus echinata* seedlings grown under CO_2 enrichment. *Tree Physiology* **3**, 203–210.

Oberbauer, S.F., Sionit, N., Hastings, S.J. and Oechel, W.R. (1986) Effects of CO_2 enrichment and nutrition on growth, photosynthesis and nutrient concentrations of Alaskan tundra species. *Canadian Journal of Botany* **64**, 2993–2998.

Oberbauer, S.F., Strain, B.R. and Fetcher, N. (1985) Effect of CO_2 enrichment on seedling physiology and growth of two tropical species. *Physiologia Plantarum* **65**, 352–356.

O'Neill, E.G., Luxmore, R.J. and Norby, R.J. (1987a) Elevated atmospheric CO_2 effects on seedling growth, nutrient uptake, and rhizosphere bacterial populations of *Liriodendron tulipifera* L. *Plant and Soil* **104**, 3–11.

O'Neill, E.G., Luxmore, R.J. and Norby, R.J. (1987b) Increases in mycorrhizal colonization and seedling growth in *Pinus echinata* and *Quercus alba* in an enriched CO_2 atmosphere. *Canadian Journal of Forest Research* **17**, 878–883.

Pearcy, R.W. and Björkman, O. (1983) Physiological Effects. In: *CO_2 and Plants: The Response of Plants to Rising Levels of Atmospheric Carbon Dioxide* (Ed. by E.R. Lemon), AAAS Selected Symposia, Vol. 84, pp. 65–105. Westview Press, Boulder.

Pearman, G.I., Etheridge, D., de Silva, F. and Fraser, P.J. (1986) Evidence of changing concentrations of CO_2, N_2O and CH_4 from air bubbles in Antartic ice. *Nature* **320**, 248–250.

Peet, M.M., Huber, S.C. and Patterson, I.T. (1985) Acclimation to high CO_2 in monoecious cucumber. II. Alterations in gas exchange rates, enzyme activities and starch and nutrient concentrations. *Plant Physiology* **80**, 63–67.

Porter, M.A. and Grodzinsky, B. (1984) Acclimation to high CO_2 in bean. *Plant Physiology* **74**, 413–416.

Raper, P.C. and Peedin, G.F. (1978) Photosynthetic rate during steady state growth as influenced by CO_2 concentration. *Botanical Gazette* **139**, 147–149.

Regehr, D.C., Bazzaz, F.A. and Boggess, W.R. (1975) Photosynthesis, transpiration and leaf conductance of *Populus deltoides* in relation to flooding and drought. *Photosynthetica* **9**, 52–61.

Reynolds, J.F. and Thornley, J.H.M. (1982) A shoot:root partitioning model. *Annals of Botany* **49**, 585–597.

Roberts, D.R. and Dumbroff, E.F. (1986) Relationships among drought resistance, transpiration rates and abscisic acid levels in 3 northern conifers. *Tree Physiology* **1**, 161–167.

Roberts, J. (1983) Forest transpiration: a conservative process? *Journal of Hydrology* **66**, 133–141.

Rogers, H.H., Thomas, J.F. and Bingham, G.E. (1983) Response of agronomic and forest species to elevated atmospheric carbon dioxide. *Science* **220**, 428–429.

Russell, G., Jarvis, P.G. and Monteith, J.L. (1988) Absorption of Radiation by Canopies and Stand Growth. In: *Plant Canopies: their growth form and function* (Ed. by G. Russel, B. Marshall and P.G. Jarvis), pp. 21–39. Cambridge University Press, Cambridge.

Rutter, A.J. (1957) Studies in the growth of young plants of *Pinus sylvestris*. I. The annual cycle of assimilation and growth. *Annals of Botany* **21**, 399–426.

Sage, R.F. and Sharkey, T.D. (1987) The effect of temperature on the occurrence of O_2 and CO_2-insensitive photosynthesis in field grown plants. *Plant Physiology* **84**, 658–664.

Sage, R.F., Sharkey, T.D. and Seemann, J.R. (1988) The in-vivo response of the ribulose-1,5-bisphosphate carboxylase activation state and the pool sizes of photosynthetic metabolites to elevated CO_2 in *Phaseolus vulgaris* L. *Planta* **174**, 407–416.

Sage, R.F., Sharkey, T.D. and Seemann, J.R. (1989) Acclimation of photosynthesis to elevated CO_2 in five C_3 species. *Plant Physiology* **89**, 590–596.

Sasek, T.W., DeLucia, E.H. and Strain, B.R. (1985) Reversibility of photosynthetic inhibition in cotton after long-term exposure to elevated CO_2 concentrations. *Plant Physiology* **78**, 619–622.

Schulte, P.J. and Hinckley, T.M. (1987) Abscisic acid relations and the response of *Populus trichocarpa* stomata to leaf water potential. *Tree Physiology* **3**, 103–113.

Schulte, P.J., Hinckley, T.M. and Stettler, R.F. (1987) Stomatal responses of *Populus* to leaf water potential. *Canadian Journal of Botany* **65**, 255–260.

Shugart, H.H., Antonovsky, M.Ja., Jarvis, P.G. and Sandford, A.P. (1986) CO_2, climatic change and forest ecosystems. In: *The Greenhouse Effect, Climatic Change and Ecosystems*, Scope 29 (Ed. by B. Bolin, B.R. Doos, J. Jager and R.A. Warrick), pp. 475–521. Wiley, Chichester.

Simon, J.-P., Potvin, G. and Strain, B.R. (1984) Effects of temperature and CO_2 enrichment on kinetic properties of phospho-enol-pyruvate carboxylase in two ecotypes of *Enchinochloa crus-galli* (L). Beaur. a C4 weed grass species. *Oecologia* **63**, 145–152.

Sionit, N. and Kramer, P.J. (1986) Woody plant reaction to CO_2 enrichment. In: *CO_2 Enrichment and Greenhouse Crops* (Ed. by H.Z. Enoch and B.A. Timball), Vol. II, pp. 69–85. CRC Press, Boca Raton.

Sionit, N., Strain, B.R. and Hellmers, H. (1981) Effects of different concentrations of atmospheric CO_2 on growth and yield components of wheat. *Journal of Agricultural Science* **79**, 335–339.

Sionit, N., Strain, B.R., Hellmers, H., Riechers, G.H. and Jaeger, C.H. (1985) Long-term atmospheric CO_2 enrichment affects the growth and development of *Liquidambar styraciflua* and *Pinus taeda* seedlings. *Canadian Journal of Forest Research* **15**, 468–471.

Siren, G. and Alden, T. (1972) *CO_2 supply and its effect on the growth of conifer seedlings grown in plastic greenhouses.* Research Note 37, p. 15. Department of Reforestation, Royal College of Foresty, Sweden.

Smagorinsky, J. (1983) Effects of Climate. In: *Changing Climate*, Report of the CO_2 Assessment Committee. National Research Council, USA, National Academy Press, USA.

Strain, B.R. (1985) Physiological and ecological control on carbon sequestering in ecosystems. *Biogeochemistry* **1**, 219–232.

Strain, B.R. and Cure, J.D. (Eds.) (1986) Direct Effects of atmospheric CO_2 on plants and ecosystems: a bibliography with abstracts. ORNL/CDIC-13. NTIS, US Department of Commerce, Springfield.

Strain, B.R. and Bazzaz, F. (1983) Terrestrial plant communities. In: *CO_2 and Plants: The response of Plants to Rising Levels of Atmospheric Carbon Dioxide* (Ed. by E.R. Lomon), AAAS Selected Symposium Vol. 34, pp. 177–222. Westview Press: Boulder.

Strand, M. and Öquist, G. (1985) Inhibition of photosynthesis by freezing temperatures and high light levels in cold acclimated seedlings of Scots pine (*Pinus sylvestris*). *Physiologia Plantarum* **64**, 425–430.

Surano, K.A., Daley, P.F., Houpis, J.L.J. *et al.* (1986) Growth and physiological responses of *Pinus ponderosa* Dougl. ex P. Laws. to long-term elevated CO_2 concentrations. *Tree Physiology* **2**, 243–259.

Telewski, F.W. and Strain, B.R. (1987) Densiometric and ring width analysis of 3-year-old *Pinus taeda* L. and *Liquidambar styraciflua* L. grown under three levels of CO_2 and two water regimes. In: *Proceedings of the International Symposium on Ecological Aspects of Tree Ring Analysis* (Ed. by G.C. Jacoby and J.W. Hornbeck). DoE CONF-8608144. NTIS, Springfield, Virginia.

Tenhunen, J.D., Lange, O.L., Gebel, J., Beyschlag, W. and Weber, J.A. (1984) Changes in photosynthetic capacity, carboxylation efficiency, and CO_2 compensation point associated with midday stomatal closure and midday depression of net CO_2 exchange of leaves of *Quercus suber*. *Planta* **162**, 193–203.

Teskey, R.O., Fites, J.A., Samuelson, L.J. and Bongarten, B.C. (1986) Stomatal and non-stomatal limitations to net photosynthesis in *Pinus taeda* under different environmental conditions. *Tree Physiology* **2**, 131–142.

Thomas, J.F. and Harvey, C.N. (1983) Leaf anatomy of four species grown under continuous long-term CO_2 enrichment. *Botanical Gazette* **144**, 303–309.

Tinus, R.W. (1972) CO_2-enriched atmosphere speeds growth of ponderosa pine and blue spruce seedlings. *Tree Planters Notes* **23**, 12–15.

Tolbert, N.E. and Zeltich, I. (1983) Carbon metabolism. In: *CO_2 and Plants: The Response of Plants to Rising Levels of Atmospheric Carbon Dioxide* (Ed. by E.R. Lemon), AAAS Selected Symposia Vol. 84, pp. 21–64. Westview Press, Boulder.

Tolley, L.C. and Strain, B.R. (1984a) Effects of CO_2 enrichment on growth of *Liquidambar styraciflua* and *Pinus taeda* seedlings under different irradiance levels. *Canadian Journal of Forest Research* **14**, 343–350.

Tolley, L.C. and Strain, B.R. (1984b) Effects of CO_2 enrichment and water stress on growth of *Liquidambar styraciflua* and *Pinus taeda* seedlings. *Canadian Journal of Botany* **62**, 2135–2139.

Tolley, L.C. and Strain, B.R. (1985) Effects of CO_2 enrichment and water stress on gas exchange of *Liquidambar styraciflua* and *Pinus taeda* seedlings grown under different irradiance levels. *Oecologia* **65**, 166–172.

Verma, S.B., Baldocchi, D.D., Anderson, D.E., Matt, O.R. and Clement, R.J. (1986) Eddy fluxes of CO_2, water vapour and sensible heat over a deciduous forest. *Boundary-Layer Meteorology* **367**, 71–91.

Wang, Y.-P. (1988) *Crown structure, radiation absorption, photosynthesis and transpiration.* PhD Thesis, University of Edinburgh.

Wang, Y.-P. and Jarvis, P.G. (1989) Description and analysis of an array model—MAESTRO. *Agricultural and Forest Meteorology* (submitted).

Warrick, R.A., Gifford, R.M. and Parry, M.L. (1986) Assessing the response of food crops to the direct effects of increased CO_2 and climatic change. In: *The Greenhouse Effect, Climatic Change and Ecosystems*, Scope 29 (Ed. by B. Bolin, B.R. Doos, J. Jaeger and R.A. Warrick), pp. 393–473. Wiley, Chichester.

Warrit, B., Landsberg, J.J. and Thorpe, M.R. (1980) Responses of apple leaf stomata to environmental factors. *Plant, Cell and Environment* **3**, 13–22.

Watson, R.L., Landsberg, J.J. and Thorpe, M.R. (1978) Photosynthetic characteristics of the leaves of golden delicious apple trees. *Plant, Cell and Environment* **1**, 51–58.

Watts, W.R. and Neilson, R.E. (1978) Photosynthesis in Sitka spruce (*Picea sitchensis* (Bong.) Carr.) VIII. Measurements of stomatal conductance and $^{14}CO_2$ uptake in controlled environments. *Journal of Applied Ecology* **15**, 245–255.

Wigley, T.M.L., Jones, P.O. and Kelly, P.M. (1986) Empirical climate studies. In: *The Greenhouse Effect, Climatic Change and Ecosystems*, Scope 29 (Ed. by B. Bolin, B.R. Doos, J. Jager and R.A. Warrick), pp. 271–322. Wiley, Chichester.

Williams, W.E., Garbutt, K., Bazzaz, F.A. and Vitousek, P.M. (1986) The response of plants to elevated CO_2 IV. Two deciduous forest communities. *Oecologia* **69**, 454–459.

Wong, S.C., Cowan, I.R. and Farquhar, G.D. (1978) Leaf conductance in relation to assimilation in *Eucalyptus pauciflora* Sieb. ex Spreng. Influence of irradiance and partial pressure of carbon dioxide. *Plant Physiology* **62**, 670–674.

Wong, S.C. and Dunin, F.X. (1987) Photosynthesis and transpiration of trees in a eucalypt forest stand: CO_2, light and humidity responses. *Australian Journal of Plant Physiology* **14**, 619–632.

Woodward, F.I. (1987) Stomatal numbers are sensitive to increase in CO_2 from pre-industrial levels. *Nature* **327**, 617–618.

Woodward, F.I. and Bazzaz, F.A. (1988) The responses of stomatal density to CO_2 partial pressure. *Journal of Experimental Botany* **39**, 1771–1781.

Woodwell, G.M., Whittaker, R.M., Reiners, W.A., Likens, G.E., Delwiche, C.C. and Botkin, D.B. (1978) The biota and the world carbon budget. *Science* **199**, 141–146.

Wulff, R.D. and Strain, B.R. (1981) Effects of CO_2 enrichment on growth and photosynthesis in *Desmodium paniculatum*. *Canadian Journal of Botany* **60**, 1084–1091.

Yeatman, C.W. (1970) CO_2-enriched air increased growth of conifer seedlings. *Forestry Chronicle* **46**, 229–230.

Originally Published in Volume 15 (this series), pp 133–302, 1986

Ecology of Coarse Woody Debris in Temperate Ecosystems

M.E. HARMON, J.F. FRANKLIN, F.J. SWANSON, P. SOLLINS, S.V. GREGORY, J.D. LATTIN, N.H. ANDERSON, S.P. CLINE, N.G. AUMEN, J.R. SEDELL, G.W. LIENKAEMPER, K. CROMACK, JR. AND K.W. CUMMINS

ADVANCES IN ECOLOGICAL RESEARCH VOL. 34
0065-2504/04 $35.00 DOI 10.1016/S0065-2504(03)34002-4

I. INTRODUCTION

Woody debris is an important, but often neglected component of many terrestrial and aquatic ecosystems. Coarse woody debris (CWD), primarily in the form of standing dead trees and downed boles and large branches, is abundant in many natural forest and stream ecosystems, forming major structural features with many crucial ecological functions—as habitat for organisms, in energy flow and nutrient cycling, and by influencing soil and sediment transport and storage.

CWD includes a wide variety of types and sizes of materials. Types of CWD include snags, logs, chunks of wood (which result from disintegration of larger snags and logs), large branches, and coarse roots. The size used to define CWD has varied widely among studies, making exact comparisons difficult. Typical minimum diameters are 7.5–15 cm in western North American studies and 2.5–7.5 cm elsewhere. Some ecologists (e.g., Christensen, 1977) make no distinction between coarse and fine woody debris. For the purposes of this article, we consider CWD as being any woody material >2.5 cm in diameter, although many of the studies that we review consider only larger material.

A brief review of the roles of CWD highlights its importance. CWD is habitat for many species, including autotrophs and heterotrophs. "Nurse logs" are a widely recognized example of logs acting as habitat for autotrophs (Kirk, 1966). Indeed, in some environments, such as the rain forests of the Pacific Northwest, logs are the major site of tree seedling establishment (McKee *et al.*, 1982). Snags (also called standing dead trees) are used by many animal taxa, particularly birds; snags are of emerging concern in wildlife research and management (e.g., Davis *et al.*, 1983). In contrast, the use of logs by forest vertebrates has not been as widely appreciated (e.g., Maser *et al.*, 1979), nor has the dependence of some vertebrates, such as salamanders, on CWD (Maser and Trappe, 1984). Invertebrate use of and dependence on both snags and logs is well known to entomologists. Finally, many decomposer bacteria and fungi utilize CWD as an energy and nutrient source as well as a habitat (Frankland *et al.*, 1982; Swift, 1977a).

The importance of CWD in energy flow and nutrient cycles of ecosystems has not always been appreciated by ecologists. Low in nutrient concentration and slow to decompose, these materials are ignored in many ecological studies despite the large amounts of organic matter represented. Although CWD may be a nutrient sink in the short run, these materials can be a major long-term source of both energy and nutrients in many ecosystems (Larsen *et al.*, 1978; McFee and Stone, 1966; Triska and Cromack, 1980; Triska *et al.*, 1984). Furthermore, although intrinsically poor in nitrogen (Merrill and Cowling, 1966), nitrogen fixation in CWD is an important source of this limiting element in both terrestrial and aquatic ecosystems (Cornaby and Waide, 1973; Sharp and Milbank, 1973).

In stream and river systems, CAD has a major influence on geomorphic processes (Swanson *et al.*, 1982a,b). Woody structures are critical in regulating sediment transport and storage. Debris accumulations in small and moderate-sized streams dissipate energy and store organic and inorganic sediment. In large streams and rivers, CWD provides a diverse array of habitats that significantly influence biological productivity (e.g., Franklin *et al.*, 1982; Sedell and Frogatt, 1984; Sedell *et al.*, 1982).

CWD is biologically important to the freshwater and estuarine ecosystems in which it occurs. In forested streams, CWD serves as an energy and nutrient source, a site for nitrogen fixation, and habitat for organisms (Triska *et al.*, 1982). CWD influences the physical structure of forest streams, creating a variety of habitats—debris dams, plunge pools, and gravel and sandbars; as much as 50% of the habitat in small forested streams of the Pacific Northwest may be provided or controlled by CWD (Swanson and Lienkaemper, 1978). CWD is instrumental in retaining organic material, particularly forest litter, in small streams until it can be utilized by aquatic organisms (Bilby and Likens, 1980; Cummins, 1979).

Despite the role of CWD as habitat, an influence on geomorphic processes, and an integral component of energy flow and nutrient cycles, this material has often been ignored. Some of this "neglect" may have arisen because of the many problems associated with the study of CWD; it varies widely in space and time, which creates sampling difficulties, and is massive, making manipulative experiments difficult. The rate of CWD production is difficult to measure, requiring long periods of observation over large areas. Decomposition is slow and also requires long periods of time for investigation.

This article partially parallels the structure of a forest-ecosystem compartment model (Figure 1). We emphasize CWD in temperate forest and stream ecosystems and include only selected references on CWD in tropical and boreal ecosystems for comparative purposes. In Section II, the rate at which CWD is added to ecosystems via tree mortality and breakage of stems and branches, as well as by physical transport from adjacent ecosystems, is considered. As soon as CWD is created, biological and physical processes begin to degrade it. Section III examines the processes involved in decay and decomposition rates of CAD. Of these, respiration and fragmentation have received most study, whereas leaching, burial, and physical transport have received least. In Section IV, we examine how rates of tree death and decomposition as well as disturbances to forests control the amounts of CAD in temperate ecosystems. The functional importance of CWD depends not only on the amount of CWD, but also on its distribution in terms of size, spatial arrangement, degree of decay, species, and position (i.e., snags

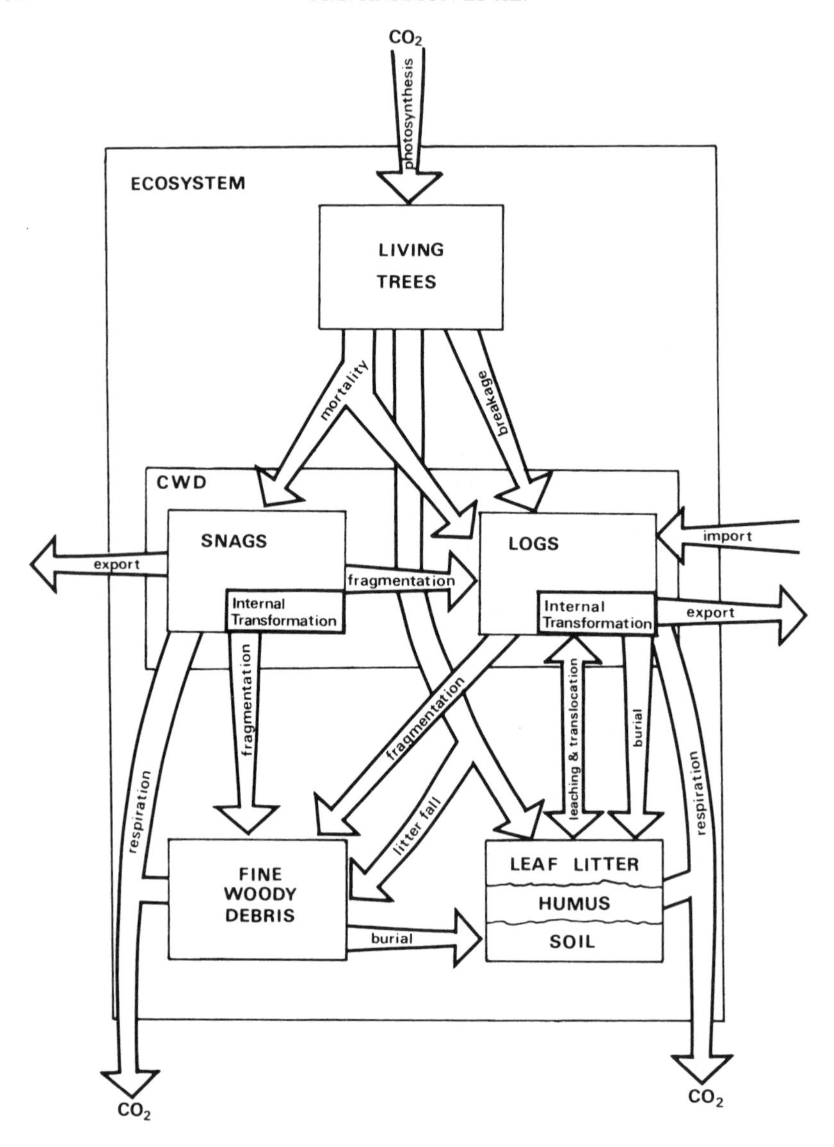

Figure 1 The flow of CWD into, within, and from an ecosystem. Physical transport (e.g., floods) adds CWD to an ecosystem, while respiration and physical transport are the major pathways of loss from an ecosystem. Within an ecosystem, CWD is added from live trees by mortality and breakage. Fragmentation and, to a far lesser extent, burial and leaching are the most important losses from CWD and result in a transfer of material to the fine woody debris and soil compartments. Fragmentation of snags also changes the form of CWD from the standing position to downed logs and large chunks of wood. Finally, numerous internal decay processes transform woody material within the snag and log compartments.

versus logs). In Section V, we focus on the influence that CWD exerts on organisms and processes within ecosystems, as a plant, vertebrate, and invertebrate habitat, as well as its role in nutrient cycles and in controlling geomorphic processes. The ecological behavior and significance of CWD varies markedly between aquatic and terrestrial environments and between managed and "natural" lands. The article concludes with a comparison of these environments.

II. INPUT OF COARSE WOODY DEBRIS

Addition or input of CWD can be considered on three levels of the ecosystem hierarchy shown in Figure 1. However, whether these processes are considered inputs or transformations depends upon the level of the hierarchy being examined. When only logs are considered, snag fragmentation is an input to logs. On the other hand, if both snags and logs are considered, then snag fragmentation is a transformation from one form of CWD to another. Similarly, tree death and breakage is an input when only CWD is considered, but is a transformation when the entire forest is examined. Finally, CWD can be moved from one part of a watershed to another, and this process may also be viewed as an input. The movement of CWD to streams from the surrounding forests is an example.

In this section, we consider the input of CWD on two levels. In forest studies, the transfer of living to dead wood is considered an input to the CWD components. In studies of streams, the transfer of CWD and living trees from streamside forests to the stream channel has also been considered an input, but obviously on, a different level than in forest studies.

A. Methods

The simplest and most direct method to assess the input of CWD from living trees is to determine tree mortality within permanently marked plots or of tagged trees (Franklin *et al.*, 1984; Grier, 1978; Harcombe, 1984; Harcombe and Marks, 1983; Harris *et al.*, 1973; MacMillan, 1981; Sollins, 1982). However, this technique alone underestimates input to CWD because large branches and broken taps of boles are not included.

Input to CWD can be measured on cleared plots. This method has been used most commonly in studying input of fine woody debris such as branches, but it is occasionally used for CWD as well (e.g., Gentry and Whitford, 1982). For large material, it is more practical to mark or map the pieces present at the start of the observation period than to remove them

(Gosz *et al.*, 1972; Swanson and Lienkaemper, 1978). Modern forest-management practices that remove all or most CWD may be potential study areas. Instead of using plots, Tritton (1980) measured the input rate of logs along line transects on which the original CWD had been removed.

These methods are designed primarily to provide estimates of input in intact stands. Stand reconstructions (Henry and Swan, 1974; Oliver and Stephens, 1977) can be used to assess input from catastrophes such as wildfires, windstorms, floods, avalanches, and insect epidemics that kill complete or major portions of stands.

B. Rates of Input

Measured input rates of CWD in forest ecosystems range from 0.12 to 30 Mg ha^{-1} $year^{-1}$ (Table 1). Input rate varies primarily with the productivity and massiveness of the trees in the ecosystem, although disturbances during the observation period can increase overall rates markedly (e.g., Wright and Lauterbach, 1958). Factors influencing measurement of input rate are the size used to define CWD, the length of the study, and the area observed. Generally, precision of measurement improves with increasing the length of the study and the size of the sample area.

Smallest input rates were observed in a scrub *Quercus nigra* stand (Gentry and Whitford, 1982) and the largest in undisturbed, old-growth coniferous stands in northwestern North America. Generally, deciduous forests appear to produce less CWD than conifer forests, although some undisturbed conifer forests such as *Pinus contorta* (Alexander, 1954) and *Pinus ponderosa* (Avery *et al.*, 1976) also have low rates of input.

Rates of CWD input to streams have been measured directly in only a few cases, all in western North America and over periods of <9 years. Lienkaemper (unpublished) monitored CWD input for 8 years for five streams ranging in size from first to fifth order in the H. J. Andrews Experimental Forest in Oregon. Input rates averaged 5.2 Mg ha^{-1} $year^{-1}$ and ranged from 2.0 to 8.8 Mg ha^{-1} $year^{-1}$, which is similar to old-growth forests for this region. However, there are situations where the input rate to streams may be larger or smaller than the transfer of live to dead wood found in old-growth forests. Earthflows and other mass-movement processes may import CWD to stream channels at rates that greatly exceed the normal transfer rate of living trees to CWD found in forests. When stream channels are wider than the canopy height, input to streams may be smaller than the transfer of living to dead wood found in adjacent forests.

All of the data reported on input of CWD, whether for forests or streams, are based on short observation periods. Although a decade is long in terms of the life of most research programs, it represents only a fraction of the

Table 1 Aboveground Input of Coarse Woody Debris for Various Temperate Ecosystems

Ecosystem	Location	Sample period (year)	Sample area (ha)	Biomass input (Mg ha^{-1} year^{-1})	Refs.[a]
		Coniferous forests			
Abies amabilis (second growth)	Washington	5	0.13	0.3	14
Picea engelmanii–Abies lasiocarpa	Colorado	11	12.8	0.18	2
Picea rubens–Abies balsamea	Maine	20	—[b]	1.45[c]	7
Picea sitchensis–Tsuga heterophylla	Oregon	40	0.4	2.8	12
Picea sitchensis–Tsuga heterophylla	Oregon	43	4.5	3.11	20
Picea sitchensis–Tsuga heterophylla	Washington	5	4	4.1	15
Picea abies–Carpinus betulus	Poland	10	1	~1.6[c]	6
Pinus banksiana	Minnesota	30	—	~2.3[c]	17
Pinus contorta	Colorado	12	8	0.17[c]	1
Pinus palustris	Georgia	1	0.2	0.79	8
Pinus ponderosa	Arizona	50	15.8	0.25	3
Pinus strobus–Acer saccharum	Minnesota	26	0.48	1.1[d]	19
Pinus echinata–Pinus virginiana	Tennessee	5	3.52	1.9	16
Pseudotsuga menziesii (old)	Oregon	2	10.2	7.0	13
Pseudotsuga menziesii (old)	Washington	29	41.6	4.54	20
Pseudotsuga menziesii (mature)	Oregon	10	80–265	0.5–30	23
Pseudotsuga menziesii (mature)	Oregon and Washington	16–46	0.2–2.8	1.55–4.25	20
Pseudotsuga–Abies–Picea	Arizona	5	—	~3.8[c]	11
		Deciduous forests			
Acer saccharum	Michigan	9	6.3	0.42[e]	5
Acer saccharum	Michigan	12	4	0.35–0.54[e]	5
Acer saccharum	Michigan	6	16	0.56[e]	5
Acer–Betula–Fagus	New Hampshire	2	—	0–14.5	22
Fagus–Acer–Betula	New Hampshire	1	13.2	1.0	10

(Continued)

Table 1 Continued

Ecosystem	Location	Sample period (year)	Sample area (ha)	Biomass input (Mg ha^{-1} year^{-1})	Refs.[a]
Liriodendron tulipifera	Tennessee	4	0.15	1.18	16
Liriodendron tulipifera	Tennessee	8	0.04	1.1	21
Populus tremuloides	New Mexico	5	3.4	0.45	9
Quercus–Carya	Tennessee	4	2.56	1.18	16
Quercus nigra	Georgia	1	0.2	0.12	8
Quercus prinus	Tennessee	4	2.88	0.55	16
Quercus robur	Denmark	2	—	0.77	4
Quercus mixed	Indiana	20	3.35	~0.64[f]	18
Quercus–Liquidambar–Liriodendron	Georgia	1	0.2	2.87	8

[a] (1) Alexander (1954), (2) Alexander (1956), (3) Avery *et al.* (1976), (4) Christensen (1977), (5) Eyre and Longwood (1951), (6) Falinski (1978), (7) Frank and Blum (1978), (8) Gentry and Whitford (1982), (9) Gosz (1980), (10) Gosz *et al.* (1972), (11) Gottfried (1978), (12) Grier (1978), (13) Grier and Logan (1977), (14) Grier *et al.* (1981), (15) Harmon (unpublished), (16) Harris *et al.* (1973), (17) Jensen and Zasada (1977), (18) MacMillan (1981), (19) Peet (1984), (20) Sollins (1982), (21) Sollins *et al.* (1973), (22) Tritton (1980), (23) Wright and Lauterbach (1958).

[b] Not available.

[c] Assumes density of 0.4 Mg m^{-3}.

[d] Converted data from Figure 1 to biomass using equations in Ker (1980) and Tritton and Hornbeck (1982).

[e] Assumes wood density of 0.6 Mg m^{-3}.

[f] Based on addition of 2.52 trees ha^{-1} year^{-1} with a mean volume of 0.44 m^3 and a density of 0.58 Mg m^{-3}.

turnover time for CWD and, given the episodic nature of much tree mortality, such data must be viewed conservatively.

C. Spatial Patterns

The input of CWD varies spatially on a number of scales. Within a stand, mortality may be aggregated or distributed randomly or regularly (see Pielou, 1977, for discussion of terms). On this scale, it is also of interest to consider the orientation (direction) and interrelationships of the pieces. Moving to a larger scale of a watershed, it is important to consider the relationship between source areas and zones of deposition. This is very important when considering input to streams from forests, because the area of initial input does not always correspond to the final resting site.

Finally, variations in input on the regional scale may be considered. This perspective is important because the cause as well as the rate of input is apt to change from region to region.

1. Patterns within Stands

Although comprehensive studies of the spatial pattern of CWD input are lacking, input is probably not distributed randomly over a forest stand or stream reach. Many causes of mortality, such as windthrow, insects, and diseases, affect patches of trees and exhibit highly contagious spatial patterns (aggregated or clumped). Therefore, CWD generated by these agents can be expected to be aggregated. Wind often uproots several nearby trees, for example. Other examples of patches of mortality include *Phellinus weirii* root rot centers (McCauley and Cook, 1980) and fir waves observed in the northeastern United States (Sprugel, 1976). Inputs of CWD due to suppression mortality are likely to be among the most evenly or regularly distributed.

2. Patterns within Watersheds

Considerable redistribution of CWD can occur on steep slopes or in streams and rivers. Substantial transfers of CWD can also be associated with catastrophic events, such as mass soil movements, floods, snow avalanches, and volcanic eruptions.

No published data exist on the effect of slope position and steepness on input and accumulation of CWD within ecosystems. Stands located on steep slopes should tend to lose material to downslope areas and accumulate CWD from upslope areas. Stands lacking source areas for CWD (upper slope or ridgetop positions or stands with nonforested areas upslope) should tend to have low accumulations because of the net downslope movement of CWD. Conversely, stands on lower slopes retain as well as receive CWD from upslope stands and therefore have high accumulations.

The source area of CWD input to streams extends substantial distances from stream channels in mountainous areas. In third-order streams flowing through old-growth *Pseudotsuga menziesii* forests, CWD pieces came from as far as 45 m from the channel, median distances of movement were ~15 m (M. H. McDade, unpublished). The probability that a falling tree or snag will be added to a stream was found to be inversely related to the distance between the channel and the site of CWD origin. Assuming uniform tree height and a random orientation of falling, those trees or snags closest to streams are most likely to contribute CWD to the channel. Stands with shorter trees would be expected to have a narrower source area of CWD for streams, while those with taller trees would contribute from a greater distance.

3. Regional Patterns

Patterns of tree mortality vary regionally because the relative importance of catastrophic and noncatastrophic agents varies widely with forest type (White, 1979). In northwestern North America, the importance of mortality agents varies along a transect from the Pacific Ocean across the Coast and Cascade Ranges into more continental regions (Franklin *et al.*, 1984). In coastal *Picea sitchensis–Tsuga heterophylla* forests, wind-related mortality provides 70% of the stem input (Greene, 1984). Forests of *P. menziesii–T. heterophylla* in the Cascade Range have 17–47% of the stems killed by wind-related injuries. Insects and diseases are rated as proximate causes of death for less than 1% of the stems in these two environments. In Rocky Mountain *P. ponderosa* (Avery *et al.*, 1976) and mixed-conifer forests (Gottfried, 1978), wind-related death amounted to only 18 and 17% of the stems added, respectively. Insects and diseases accounted for 40% of the stems killed in the *P. ponderosa* stand (Avery *et al.*, 1976). Most CWD input in Rocky Mountain forests is as snags; most CWD in the coastal and Cascade forests is as logs.

D. Temporal Patterns

Important temporal variations in CWD input are associated with seasonal, annual, and successional time scales (Figure 2). The variation within each of these scales is probably large. A clearer understanding of variation associated with annual and successional time scales will be required before comprehensive comparisons of CWD input rates for various ecosystem types can be made.

1. Seasonal

The few studies that document seasonal patterns of CWD input show wide variation between seasons. A 16-month study of seasonal input patterns in three forest types in Georgia showed a peak in late summer and fail for *Pinus and Liriodendron* stands that was caused by hurricane winds (Gentry and Whitford, 1982); in a *Q. nigra* stand in the same area, however, there was little consistent variation from season to season.

In a tropical forest in Panama, there was a distinctive seasonal cycle of tree falls, with a maximum in August to September and a minimum in

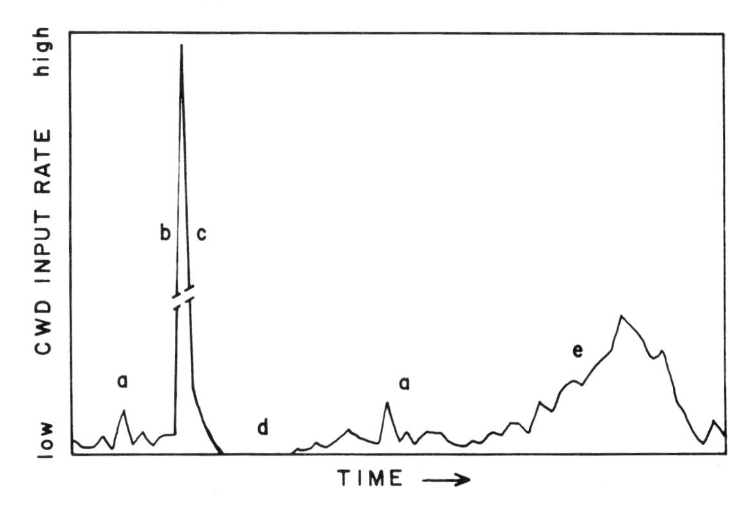

Figure 2 The hypothetical types of changes in CWD input rate over time. (a) Represents minor annual fluctuation of the input rate (10^{-1} to 10^{1} Mg ha^{-1} year^{-1}) around a long-term mean. (b) A large, rapid, but temporary increase in CWD input rate caused by a sudden, catastrophic disturbance that adds 10^{2} to 10^{3} Mg ha^{-1} during a few years. (c) Declining input rate as injured trees continue to die for a period after the disturbance. (d) Initial lack of input and then increase in input rate to long-term mean during the course of forest succession. (e) Gradual increase of CWD input rate associated with a gradually increasing disturbance (such as a pathogen).

December to March (Brokaw, 1982). The maximum rate of treefall corresponded to the middle of the wet season. Other agents of mortality that can cause seasonal variations in input rate include snow and ice, fire, insects, and floods.

2. *Annual*

Annual variation in input of CWD can be very large, even excluding catastrophic events that destroy the whole stand. Again, data are very limited because few studies of mortality have involved yearly observations. Annual studies are in progress in several localities, however (Franklin *et al.*, 1984; Harcombe and Marks, 1983), and suggest low background levels of input, with occasional spikes of mortality associated with a specific event, such as a windstorm (Harcombe, 1984) or a bark beetle epidemic (Wright and Lauterbach, 1958). In a 180-year-old *P. menziesii* stand in coastal Oregon, input, reached 69.8 m^{3} ha^{-1} year^{-1} (30 Mg ha^{-1} year^{-1}) during a bark beetle epidemic, with an average input of 39.2 m^{3} ha^{-1} year^{-1} (17 Mg ha^{-1} year^{-1})

over a 10-year period (Wright and Lauterbach, 1958). However, prior to the epidemic, the rate of input averaged ~ 0.5 Mg ha^{-1} year^{-1}.

3. *Long Term*

The amount and type (i.e., snags versus logs) of CWD input change with succession (see also Section IV.C). Many years may be required for the new stand to become fully stocked with trees and to grow boles large enough to exceed the minimum CWD size limit. These two factors markedly reduce the input rate of CWD for many years following catastrophic disturbance. The input rate and average size of pieces added to CWD generally increase with succession. However, the development of a reverse "J"-shaped size structure in older forests may cause a reduction in the size of pieces and the input rate. The type of CVWD added, either as snags or logs, and the decomposability of this material also change with succession. In young stands, suppression is a major cause of mortality and many pieces are therefore added as snags. In contrast, wind-related mortality is probably more important in older forests, shifting input to fallen boles and broken branches.

Tritton's (1980) study of succession in northern hardwood forests following clear-cutting illustrates these patterns. No CWD was input to the 10-year-old stand during the 2 years of observation. CWD accrued at a rate of 0.4 Mg ha^{-1} year^{-1} in a 20-year-old stand, and this rate increased 10-fold over the next 20 years to 4.1 Mg ha^{-1} year^{-1} in a 40-year-old stand. In stands older than 40 years, the input rate of CWD varied considerably from a minimum of 1.3 Mg ha^{-1} year^{-1} at 60 years to a maximum of 14.5 Mg ha^{-1} year^{-1} in old-growth forests. These variations may have been due, in part, to the relatively small plots and short observation period. However, as Tritton (1980) points out, the peak at 30–40 years is due to the death of large numbers of *Prunus pensylvanica*, an early successional species with a short life span. The form of input also varied with succession. During the period between 30 and 40 years, snags comprised 78–87% of the CWD input, but at other times, snags comprised $<61\%$ of the input. The peak in snag production observed by Tritton (1980) reflects the senescence of *P. pensylvanica*.

E. Agents of Mortality

1. *Wind*

Wind is an agent of mortality throughout temperate forests, generating CWD by uprooting and snapping trees and breaking branches. Trees may also be crushed and broken by pieces generated by wind.

Catastrophic windstorms, common in many temperate and tropical regions (White, 1979), result in very large inputs of CWD at irregular intervals. Effects of hurricanes have been studied in New England (Henry and Swan, 1974; Stephens, 1955) as well as in other locales in eastern North America (e.g., Reiners and Reiners, 1965) and the Carribean (e.g., Wadsworth and Englerth, 1959). Holtam (1971) describes extensive wind damage in forests of Scotland. Typhoons are characteristic of eastern Asia; the Japanese Forestry Agency (1955) describes loss of around 5×10^6 m^3 of growing stock in the Ishikari River region of Hokkaido in two typhoons in 1954. A severe windstorm on October 12, 1962 blew down an estimated 2.6×10^7 m^3 of timber in northwestern North America (Orr, 1963).

Wind also creates CWD on a much smaller scale by killing single or small clusters of trees. Chronic wind-caused mortality in many temperate forests has been reported (e.g., Falinski, 1978; Gentry and Whitford, 1982; Grier, 1978; Harcombe and Marks, 1983; Sollins, 1982). The importance of wind as an agent of mortality varies among forest types; for example, in northwestern North America, wind kills 70% of the stems in coastal *P. sitchensis*–*T. heterophylla* forests (Greene, 1984), but only 15–20% of the stems in interior *P. ponderosa* forests (Avery *et al.*, 1976). However, in other interior forests, such as *P. contorta* forests of Colorado, wind can cause up to 70% of the stem mortality (Alexander, 1954). The importance of wind also varies considerably within deciduous forests. Eyre and Longwood (1951) observed that wind-related mortality produced 87% of the CWD volume added to an *Acer saccharum* forest in Michigan. In a *Fagus*–*Magnolia* forest in Texas, wind accounted for 10% of tree death over a 5-year period (Harcombe and Marks, 1983). In part, these differences are related to differences in the variables measured: Eyre and Longwood (1951) considered volume, while Harcombe and Marks (1983) considered number of stems. The latter study found larger trees were more prone to windthrow, indicating more than 10% of the volume was input by wind.

The importance of wind also varies with topographic and edaphic conditions (Alexander, 1954; Gratkowski, 1956; Ruth and Yoder, 1953). Trees growing on wet sites are generally more susceptible to windthrow than those on dry sites because of shallow rooting. Rooting can also be shallow on drier ridgetop sites because of shallow, rocky soils, increasing the chances of windthrow, although consistent wind exposure on ridges may produce more windfirm trees (Gratkowski, 1956). Topographic position and valley configuration can also affect the chances of windthrow. For example, trees growing in saddles of main ridges or in areas of valleys that are constricted have a high probability of wind damage (Gratkowski, 1956). Heavy mortality has been observed on lee slopes with moderate slope gradient during strong windstorms (Ruth and Yoder, 1953).

Species also differ in wind firmness, although these patterns tend to be confounded with size, age, habitat, and position within a stand. Species with open-grown crowns appear to be less prone to wind damage than those with dense crowns (Boyce, 1929; Curtis, 1943; Gratkowski, 1956). The strength properties of wood are also related to susceptibility to wind breakage (Putz *et al.*, 1983). Root and heart rot structurally weaken trees and predispose them to wind damage.

2. *Fire*

Trees are killed directly by fire by stem girdling, scorching of crowns, and burning of root systems. In addition, fire indirectly contributes to other causes of mortality. By causing basal wounds, fire allows decay organisms to weaken the stem, contributing to windthrow. Survivors of severe fires experience greater exposure to winds than in the original stand and may have an increased risk of wind damage. Removal of part of the crown or phloem by heat weakens trees and makes them susceptible to insect attack. Healthy trees can also be killed by insects that came from adjacent fire-killed trees (Furniss, 1936).

The amount of mortality caused by fire depends on the type of fire (i.e., ground, surface, and crown) and its intensity, as well as the species and size structure of the forest. Species vary in their tolerance to fire (Starker, 1934). However, this tolerance changes with tree size. Increasing fire intensity increases the size of the trees killed by fire (Van Wagner, 1973). The frequency and average intensity of fires varies between regions and forest types. Such differences, recently considered by Mooney *et al.* (1981) and Wright and Bailey (1982), are beyond the scope of this article.

Although fire is typically less frequent as a cause of mortality than wind, insects, and competition, it is important over the long term; the input from a single fire can be equivalent to centuries of "normal" input. For example, normal input of CWD in the Pacific Northwest ranges from 1 to 5 Mg ha^{-1} $year^{-1}$ (Sollins, 1982). In contrast, biomass of living tree stems may range from 447 to 892 Mg ha^{-1} (Grier and Logan, 1977), and if this is the assumed range of fire-caused input, then between 105 and 575 years of "normal" input could be added by a single intense fire.

3. *Insects*

Snags are input when insects kill trees. Coleoptera (beetles) larvae and adults kill trees by girdling the phloem. In the case of angiosperms, repeated defoliation is required to cause mortality (Churchill *et al.*, 1964). Even in the case of gymnosperms, repeated and rather complete defoliation must occur before mortality is assured (Wickman, 1978). An initial attack by defoliators may weaken a tree and allow a successful attack by a secondary insect, often a bark beetle (McMullen *et al.*, 1981; Wickman, 1978). Insects also cause mortality when they introduce pathogens, as in the case of Dutch elm disease [*Ceratocystis ulmi* (Buisman) C. Moreau] being spread by *Scolytus multistriatus* (Marsham). Although sucking insects usually do not cause mortality, some such as the balsam woolly aphid [*Adelges picea* (Ratz)] may alter the host tree's anatomy to the point of causing death (Balch *et al.*, 1964).

Insects periodically reach epidemic proportions and may add a large amount of CWD over an extensive area. Outbreaks covering areas up to 1.9×10^7 ha have been observed in western North America (Furniss and Carolin, 1977). Various causes of outbreaks have been proposed. In the case of bark beetles, stress induced by drought, flooding, excessive competition, and mechanical damage reduces the ability of host trees to repel attacking insects (Berryman, 1982; Larsson *et al.*, 1983; Mitchell *et al.*, 1983).

4. Diseases

There are an exceedingly large number of biotic diseases that generate CWD within forests and most cannot be discussed here (see Hepting, 1971, for an extensive listing of North American diseases). The majority are caused by fungi, although parasitic vascular plants, e.g., *Arceuthobium* spp., also cause disease. In many ecosystems, diseases generate small amounts of CWD and are a contributing factor to wind and insect-caused mortality. However, diseases—especially those newly introduced to a host—may generate large amounts of CWD over extensive areas. *Endothia parasitica*, which virtually eliminated its host, *Castanea dentata*, from the eastern United States between 1904 and 1935, is a dramatic case in point (Beattie and Diller, 1954). Other examples of newly introduced biotic diseases that are greatly increasing the rate of CWD generation in forests include Dutch elm disease, beech bark disease, white pine blister rust, and *Phytophthora lateralis* Tucker and J. Milb., which is currently decimating *Chamaecyparis lawsoniana*. Abiotic diseases caused by industrial and other forms of pollution may also generate considerable amounts of CWD (Tomlinson, 1973). An example is the recent decline in *Picea rubens* noted in the northeastern United States (Scott *et al.*, 1984; Siccama *et al.*, 1982), which may be related to acid deposition.

5. Suppression and Competition

Suppression mortality is defined as the death of trees because of slow growth caused by competition. Suppressed trees die standing and often exhibit slow radial or height growth and often lack apical dominance. The smaller trees in a stand often die from suppression (Harper, 1977). Low vigor may predispose trees to fatal attack by insects and pathogens; it is difficult to determine, for instance, whether bark beetles attacked a tree after the tree's death or killed a weakened one.

Suppression mortality occurs in stands of all ages, but probably is most important as a source of CWD during succession in mature forests with closed canopies because of the combination of high competitive stress coupled with large tree size. Forest managers have used the −3/2 power model (Yoda *et al.*, 1963) and yield tables to estimate suppression mortality. Statistical models also exist relating the probability of survival to relative diameter, crown class, crown ratio, growth rate, or other indices of tree vigor (Monserud, 1976). These models have generally not been applied to natural stands.

6. *Agents Adding Coarse Woody Debris to Streams*

All agents creating CWD discussed above apply to aquatic as well as terrestrial ecosystems. Three additional processes add CWD to streams and rivers: stream and river undercutting of banks, mass movement of soil on hillslopes, and transport by floods. Slow mass-movement processes, such as soil creep, slumping, and earthflows on hillslopes, add CWD to the stream by moving forests toward streams and tipping trees into the channel. Very rapid mass movements, such as landslides, debris flows, and snow avalanches, produce CWD by knocking over and uprooting trees and then depositing them into stream channels. Deposition of soil and physical damage to live trees by these processes can also lead to mortality. Flooding and movement of ice add CWD to streams by tearing trees from the bank. Floods also remove CWD from floodplains, adding it to stream ecosystems. Bank undercutting is a very important process that adds material along most forested reaches; it is particularly significant along meandering rivers with high rates of lateral migration.

Generalizations about the relative importance of mechanisms have been made for particular landscapes and ecosystems by Keller and Swanson (1979) (Figure 3). At sites of earthflows and tributary streams that have frequent debris-torrent activity, CWD is added to streams at a very high rate. There are no reports quantifying the relative importance of mechanisms by which CWD is delivered to streams and rivers, but several studies suggest major differences exist between landscapes. In Soguel Creek, California, mass movement from heavily forested areas adjacent to hillslopes is the most important source of large trees for streams (Singer and Swanson, 1983). In 7 years of observation along 520 m of third-order streams flowing through old-growth *P. menziesii* in Oregon, undercutting

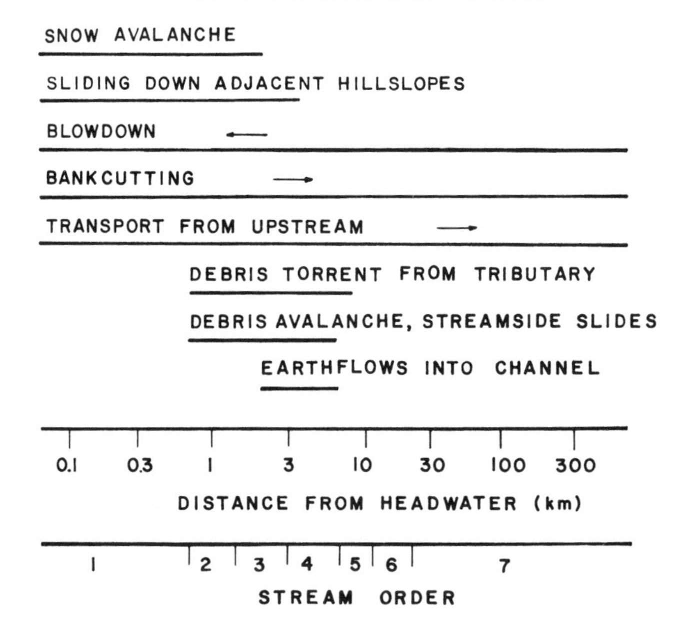

Figure 3 Stream distribution of input–output processes associated with organic debris in the Lookout Creek–McKenzie River System, Oregon. Arrows indicate direction of increasing importance. From Keller and Swanson (1979). Reprinted by permission of John Wiley & Sons, Ltd.

may have contributed 25% of the trees added (Lienkaemper, unpublished). Bank undercutting and CWD entrainment during floods adds much of the CWD in sandbed channels where lateral migration is rapid, such as the Little Missouri River (Everitt, 1968).

III. DECOMPOSITION OF COARSE WOODY DEBRIS

The following section considers the many ways CWD is decomposed in ecosystems, the rates at which these processes have been observed to operate, and, finally, the factors that control these rates.

A. Types of Decay Processes

1. Leaching

Water percolating through CWD dissolves some materials and results in a weight loss. Leaching is a very important process in the loss of mass from decomposing leaf litter in terrestrial (Singh and Gutpka, 1977; Swift *et al.*, 1979) and aquatic ecosystems (Petersen and Cummins, 1974) as well as from living plants (Tukey, 1970). However, little work has been published on leaching losses from CWD (Matson and Swank, 1984). Leaching is probably of minor importance initially, largely because CWD is high in polymeric material and low in soluble substances. As microbes transform these polymers to soluble material, leaching may increase. Another consideration is the low surface-to-volume ratio of CWD, which may reduce leaching losses relative to that found in leaf litter. Fragmentation may increase the importance of leaching as a loss of CWD because it increases the surface-to-volume ratio.

2. *Fragmentation*

Fragmentation or communition of CWD takes many forms. Physical fragmentation, caused by gravity and flowing water, is accelerated by decay organisms that weaken wood and bark. In terrestrial ecosystems, the fragmentation of snags creates a range of sizes from entire logs to chunks to finer woody debris. The latter two size classes may fall directly from snags or may be created when extremely decayed snags hit the forest floor and shatter into many fine particles. The range in particle size created by log fragmentation is probably smaller. Perhaps the smallest particles are created by the abrasion of the decayed surface layers of logs in streams by flowing water. Separation of individual wood cells (i.e., tracheids) occurs to a minor extent in terrestrial systems where tracheids are washed off the upper portions of sun-exposed logs.

Given the slow rates of microbial decay in streams, one would expect that CWD would remain longer in aquatic than in terrestrial environments. However, erosion of decayed wood surfaces by flowing water accelerates the overall rate of CWD loss in streams (Aumen, 1985). Of the five major decay classes used to classify wood debris in the Pacific Northwest (see Section IV.A), only the three least decayed stages are commonly found in streams (Triska and Cromack, 1980). CWD in the two most decayed stages appears to be fragmented and transported by flowing water and rarely accumulates. The erosive effects of current are probably more pronounced in larger streams than in small headwater tributaries.

Biological fragmentation of CWD is caused by both plants and animals, with invertebrates probably the most important. By chewing, ingesting, and excavating, invertebrates create a dust that decays more rapidly than the original CWD because of increased surface-to-volume ratio. Depending on the species, invertebrates may either transport these particles from the log or snag or leave them inside. The galleries made by invertebrates allow microbes to colonize CWD more rapidly (Ausmus, 1977; Leach *et al.*, 1934, 1937). Invertebrates such as bark and ambrosia beetles may actively bring microbe symbionts into CWD. Invertebrates in CWD are an important food source for vertebrates, such as bears and birds, that fragment the material while foraging. By growing on CWD and then subsequently falling off, trees and shrubs can cause CWD to fragment (see Section V.A). Finally, when trees and snags fall over, they often fragment. CWD by knocking over other snags or crushing parts of logs.

Material exported from CWD as fragments has important influences on other ecosystem components. Fragmentation may be a large source of fine-particulate organic matter in stream sediments. Addition of ground bark and heartwood of *Alnus* and *Pseudotsuga* to muddy sediments in a small stream in Oregon increased rates of respiration and methane production, with *Alnus* wood causing the greatest increase (Baker *et al.*, 1983). Rates of nitrogen fixation associated with the sediments were unaffected by increases of ground bark and wood, except that *Alnus* wood enhanced nitrogen fixation. Sugars with low molecular weight were identified by Baker *et al.* (1983) as a possible factor in the stimulation of nitrogen fixation in *Alnus* wood.

3. *Transport*

Transport rates of CWD within stream channels appear to be larger than those of terrestrial ecosystems, although there are no published studies that quantify the role of transport in either type of ecosystem. The transport of CWD in stream systems is a function of stream size; the wider the stream relative to a log's length, the more likely the log will be moved (see Section V.D). Trees, large rocks, and other pieces of CWD anchor logs and reduce movement within stream channels. Trees and rock outcrops are also probably important in reducing downslope movement of CWD on steep hillslopes.

4. *Collapse and Settling*

As decay proceeds and structural strength declines (Hartley, 1958; Toole, 1969), logs are unable to support their own weight and settle to the ground. Sollins (1982) found settling coincided with structural weakening of the heartwood to the point that branch stubs can be pulled from logs by hand. Settling increases the degree of contact between soil and log and changes the suitability of CWD as microbial, vertebrate, and invertebrate habitat (Maser and Trappe, 1984). During settling, the cross-sectional profile of logs changes from circular to elliptical and the contact between soil and log increases.

5. *Seasoning*

As CWD ages in dry environments, it undergoes a series of changes known as seasoning (Panshin and deZeeuw, 1980). This comprises a decrease in moisture, shrinkage, and formation of checks or cracks that increase access to microbes. Case hardening is a form of seasoning in which the outer rind of wood a few centimeters thick becomes sun bleached and dried, which may initially protect CWD from fragmentation losses and reduce loss of moisture from the interior.

6. *Respiration*

Respiration, primarily by microbes, removes matter from CWD. In terrestrial ecosystems, basidiomycetes are responsible for the majority of respirational loss (Käärik, 1974; Swift, 1977a), but in aquatic ecosystems, bacteria, including actinomycetes, are most important (Crawford and Sutherland, 1979).

Field studies indicate that respiration rates (rate of CO_2 evolution) of CWD increase as wood density decreases (Ausmus, 1977; Yoneda, 1975; Yoneda *et al.*, 1977). To calculate a combined respiration and leaching decay rate, the density or specific gravity of CWD can be plotted against the time CWD has been exposed to decay. Assuming volume remains constant, any decreases in density with time will reflect losses in mass (Christensen, 1984). However, a number of problems can arise when using this method.

Measuring volume is difficult. In regularly shaped solids, such as cylinders and cubes, volume can be calculated from external measurements of lengths and diameters. For very decayed wood, it is important to measure these dimensions in the field because samples often compress during transport. The volume of irregularly shaped pieces can be estimated by displacement, usually of water.

The dependence of density on moisture below the fiber saturation point causes errors in estimating decay rate. In undecayed wood, density at 30% moisture content is typically 10% less than that at 0% moisture (U.S. Forest Products Laboratory, 1976). It is therefore important to control moisture content when volumes are measured below the fiber saturation point and to report moisture content so valid comparisons can be made between studies.

Variations in the initial density of wood also cause problems in estimating decay rates. Density in living trees is influenced by species, age, and position along radial and longitudinal axes (Spurr and Hsuing, 1954). In many gymnosperms, density declines with height along the bole, although in some species density remains constant (Heger, 1974; Okkonen *et al.*, 1972). Density also increases in young trees from the pith outward, while in mature to overmature trees, the outermost wood tends to have lower density (Spurr and Hsuing, 1954). Growth rate apparently has little influence on density (Spurr and Hsuing, 1954). There also appears to be little difference between open- versus forest-grown trees in the case of *Abies balsamea* (Heger, 1974). The wood density of some species has been studied in detail (U.S. Forest Service, 1965a, U.S. Forest Service, 1965b; Wahlgren and Fassnacht, 1959; Wahlgren *et al.*, 1968).

7. *Biological Transformation*

CWD is metabolized by microbes and, to a lesser degree, by invertebrates (Käärik, 1974; Swift, 1977a). The cell walls of basidiomycetes contain chitin, and because chitin is not present in undecayed wood, fungal biomass can be estimated if a chitin–biomass conversion factor is known. Using this technique, Swift (1973) calculated that when 39% of the original wood mass was lost, an additional 35% had been converted to fungal biomass, demonstrating that much heterotrophic activity is overlooked when only respiration is considered.

Lignin decays more slowly than cellulose and hemicelluloses (Crawford, 1981), leading to an increase in the lignin-to-cellulose ratio as decay proceeds. This pattern has been observed for fine litter (see Swift *et al.*, 1979), but has received much less attention in the case of CWD. In undecayed wood, the lignin-to-cellulose ratio ranges from 0.6 to 1.2 for angiosperms and from 0.5 to 0.9 for gymnosperms (Table 2). Means *et al.* (unpublished) measured lignin and cellulose in a chronosequence of *Pseudotsuga* logs. In undecayed wood, the lignin-to-cellulose ratio was ~0.9, increasing to ~1.3, 1.9, and 2.7 after 50, 100, and 150 years of decay, respectively. The changes in this ratio may be affected by the type of decomposers present. White-rot fungi can degrade both lignin and cellulose and therefore may not increase the ratio. Conversely, soft-rot and brown-rot fungi only degrade cellulose, and the proportion of lignin should increase markedly in wood decayed by these organisms.

Many nutrients occur in fresh wood in very low concentrations. However, as decay proceeds and carbon is lost via respiration, the

Table 2 Cellulose, Hemicellulose, and Lignin Content of Wood from Selected Temperate Tree Species[a]

Species	Percentage of dry weight			Refs.[b]
	Cellulose	Hemicellulose	Lignin	
	Angiosperms			
Acer rubrum	44–45	29–32	24	2,5
Acer saccharum	40	35	25	3
Betula lutea	43	31	26	3
Betula papyrifera	41	40	19	6,5
Betula verrucosa	39	39	21	4
Fagus grandifolia	39–42	31–36	22–26	5,3
Populus tremuloides	43–53	27–34	16–23	2,5,1,3
Quercus rubra	44	31	25	7
Robinea pseudoacacia	40	28	32	3
Ulmus americana	51	23	24	6
Mean[c]	43	32	24	
Range	39–53	23–40	16–32	
	Gymnosperms			
Abies balsamea	42–43	22–27	29–30	2,5,1,3
Larix larcina	42–44	27–31	27–29	5,3
Picea abies	43	27	30	4
Picea glauca	40–50	20–31	27–30	6,5,1,3
Picea mariana	48–51	22–24	27–28	1,7
Pinus banksiana	42–47	26–29	27–29	5,1
Pinus elliotii	39	31	30	3
Pinus strobus	41–48	26–30	26–29	6,1,3
Pinus sylvestris	45	25	30	4
Pinus taeda	47	24	29	7
Pseudotsuga menziesii	43–48	24–31	26–28	3,7
Sequoia sempervirens	38	28	34	3
Thuja occidentalis	41–45	24–26	31–33	2,5,3
Thuja plicata	47	21	32	7
Tsuga canadensis	40–53	17–26	30–34	6,5,1,3
Tsuga heterophylla	44	26	30	7
Mean[c]	44	26	30	
Range	38–53	17–31	26–34	

[a] Values are based on extractive free wood.
[b] (1) Clermont and Schwartz (1951), (2) Côté (1977), (3) Panshin and deZeeuw (1980), (4) Rydholm (1965), (5) Timell (1957), (6) Timell (1967), (7) Wise and Jahn (1952).
[c] The overall mean was based on the mean for each species.

concentration of nutrients may increase; other mechanisms, such as N fixation, leaching, and fragmentation, also contribute to increased nutrient concentrations (see Section V.C). Nitrogen content of *Pseudotsuga* wood increased ~3.5-fold from the initial to the final stages of decay (Sollins *et al.*, unpublished). Other nutrients also increased in *Pseudotsuga* wood, with P, K, Ca, Mg, Mn, and Na exhibiting 5.3-, 2.2-, 3.0-, 6.8-, 1.6-, and 2.2-fold increases, respectively, as wood proceeded from the least to the most decayed stages. In many cases, the C:N ratio decreases as decay proceeds (Foster and Lang, 1982; Grier, 1978; Harris, 1978; Lambert *et al.*, 1980), although an exception has been found in a tropical forest (Yoneda *et al.*, 1977).

B. Decay Models

Many of the models proposed for leaf decomposition have been used in CWD studies. Wieder and Lang (1982) examined the models commonly used in litterbag experiments and concluded that the single-exponential (Jenny *et al.*, 1949; Olson, 1963) and the double-exponential models were the most useful and realistic biologically, whereas the asymptotic, linear, quadratic, and power models of decay were less useful. Their conclusions might also reasonably apply to CWD.

The most commonly used model in CWD studies has been the single-exponential model discussed in detail by Olson (1963). The assumption that decay is proportional to the amount of material remaining leads to the model

$$Y_t = Y_0 e^{-kt}$$

where Y_0 is the initial quantity of material, Y_t is the amount left at time t, and k is the decay rate constant. In most CWD studies, wood density is used as the Y variable, although mass, volume, or cover of bark could also be used. The times to decompose one-half ($t_{0.50}$) and 95% ($t_{0.95}$) of the material are often reported, where $t_{0.50} = 0.693/k$ and $t_{0.95} = 3/k$. It is also common practice to report the turnover time, which is equal to the reciprocal of k.

Minderman (1968) objected to the single-exponential model because most substrates are not homogeneous, but contain substances decaying at different rates. If some substances are labile and others are recalcitrant, actual decay curves can depart markedly from the single-exponential model (Figure 4). This problem has led to the use of double-exponential

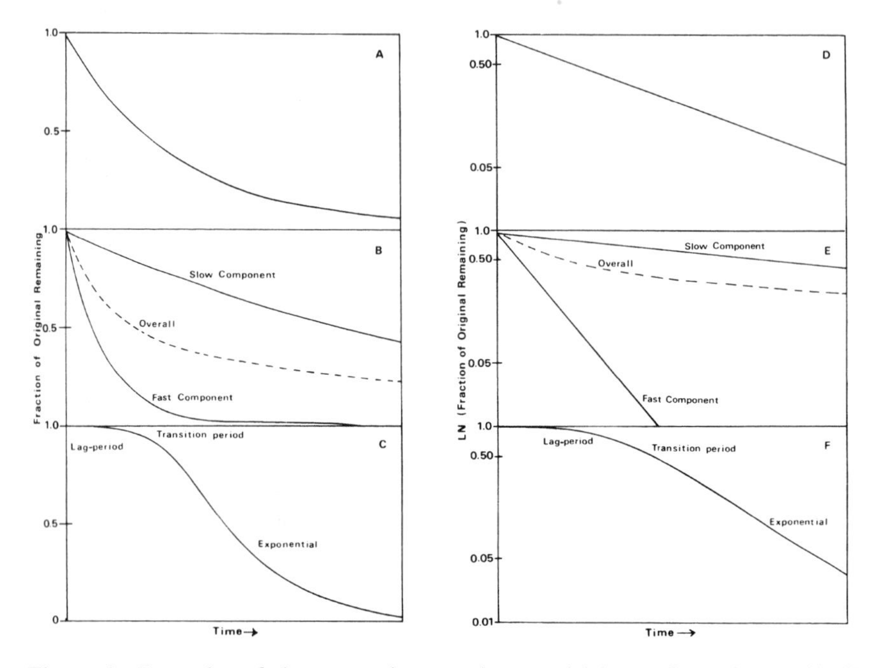

Figure 4 Examples of three equations used to model losses from CWD. (A–C) Examples of single-exponential, multiple-exponential, and lag-time models, respectively, plotted on arithmetic scale. (D–F) The same models plotted when the fraction remaining has been. Transformed to natural logarithmic scale. The multiple-exponential model used here has fast and slow components, each composing half of the original mass and having rate constants differing by an order of magnitude.

models where the substrate is partitioned into two fractions, each of which is characterized by its own decay rate constant.

In the case of CWD, where decomposition is complex, it might be considered useful to expand the double-exponential into a multiple-exponential model. Multiple-exponential models may prove quite useful in understanding how components such as bark, sapwood, and heartwood contribute to the overall decay of CWD and how they might control differences between species and sizes of material. Means *et al.* (unpublished) adapted this model to examine the decay of *Pseudotsuga* boles as a function of three components: cellulose, lignin, and an acid detergent-soluble fraction. Although the lignin decayed at one-third the rate of the other fractions, the single-exponential model fit the data as well as the multiple-exponential one. This result was caused, in part, by the low coefficients of variation associated with both models. Other factors that might influence the comparison are the proportions of the components and the differences in decay rates of the components. The predicted difference between the single- and multiple-exponential models will increase as the differences between the decay rate of components increases. The differences between models also increase as the proportion of components becomes equal.

In the single-exponential model, material is assumed to be homogeneous, and the multiple-exponential model was developed to address this problem. However, an assumption of both models is that detritus is not transformed into more or less decomposable forms. This is an important shortcoming given the large amount of biological transformation occurring during decay. Carpenter (1981) developed a more general decay model with three parameters that, under special conditions, simplifies to either of the models presented above. Carpenter's general model incorporates the notion that more or less decomposable substances may be created during decay and deserves further examination.

All of the models described above are primarily concerned with the loss of mass via respiration and leaching. However, CWD is also lost and transformed via fragmentation (Lambert *et al.*, 1980; Sollins, 1982). To address this problem, the decay rate constant can be divided into two parts so that $k = k_m + k_f$, where k is the overall rate constant, k_m is the rate constant for mineralization losses due to respiration and leaching, and k_f is the decay rate constant for losses due to fragmentation (Lambert *et al.*, 1980; Sollins *et al.*, 1979). In this approach, k_m and k_f are constant with time; however, there is a time lag before fragmentation begins. During this period, k_f is near zero. Harmon (1985) adapted the exponential model to incorporate a lag time. This model is $Y_t = 1 - (1 - \exp[-k_f t])^N$, where Y_t, is the amount left at time t, k_f is the fragmentation rate constant, and N is a constant related to the lag time (Figure 4). This model may also be useful in examining the effect of fungal colonization and the lag time it introduces into mineralization losses.

C. Methods to Determine Decay Rates

1. Chronosequences

Decomposition of CWD is slow, necessitating long time periods to accurately measure the decay rate of individual pieces. An alternative, short-term method is to determine the length of time snags or logs have been dead and examine how volume, density, or other characteristics change with time. This array of aged pieces forms a chronosequence.

Dating the age of logs and snags requires imaginative detective work. More methods exist for dating fallen boles than standing dead trees. Logs can be aged by aging scars left on live trees adjacent to the fallen boles. In moist climates where tree seedlings can grow on logs, the age of the oldest seedlings gives a minimal estimate of log age (Triska and Cromack, 1980), but the time between tree fall and seedling establishment must be estimated to calculate log age more accurately. Bark sloughing occurs as logs decay, and it is therefore important to note if the aged seedlings are growing on bark; if sloughing has occurred, additional years should be added to the log age. Living stumps can be analyzed to determine the ages of trees snapped by wind (Harmon and Cromack, unpublished).

Historical records have also been used to provide ages of CWD. Permanent-plot records of tagged trees were used by Grier (1978) to date logs in *T. heterophylla* forests in Oregon. MacMillan (1981) used maps of a mixed mesophytic forest to age logs. By comparing maps made a decade apart, Falinski (1978) estimated rates of log disappearance in mixed hardwood and *Picea* forests in Poland. Records of logging and thinning operations were used by Foster and Lang (1982) and Savely (1939).

Records of natural disturbances have also been used to age CWD. Dates of fires have been extensively used to age snags (Harmon, 1982; Kimmey, 1955; Kimmey and Furniss, 1943), as have records of insect outbreaks (Keen, 1929, 1955; Wright and Harvey, 1967) and catastrophic windthrow (Buchanan and Englerth, 1940).

Historical reconstructions of stands have also been used to estimate the age of CWD (McFee and Stone, 1966). Lambert *et al.* (1980) aged dead boles by aging tree seedlings in *A. balsamea* stands. Fahey (1983) assumed all *P. contorta* trees established the same year and then aged snags by noting the difference in age between living and dead trees.

Some studies have monitored the decomposition of individuals and cohorts over short periods of time. Most of these investigations have monitored snags created by fires (Dahms, 1949; Lyon, 1977) or insect attacks (Bull, 1983; Hinds *et al.*, 1965; Keen, 1955), although Buchanan and Englerth (1940) followed the deterioration of windthrown trees. Trees have also been deliberately felled for later sampling (Gosz, 1980; Harris *et al.*, 1972; Miller, 1983).

2. *Input/Biomass Ratio*

It is possible to estimate decay rates from the ratio of CWD input to biomass, assuming the biomass of CWD is in steady state. Christensen (1977) estimated wood turnover rate in a Danish *Quercus robur* forest. Sollins (1982) combined input data derived from long-term mortality records and current biomass of CWD to estimate the decay rate constant of an old-growth *Pseudotsuga* forest and found that this method yielded a decay rate constant of 0.03 $year^{-1}$, whereas those based on wood-density change ranged from 0.01 to 0.02 $year^{-1}$ (Graham, 1982).

D. Decay Rates

In this section, we review the rates at which some of these processes have been observed to operate. Because of the limited number of observations for some processes, we have chosen to emphasize fragmentation and mineralization rates of snags and logs.

1. Snag Fragmentation Rates

The rates snags fall and/or fragment have been studied for many coniferous species in the western United States. Less information is available for other forested ecosystems. Most snag studies have used chronosequences, although some (e.g., Bull, 1983; Lyon, 1977) have observed a single population through time. In most studies, proportions of snags standing were noted, but in some the volume standing was estimated. Published data were used to calculate fragmentation rate constants, k_f, of snags and the length of the lag time before fragmentation (Table 3).

The lag time between tree death and the onset of fragmentation is influenced by species, size, microclimate, and type of mortality. Lag times required for snags to begin falling appear to be <20 years for all the species examined. *Pseudotsuga menziesii* snags have the longest lag time, although considerable variation exists within this species. For example, Cline *et al.* (1980) observed that as size increased, the lag time also increased. However, Graham (1982) indicated both large and small *P. menziesii* had lag times of <6 years. Most other species appear to have lag times of <3 years. The increase in lag time with increasing size also appears to occur in the case of *P. contorta* and *P. ponderosa* (Bull, 1983), but these differences were minor (i.e., 1–2 years).

After the initial period without breakage, snag volume declines exponentially and can be described by a rate constant (see Section III.B). The rate of volume decline varies considerably between studies, species, and sizes. The fastest rates of fragmentation, k_f, have been observed for *P. menziesii*, 10–18 cm diameter at breast height (dbh). After standing unbroken for 4 years, one-half of the average snag would remain standing at 6 years ($k_f=0.354$ $year^{-1}$; Cline *et al.*, 1980) and similar values of 0.317 $year^{-1}$ were observed for *Abies lasiocarpa* (Lyon, 1977) and

Table 3 Decomposition Rates of Snags and Logs in Temperate Forest Ecosystems

Species	DBH (cm)	Study length (year)	Cause of death[a]	Lag time (year)	Decay rate constant	Half-time (year)	Refs.[b]
			Snag-bole fragmentation				
Abies balsamea	—[c]	5	D	3	0.076	12	3[d]
Abies balsamea	>8	80	FW	<5	~0.085	~13	19[e]
Abies lasiocarpa	<7.5	15	F	2	0.317	4	20[d]
Picea engelmannii	7.5–24	25	BB	10	0.015	56	24[f]
Picea engelmannii	25–39	25	BB	10	0.012	67	24[f]
Picea engelmannii	>40	25	BB	10	0.009	87	24[f]
Picea engelmannii	>12	22	BB	4	0.023	34	16[e]
Picea glauca	—	7	D	0	0.012	57	26[d]
Pinus contorta	7.5–30	15	F	2	0.089[g]	10	20[d]
Pinus contorta	<25	8	BB	2	0.318[g]	4	4[d]
Pinus contorta	>25	8	BB	3	0.133[g]	8	4[d]
Pinus contorta	—	17	BB	<4	0.021	~37	16[e]
Pinus ponderosa	<25	8	BB	3	0.283	5	4[d]
Pinus ponderosa	25–49	8	BB	3	0.113	9	4[d]
Pinus ponderosa	>50	8	BB	5	0.161	9	4[d]
Pinus ponderosa	>20	22	F	<9	0.073[g]	~18	6[d]
Pinus ponderosa	>25	29	BB	1	0.197[g]	4.5	18[e]
Pinus ponderosa	>25	29	BB	2	0.112[g]	8	18[e]
Pinus ponderosa	>25	9	BB	1	0.189[g]	4.6	17[e]
Pseduotsuga menziesii	10–18	25	U	4	0.354[g]	6	5[e]
Pseduotsuga menziesii	29–31	60	U	6	0.109[g]	12	5[e]
Pseduotsuga menziesii	32–46	40	U	11	0.033[g]	32	5[e]
Pseduotsuga menziesii	47–71	45	U	17	0.055[g]	29	5[e]
Pseduotsuga menziesii	<40	50	U	<5	0.026	~31	10[e]

Pseduotsuga menziesii	>65	30	U	<6	0.014	~55	10[e]
Pseduotsuga/Tsuga	—	140	F	<20	0.096	~27	22[e]
Tsuga heterophylla	>25	30	U	<2	0.067	~12	10[e]
			Log-bole fragmentation				
Abies balsamea	>8	75	FW	<35	0.019	~71	19[e]
Abies concolor	>15	60	W	25	0.060	36	14[e]
Pseudotsuga menziesii	>20	250	W	80	0.008	166	10[e]
Tsuga heterophylla	>15	50	W	>50	0	—	10[e]
			Snag-bark fragmentation[h]				
Picea engelmannii	>7.5	25	BB	—	0.012	57	24[f]
Pinus contorta	>1	20	U	<12	0.231	15	7[e]
Pinus contorta	>1	8	BB	—	0.015	46	4[d]
Pinus ponderosa	>1	8	BB	—	0.005	138	4[d]
Pseudotsuga menziesii	<40	50	U	<5	0.110	~11	10[e]
Pseudotsuga menziesii	>65	30	U	<6	0.038	~24	10[e]
Tsuga heterophylla	>25	30	U	<5	0.140	~10	10[e]
Tsuga heterophylla	>25	30	U	<5	0.096	~12	10[e]
			Log-bark fragmentation				
Abies concolor	>15	60	W	7	0.125	12	14[e]
Populus tremuloides	>2	5	C	1	0.145	6	25[d]
Pseudotsgua menziesii	<40	200	W	<30	0.039	~47	10[e]
Pseudotsuga menziesii	40–65	200	W	<30	0.018	~68	10[e]
Pseudotsuga menziesii	>65	250	W	<30	0.021	~63	10[e]
Tsuga heterophylla	<25	50	W	< 10	0.026	~ 36	10[e]
Tsuga heterophylla	>25	50	W	< 20	0.019	~ 56	10[e]

(Continued)

Table 3 Continued

Species	DBH (cm)	Study length (year)	Cause of death[a]	Lag time (year)	Decay rate constant	Half-time (year)	Refs.[b]
Snag-bole mineralization							
Abies balsamea	—	2	D	~1	0.062	12	2[d]
Acer rubrum	<15	6	F	~1	0.08	10	12[e]
Cornus florida	<15	6	F	~1	0.05	15	13[e]
Carya spp.	<15	6	F	~1	0.08	10	13[e]
Nyssa sylvatica	<15	6	F	~1	0.20	4.4	13[e]
Oxydendrum arboreum	<15	6	F	~1	0.05	15	13[e]
Pinus contorta	>1	20	U	12	0.006	127	7[e]
Pinus rigida	<15	12	F	~1	0.06	13	13[e]
Pinus virginiana	<15	15	F	~1	0.04	18	13[e]
Pseudotsuga menziesii	<40	50	U	—	0.027	25	10[e]
Pseudotsuga menziesii	40–65	30	U	—	0.013	53	10[e]
Pseudotsuga menziesii	>65	30	U	—	0.10	230	10[e]
Quercus coccinea	<15	6	F	~1	0.18	8	13[e]
Quercus prinusl	<15	6	F	~1	0.04	5	13[e]
Tsuga canadenis	<15	6	F	~1	0.04	18	13[e]
Tsuga heterophylla	<25	30	U	—	0.017	40	10[e]
Tsuga heterophylla	>25	30	U	—	0.016	43	10[e]
Log-bole mineralization							
Abies balsamea	>8	80	FW	—	0.011	63	19[e]
Abies balsamea	10—15	29	C	—	0.029	24	8[e]
Abies concolor	>15	60	W	—	0.05	14	14[e]
Carya spp.	>2.5	4	C	1	0.305	3.3	15[d]
Liriodendron tulipifea	>2.5	4	C	1	0.520	2.3	15[d]

Picea rubens	>10	64	C	—	0.033	21	8[e]
Picea sitchensis	>15	100	W	—	0.011	63	11[e]
Pinus contorta	>1	40	C	—	0.012	57	7[e]
Pinus taeda	>2.5	4	C	1	0.274	3.5	15[d]
Populus tremuloides	>2	5	C	0	0.049	14	25[d]
Populus tremuloides	17	5	C	0	0.070	9.8	9[d]
Pseudotsuga menziesii	<40	200	W	—	0.004	172	10[e]
Pseudotsuga menziesii	40—65	200	W	—	0.004	172	10[e]
Pseudotsuga menziesii	>65	250	W	—	0.006	115	10[e]
Pseudotsuga menziesii	>15	320	W	—	0.007	98	23[e]
Quercus prinus	>2.5	4	C	1	0.266	3.6	15[d]
Quercus prinus	3—5	1	C	0	0.098	7	1[d]
Quercus spp.	>5	23	U	—	0.029	24	21[e]
Tsuga heterophylla	<25	50	W	—	0.012	57	10[e]
Tsuga heterophylla	>25	50	W	—	0.024	29	10[e]
Tsuga heterophylla	>25	38	U	2	0.008	88	12[e]
Tsuga heterophylla	>15	50	W	—	0.010	69	11[e]

[a] BB, bark beetless; C, cutting; D, defoliated; F, fire; FW, fir-ware; U, undetermined; W, windthrow.

[b] (1) Abbott and Crossley (1982), (2) Barnes and Sinclair (1983), (3) Basham (1951), (4) Bull (1983), (5) Cline *et al.* (1980), (6) Dahms (1949), (7) Fahey (1983), (8) Foster and Lang (1982), (9) Gosz (1980), (10) Graham (1982), (11) Graham and Cromack (1982), (12) Grier (1978), (13) Harmon (1982), (14) Harmon and Cromack (unpublished), (15) Harris (1976) (16) Hinds *et al.* (1965), (17) Keen (1929), (18) Keen (1955), (19) Lambert *et al.* (1980), (20) Lyon (1977), (21) MacMillan (1981), (22) McArdle (1931), (23) Means *et al.* (unpublished), (24) Mielke (1950), (25) Miller (1983), (26) Riley and Skolko (1942).

[c] Not available.

[d] Monitoring population.

[e] Chronosequence.

[f] One-time sample.

[g] Assumes entire snag falls.

[h] Transfer from standing material only and does not include transfer with stem breakage.

0.318 year^{-1} for *P. contorta* (Bull, 1983). Smallest fragmentation rate constants (k_f=0.009–0.012 year^{-1}) were observed for *Picea engelmannii* in Utah (Mielke, 1950), indicating a half-time of 67–87 years when a 10-year lag time is included. *Picea glauca* (k_f=0.012 year^{-1}; Riley and Skolko, 1942) and large *P. menziesii* (k_f=0.014 year^{-1}; Graham, 1982) also appear to fragment at slow rates. Snags of both species would have, on the average, 50% of their volume standing 50–57 years after the onset of fragmentation. Fragmentation rate constants appear to decline as snag size increases. Data from Bull (1983), Cline *et al.* (1980), Graham (1982), and Mielke (1950) all confirm this general pattern. However, the data from Cline *et al.* (1980) for *P. menziesii* and Bull (1983) for *P. ponderosa* indicate that above a certain size the fragmentation rate constant remains constant. For *Pseudotsuga*, k_f declines from 0.345 to 0.109, 0.033, and 0.055 as dbh increases from 10–18 to 29–31, 32–46, and 47–71 cm, respectively.

Even within a species and size class, k_f can vary considerably. The most frequently studied species is *P. ponderosa*. For snags >25 cm dbh, k_f ranges from 0.073 to 0.197 year^{-1}, which means that 50% of the population would be standing 3.5–9 years after the onset of fragmentation. Keen (1955) observed that snags fell at slower rates on drier pumice soils (k_f=0.112 year^{-1}) than on moister loam soils (k_f=0.117 year^{-1}). The cause of death also appears to influence k_f, with fire-killed trees (Dahms, 1949) falling slower than trees killed by bark beetles (Keen, 1955; k_f=0.073 year^{-1} versus 0.112 year^{-1}).

All the studies mentioned above examined snag fragmentation on a population and not on an individual basis. On an individual basis, fragmentation is highly variable and therefore harder to predict (Graham, 1982). Until the lag time is exceeded, a population of snags is composed of individuals with no breakage (Figure 5). Once fragmentation begins, the frequency distribution of the snag population has a strong positive skew; that is, only a few individuals are broken and most of the population is intact. As time progresses, more breakage occurs and the frequency distribution of the population is more positively skewed. When most of the snags have fallen, a few individuals remain standing, which causes the frequency distribution of the population to be negatively skewed.

Exponential fragmentation curves indicate continuous loss; in fact, many boles may break at the same time, and this causes discrete steps in the fragmentation curve. For example, Lyon (1977) observed trees killed by a fire and found years with breakage followed by years with none.

Loss of bark is an important feature in creating animal habitats and transferring nutrients to the forest floor, but has received little study. Bark moves to the forest floor either on logs broken from snags or by falling off snags (Graham, 1982). The rate constants presented in Table 3 represent

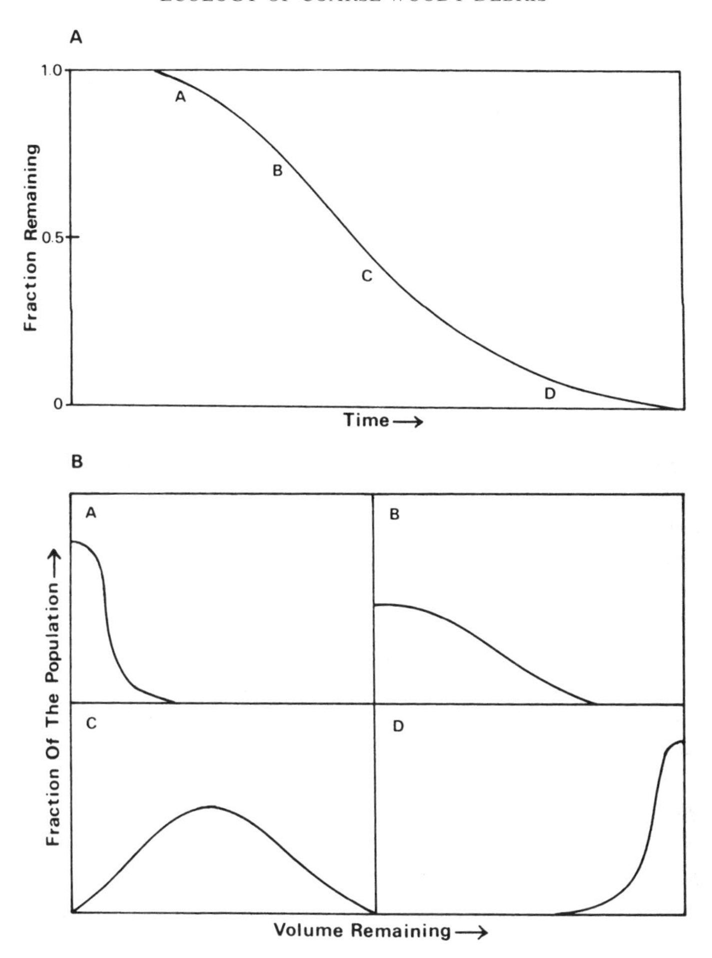

Figure 5 The loss of volume over time caused by fragmentation from a hypothetical cohort of snags. (A) The mean fraction of volume remaining in the cohort as a function of time. Letters in this part of the figure indicate points along the curve which correspond to the distributions of individuals (B). As time progresses, both the mean, variance, skewness, and kurtosis of the cohort is expected to change.

changes in bark cover. Lag times for bark to fall from snags is known with less certainty than for bole volume losses. Most species begin to lose bark < 6 years after tree death. The longest lag time was < 12 years and observed for *P. contorta* (Fahey, 1983). Woodpeckers have a profound effect on bark loss (Bull, 1983) and may begin to remove bark within the first year of tree death.

Rate constants of bark loss are highly variable between species. The smallest ($k_f = 0.005$ year^{-1}) was observed for *P. ponderosa* (Bull, 1983), and because this is much smaller than the bole fragmentation rate constant, most of the bark of this species is probably transferred to the forest floor via bole fragmentation. The largest bark k_f (0.231 year^{-1}) was observed for *P. contorta* in Wyoming (Fahey, 1983), which lost 50% of its bark 3 years after a 12-year period with no toss. Because bole fragmentation was not measured by Fahey (1983), the relative proportion of bark transferred to the forest floor via boles cannot be judged. In contrast to *P. ponderosa*, some species such as *Pseudotsuga* do transfer the majority of bark via direct fragmentation and not via bole fragments. Increasing tree size may cause slower bark fragmentation. Graham (1982) found that *P. menziesii* larger than 65 cm dbh had a k_f of 0.038 year^{-1}, while smaller snags had a k_f of 0.11 year^{-1}. A similar trend was noted by Graham (1982) for *T. heterophylla* snags >25 cm dbh versus those that were smaller, with k_f of 0.096 and 0.140 year^{-1}, respectively.

2. *Snag Mineralization Rates*

Changes in wood density are caused by a combination of respiration and leaching and are termed mineralization losses here (see Section III.B). Wood density of small snags in the southern Appalachians decreases after 1 year (Harmon, 1982), indicating a short lag time is associated with snag mineralization. In contrast, *P. contorta* snags in Wyoming. had no significant change in density for 12 years, and boles in these ecosystems may not be mineralized until they fall to the forest floor (Fahey, 1983).

Mineralization losses will not peak until most of the bole is colonized by microbes. Fungi begin to colonize snags within a year of death (Basham, 1951; Hinds *et al.*, 1965; Kimmey, 1955; Kimmey and Furniss, 1943). However, when the volume of respiring tissue is small compared to the total volume, total mineralization losses may be minimal. There may be an extended transitional period between the time when losses are first detected and when the snag is losing material at the maximum rate. Kimmey (1955) found that *Abies concolor*, *Pinus jefreyi*, and *P. ponderosa* snags were completely colonized by fungi: within 6 years, while *Pinus lambertiana* and *P. menziesii* snags were not colonized completely even after 10 years. This indicates that the mineralization lag time probably differs between species.

Mineralization rate constants, k_m, of snags vary with species and size. The smallest were observed for large *P. menziesii* ($k_m = 0.003$ year^{-1}; Graham, 1982) and *P. contorta* ($k_m = 0.006$ year^{-1}; Fahey, 1983). The small k_m of the former species is probably due to a combination of large size and high decay resistance of heartwood. The largest k_m for snags were observed in *Nyssa sylvatica* (0.20 year^{-1}) and *Quercus prinus* (0.18 year^{-1}) (Harmon, 1980). The large rate constants of the latter species were probably caused by a combination of small size and a high proportion of sapwood and/or non-decay-resistant heartwood.

3. Log Fragmentation Rates

In studies of log fragmentation, it has been assumed that all volume changes are associated with fragmentation (Graham, 1982; Harmon and Cromack, unpublished; Lambert *et al.*, 1980). However, volume decreases when boles collapse, and these studies probably overestimate fragmentation rates.

Lag times for log fragmentation, typically 25 years or more, are considerably longer than those for snags, which are typically less than 10 years. The shortest lag time reported for log fragmentation is 25 years for *A. concolor* (Harmon and Cromack, unpublished), while the longest was 80 years observed in *P. menziesii* (Graham, 1982).

Log fragmentation rates, k_f, vary considerably from 0 for *T. heterophylla* (Graham, 1982) to 0.06 year^{-1} for *A. concolor* (Harmon and Cromack, unpublished). The k_f of logs appear to be considerably smaller than those observed for snags for species where data exist for both forms of CWD. This indicates that snags will disappear as recognizable structures much faster than logs.

Although data are rare, there are also some general trends in the loss of bark cover from logs. For the three species that have been studied, *A. concolor*, *P. menziesii*, and *T. heterophylla*, loss of bark precedes bole fragmentation. The lag time for the onset of bark loss ranges from 1 year in *Populus tremuloides* to 30 years in *T. heterophylla*. Once initiated, bark k_f vary considerably. *Abies concolor* and *P. tremuloides* lost bark fastest, with k_f of 0.125 and 0.145 year^{-1}, respectively. *Pseudotsuga* (k_f=0.018 year^{-1}) and *Tsuga* (k_f=0.019 year^{-1}) have the slowest rates. It is not known why some species lose bark faster than others or why certain individuals vary. Trees falling during the growing season probably lose bark faster than those failing at other times. Furthermore, suspended logs are apt to lose bark faster than those resting on the ground. Thin bark is more apt to break due to shrinking, but thick bark is heavier and more apt to separate from the underlying wood.

4. Log Mineralization Rates

Most studies of log mineralization have used the single-exponential model without a lag time. However, studies conducted on beetle-killed

P. ponderosa (Boyce, 1923) and windthrown trees (Boyce, 1929; Buchanan and Englerth, 1940) indicate lag times exist for many species. Grier (1978) observed a lag of 2 years for *T. heterophylla*. Harris (1976) reported a lag of 1 year for logs of a range of species in Tennessee. Lag time associated with log mineralization should increase as log size increases.

The k_m of logs vary greatly between species. The smallest k_m, 0.004–0.007 $year^{-1}$, have been observed for *Pseudotsuga* (Graham, 1982; Means *et al.*, unpublished). *Pseudotsuga* logs would have a half-time of 98 to 172 years if only respiration losses were considered. The largest k_m, 0.52 $year^{-1}$, is for *Liriodendron tulipifera*, with a half-time of 2.3 years, even with the lag time added (Harris, 1976). Other angiosperm species have k_m an order of magnitude smaller than observed for *Liriodendron*. The half-time of *P. tremuioides* logs ranges from 10 years (Gosz, 1980) to 14 years (Miller, 1983). Large mineralization rates are not restricted to angiosperm species. For example, Harmon and Cromack (unpublished) found that *A. concolor* boles had a half-time of 14 years ($k_m = 0.05$ $year^{-1}$).

Logs entering streams remain intact for intervals ranging from a few decades to several hundred years (Anderson *et al.*, 1978; Franklin *et al.*, 1981). Decay rates of *Populus balsamifera* logs in a Canadian beaver pond indicate that ~ 250 years would be required for 95% of the mass to be lost (Hodkinson, 1975), which is considerably longer than the 43–60 years required on land (Gosz, 1980; Miller, 1983). In streams of the Pacific Northwest, dendrochronological dating indicates that pieces have been in channels for >108 years (Swanson and Lienkaemper, 1978; Swanson *et al.*, 1976). Of the species examined in these two studies, *Thuja plicata* decayed most slowly, followed in order of increasing decay rate by *P. menziesii*, *T. heterophylla*, and *Alnus rubra*. Using respiration rates, Anderson *et al.* (1978) calculated that CWD in streams of the Pacific Northwest requires 5–200 years to decay completely. Naiman and Sedell (1980) measured respiration losses from fine wool debris (<10 cm diameter) and estimated that CWD would require 200–300 years for total decomposition. Respiration measurements underestimate the time required for complete decay of CWD by microbial activity because wood surfaces in streams are covered with a film of bacteria, fungi, algae, moss, and protozoans, which also add CO_2 to the respiration losses, but do not degrade wood.

E. Factors Controlling Decomposition

Factors that can control decomposition of CWD include temperature, moisture, oxygen, carbon dioxide, substrate quality, and the organisms involved. Unfortunately, the effects of most of these factors are known only qualitatively or from laboratory tests.

1. Temperature

The role temperature plays in the activity of organisms is well studied in the laboratory. Fungi survive temperatures from below 0° up to 60°C (Deverall, 1965). However, most wood-decaying fungi are mesophilic, i.e., cannot grow above 40°C, and have a temperature optimum of 25°–30°C (Käärik, 1974). Between 13° and 30°C fungal respiration rate doubles to triples for every 10°C increase (Deverall, 1965). Variations in tolerance to temperature affect the distribution of fungi in CWD; for example, thermophilic species are most likely to occur in the outer portions of CWD exposed to direct sunlight (Käärik, 1974). Savely (1939) found that wood-inhabitating insects had upper temperature limits from 40 to 52°C. Moreover, for many species the ability to tolerate elevated temperatures decreases as relative humidity increases. Relative humidity is apt to be high in CWD, and Savely (1939) reported lethal temperatures of 41–44°C for insects. Thermophilic insect species inhabit the upper portions of logs, while those with lower tolerances inhabit the sides and lower portions (Graham, 1925). The respiration rate and feeding activity of invertebrates also increases with temperature, although at high temperatures they may become torpid. For example, fecal production by larvae of the xylophagous aquatic beetle *Lara avara* doubled as temperatures increased from 5 to 15°C (Steedman, 1983). Elevated temperatures shorten the length of the life cycle of insects (Graham, 1925), which, in turn, may increase the rate at which CWD is decayed.

Diurnal fluctuations under the bark of *Pinus* and *Quercus* logs were noted by Graham (1925) and Savely (1939). On nights and cloudy days, temperatures just below the bark closely followed the surrounding air, but under sunny conditions, temperatures on the upper side of logs exceeded air temperature, often exceeding lethal levels for insects where logs were exposed to full sunlight (Graham, 1925). However, the underside of logs tended to follow the air temperature even during sunny weather. Temperature fluctuations were increasingly dampened closer to the center of a log (Savely, 1939).

Paim and Becker (1963) monitored changes in oxygen and carbon dioxide in *Fagus grandifolia* logs and found oxygen concentration decreased with temperature increases, while CO_2 concentration increased. Savely (1939) observed similar patterns in CO_2 contents in *Pinus* and *Quercus* logs. Seasonal patterns in CO_2 concentration are probably caused by increasing respiration rates with increasing temperature. However, moisture content is also apt to be unfavorable to microbes during winter months and reduces respiration during this season.

2. *Moisture*

Both extremely low and high moisture content can limit the activity of organisms. Below 30% moisture content (the fiber saturation point), water is generally not available to microbes (Griffin, 1977; Käärik, 1974). Above 30% moisture content, water becomes available and the activity of organisms increases. However, as the pores fill with water, oxygen diffusion is reduced and aerobic activity is limited.

The moisture content of CWD can be measured a number of ways, but it is not obvious which technique gives the most meaningful measurement in terms of controlling the activity of organisms (Griffin, 1977). Moisture content is often expressed as percentage moisture on a dry weight basis, but it can represent different amounts of water if wood density varies. Water potential may be more appropriate for comparing material of various densities. Griffin (1977) defined three types of pores that store water in wood. The largest pores are cell lumina that contain water when water potential exceeds −0.3 MPa. A second set of smaller pores include pit apertures, pit-membrane pores, and other small voids that drain at water potentials of −0.3 to −14 MPa. At water potentials of <-14 MPa, water is lost from micropores in the cell walls.

The optimal amount of moisture depends, of course, on the organisms involved. A range of 30–160% moisture (dry weight basis) appears to support growth of basidiomycetes (Käärik, 1974). Griffin (1977) found that below −4 MPa basidiomycetes ceased to grow and a decrease in growth occurred below −0.1 MPa. Bacteria, ascomycetes, and fungi imperfecti can tolerate higher amounts of moisture than basidiomycetes. For example, the ascomycetes and fungi imperfecti causing soft rots can tolerate moisture contents up to 240% (Käärik, 1974). Compared to fungi, less work appears to have been done on the tolerance of insects to moisture content. Certain species such as powder post beetles (Anobiidae and Bostrichidae) can live in extremely dry wood.

The radial distribution of microbial activity in CWD differs markedly between terrestrial and aquatic ecosystems. Microbial communities in log-holding ponds are composed primarily of superficial bacterial films (Savory, 1954b). Observations of microbial discolorations and wood firmness indicate microbial decay of logs in streams is also a surface phenomenon (Anderson *et al.*, 1978; Dudley and Anderson, 1982; Triska and Cromack, 1980). Aumen (1985) investigated microbial distributions in *Pseudotsuga* logs in a third-order stream in Oregon. Aerobically incubated samples of the wood surface developed seven times more colonies than anaerobically incubated surface samples, and samples from a depth of 25 cm failed to develop colonies after either aerobic or anaerobic incubation. Cellulose and lignin decomposition was four times faster for surface samples than for samples from the interior of the log. SEM analysis revealed many single-cell bacteria and actinomycete filaments on the surface of the log, but little evidence of microbes from the core of the log. Microbial activity was restricted to the surface of the log, with bacteria and fungi virtually absent from samples taken at a depth of ~5 mm or less (Aumen, 1985).

Most work on the daily and seasonal fluctuations of wood moisture has involved wood products applications with moisture content below 30% (Skaar, 1972). Boddy (1983) observed the moisture regime of branchwood 1.5–2.5 cm in diameter over a year, and presumably some of these observations apply to larger pieces as well. Moisture content was highest during winter, and branchwood remained saturated as long as precipitation remained high and temperatures low. Minimum moisture content was observed during late summer and early autumn, but even then rarely fell below fiber saturation point. Short-term fluctuations in moisture content were highest during summer because storms and droughts caused a series of wetting–drying cycles. The response of branchwood to wetting was very rapid, with most pieces reaching saturation within 24 hours of the onset of rain. Bark cover appeared to have a minor influence on drying rates.

Seasonal variations of moisture in sound CWD have been studied by Hayes (1940) and Brackebusch (1975) in Idaho. Hayes (1940) monitored moisture content for 4 years using electrodes implanted in *Pinus monticola* logs. Drying proceeded from the outer regions toward the inside of the log and in all parts of the log except those in direct contact with the soil. Summer was a period of drying, and fall, winter, and spring were periods of recharge. During wetting cycles, the upper half of the log was the most apt to absorb water, although with longer storms the bottom half of the log was also wetted.

Brackebusch (1975) weighed debarked *T. plicata* logs at biweekly intervals to test the influence diameter, soil contact, and shade had on moisture content. The observation period was 19 years, during which the logs were replaced twice so that sound logs were observed. Unfortunately, the logs were weighed only during the fire season (May to October). However, some facets of log moisture cycles were revealed. As with the other two studies, logs started drying during spring and reached the annual low (15–30%) during August. Moisture content increased during late August and September and presumably remained high during fall, winter, and early spring. Increasing log size, contact with the soil, and shading all significantly increased moisture content.

As decay proceeds, the moisture content of CWD changes. For living trees, gymnosperms generally contain more moisture in the sapwood than angiosperms, 98–249% versus 44–146% (Peck, 1953). The heartwood of both taxa is drier than sapwood and is more similar in moisture content, with gymnosperms ranging between 31 and 121% and angiosperms ranging between 44 and 162%. In the case of undecayed wood, maximum moisture content is negatively correlated with density (Peck, 1953). A similar negative correlation between density and maximum moisture content has also been observed for decayed wood (Boddy, 1983; Yoneda, 1975), indicating that as decay proceeds, maximum moisture content also increases.

Water in CWD does not all come from external sources; for every 1 g of cellulose respired, 0.555 g of water is liberated (Griffin, 1977). However, the contribution of this respirational water to the overall moisture regime is not known.

3. Oxygen and Carbon Dioxide

The quality of the atmosphere within CWD influences the species of organisms present and the rate at which they degrade material. However, few studies have monitored either O_2 or CO_2 concentrations or the factors that control them, or examined their biological consequences.

The composition of air in logs is affected by moisture and temperature. As noted before, O_2 concentration decreases and CO_2 concentration increases as temperature increases (Paim and Becker, 1963) because of increased respiration. As moisture content rises above the fiber saturation point, respiration increases (Griffin, 1977) and causes CO_2 content to rise and O_2 content to fall. However, at very high moisture content, gas diffusion is restricted, and even slow rates of respiration deplete O_2. Gas diffusion even in relatively dry conifer wood is slow (Tarkow and Stamm, 1960) and would probably be even slower in saturated wood.

The response of wood-decaying organisms to elevated CO_2 and reduced O_2 concentrations in wood is known for few species. Oxygen is required for significant rates of microbial decomposition of lignin (Crawford, 1981; Kirk *et al.*, 1978; Zeikus, 1980). Benner *et al.* (1984b) observed fungal degradation of lignin under anaerobic conditions, but at rates far below those observed under aerobic conditions. Paim and Becker (1963) found that the cerambycid beetle *Orthosoma brunneum* inhabited decaying beech logs with O_2 concentrations as low as 2% and CO_2 concentrations as high as 15%. This species was more sensitive to low O_2 rather than high CO_2 concentrations and was not usually found in logs where O_2 concentrations dropped below 2%. The growth response of four wood-decaying fungi to O_2 and CO_2 was examined by Jensen (1967) in liquid-culture experiments where O_2 and CO_2 were varied between 0 and 40% and 0 and 30%, respectively. Fungal growth decreased with increasing CO_2 concentrations and increased with O_2 concentrations up to 20% for all four species. The effects of gaseous environments on fungi were reviewed by Tabak and Cooke (1968). They concluded that O_2 is essential for fungal growth, but the minimum concentrations for survival are very low, and that fungi are more sensitive to elevated CO_2 than low amounts of O_2.

4. Substrate Quality

CWD is a structurally and chemically heterogeneous substrate. This heterogeneity is partially responsible for the variations in CWD decay rates observed between species and within individual pieces of CWD. This section reviews how substrate quality varies from species to species and within individual boles and how these differences in turn influence decay rates.

Most ecological studies of CWD decomposition have considered entire tree boles. While this is a useful starting point, the differences between sizes of material and species will only be understood on a more detailed level. CWD can be divided into four components: outer bark; inner bark, which includes the cambium and the phloem; sapwood; and heartwood (Figure 6). This classification has proved useful in understanding which tree species are resistant to decay (Scheffer and Cowling, 1966) as well as how CWD varies internally as an invertebrate habitat (Savely, 1939). Each component can, in turn, be considered on an anatomical or chemical level. Although we will emphasize the chemical quality of the wood and bark, it should be noted that anatomical differences are often correlated to chemical differences and that anatomical structure appears to exert a great deal of control over how microbes colonize wood (Wilcox, 1973).

Each component can be divided into cell wall constituents, nutrient elements, and extractives, i.e., substances such as sugars, amino acids, fats, and waxes that can be removed using solvents. Cell walls are generally composed of cellulose, hemicelluloses, and lignin, which constitute the majority of CWD biomass. Some elements, such as Ca, may also be a part of cell walls.

a. Structural Components. The proportion of bark, sapwood, and heartwood in boles varies with species and size. The proportion of heartwood, for example, increases with diameter (Hillis, 1977), and because heartwood is usually the most resistant component, the overall bole decay rate should decrease with size. Also, decay rate should decline from the base to top of a bole because the proportion of heartwood decreases. The fraction of a

A

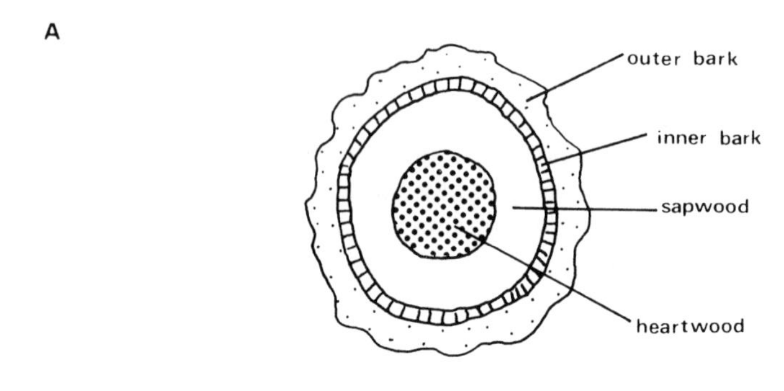

B

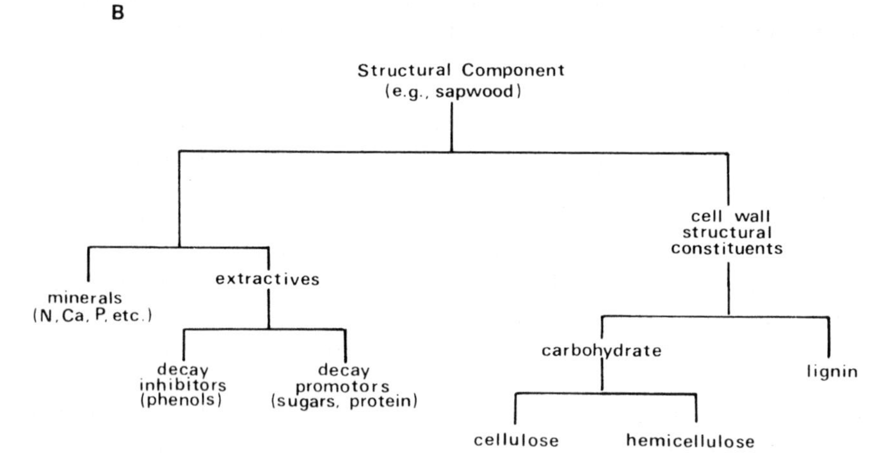

Figure 6 The structural (A) and chemical (B) division of CWD. Structurally, the cross section of a piece of CWD can be divided into outer bark, inner bark, sapwood, and heartwood. Each of these layers can in turn be broken into its chemical constituents, which influence the substrate quality of these structural layers.

bole of a given size in sapwood versus heartwood can often be predicted from diameter (Buchanan and Englerth, 1940). Likewise, bark volume can also be predicted from diameter for log segments (Pneumaticos *et al.*, 1972) or entire trees (Kozak and Yang, 1981).

b. Anatomical Structure. Gymnosperm wood is less complex and contains less living tissue than that of angiosperms. Gymnosperm sapwood contains 5–10% living tissue by volume (Panshin and deZeeuw, 1980, p. 132), while angiosperm sapwood contains 11–48% living tissues (Panshin and deZeeuw, 1980, p. 181). Increases in living cell volume probably increase decay rates because living tissue has a higher concentration of readily decomposable material such as sugars, starches, and protein. Also, nutrient content is higher in living than dead tissues. For example, Merrill and Cowling (1966) found that N content was positively correlated with parenchyma volume in angiosperm sapwood; increasing N content should speed decay. Most living-wood tissue is composed of rays that have their long axis oriented radially, thus forming pathways of colonization into CWD for microbes (Wilcox, 1973). The longitudinal orientation of water-conducting elements also influences microbial colonization patterns in that longitudinal growth occurs much more rapidly than radial growth. An anatomical feature that may cause angiosperms to decay faster than gymnosperms is the size and continuity of water-conducting elements (Wilcox, 1973). In general, vessels of angiosperms have larger diameters than gymnosperm tracheids and form a more continuous pathway, whereas tracheids are connected by pits. Both factors would tend to favor more rapid fungal colonization in angiosperms. There is considerable variation between angiosperms in the proportion of volume that is composed of vessels, e.g., 6.5% in *Carya ovata*, 50.8% in *Celtis occidentalis* (Panshin and deZeeuw, 1980, p. 181). Tracheids and vessel diameters also vary from spring to summer wood, and it is possible that colonization is faster in the larger diameter elements of the spring wood.

The size of the conducting elements also influences the depth fungal spores penetrate the exposed ends of CWD. Hintikka (1973), for example, found that spores of *Armillaria mellea* cannot penetrate *Alnus*, *Betula*, *Pinus*, or *Picea* wood in the longitudinal direction more than 1–2 cm, whereas few spores are stopped in *Fraxinus* and *Quercus* wood over a distance of 16 cm.

c. Cell Wall Chemistry. The chemical composition of wood cell walls has been studied extensively (Wenzl, 1970; Wise and Jahn, 1952) and will only be briefly reviewed here. Wood is largely composed of cellulose, the hemicelluloses, and lignin. The chemical structure of cellulose is well known, although its exact arrangement in fibrils and cell walls is quite complex and still being actively investigated. Cellulose is a linear polymer of D-glucose units joined by β-1,4 linkages and appears to be the major cell wall constituent of both gymnosperms ($\bar{x} = 44\%$) and angiosperms ($\bar{x} = 41\%$). Hemicelluloses are a diverse group of polysaccharides that compose 13–31% of gymnosperm wood ($\bar{x} = 22\%$) and 23–39% of angiosperm wood ($\bar{x} = 33\%$). These compounds are thought to be in an amorphous state surrounding the cellulose fibrils. Unlike cellulose molecules, hemicellulose molecules are branched and are composed of several sugars, including glucose, galactose, mannose, arabinose, xylose, and glucuronic acid. The terminology describing the hemicelluloses is beyond the scope of this article, but is described by Wenzl (1970) and Kirk (1973). Because hemicelluloses are more complex structurally and chemically than cellulose, microbes need a wider variety of enzymes to decompose them. Lignin is more amorphous and highly branched than are the hemicelluloses. This polymer has a highly complex structure (see Adler, 1977; and Gross, 1980, for reviews). It is basically phenylpropane units derived from *p*-coumaryl, coniferyl, and synapyl alcohols, though the proportion of these varies among taxa. This complexity of lignin makes it difficult for microbes to degrade. Moreover, microbial lignin degradation appears to be more sensitive to variations in O_2, N, and P than that of cellulose (Aumen, 1985). Of the three cell wall constituents, lignin is the most interesting ecologically because of its resistance to decay and the negative correlation between decay rates of leafy litter and increasing lignin content (Cromack and Monk, 1975; Fogel and Cromack, 1977; Melillo *et al.*, 1982). Melillo *et al.* (1983) found that decay rate constants of wood chips of different tree species were inversely related to lignin : nitrogen ratios in a first-order stream in Quebec; lignin content alone was the best predictor of wood decay in a sixth-order stream. CWD is thought to decay slower than leaf litter because of its higher lignin content (Käärik, 1974; Merrill and Cowling, 1966; Swift, 1977a).

Development of radioisotope labeling of substrates has permitted more exact measurement of decomposition rates of cellulose and lignin (Crawford and Crawford, 1976, 1978; Crawford *et al.*, 1977a,b, 1980). This approach has been modified and used to examine lignocellulose degradation in a variety of aquatic ecosystems (Aumen, 1985; Aumen *et al.*, 1983; Benner *et al.*, 1984a,b; Federle and Vestal, 1980; Maccubbin and Hodson, 1980). These studies indicate lignin breakdown is much slower than cellulose breakdown in aquatic ecosystems.

The generally slower decay rates of gymnosperms relative to angiosperms might be attributed, in part, to the higher lignin content of the latter. Gymnosperms have a higher mean lignin content (30%, range 26–34%) than angiosperms (24%, range 16–32%) (Table 2). The degree that this variation is correlated with decay rates is not clear. In many species, the heartwood is more resistant to decay than the sapwood (Scheffer and Cowling, 1966), and yet the lignin contents of extractive-free heart- and sapwood are quite similar within many species (Hawley and Wise, 1929).

The distribution of cellulose, hemicellulose, and lignin within cell walls is not uniform. Appreciation of these patterns is quite important when considering the ecological niche of microbes (Côté, 1977; Kirk, 1973; Sutherland *et al.*, 1979; Wilcox, 1973). Middle lamellae have higher concentrations of lignin (~75%) than the other cell wall layers and therefore might be expected to decompose slowest (Panshin and deZeeuw, 1980, p. 107). In contrast, cellulose composes only ~5% of the middle lamellae, but 50% of the S2 layer. The S1 layer is intermediate in composition between the middle lamellae and S2 layer, while the S2 and S3 layers are quite similar. The S2 layer dominates the overall cell wall composition because of its thickness.

d. Extractives and Heartwood Decay Resistance. Species vary considerably in the decay resistance of their heartwood. Species ratings are available because decay resistance is important in predicting wood durability in service (Reis, 1972; Scheffer and Duncan, 1947; U.S. Forest Products Laboratory, 1967). Wood extractives appear to be the principal source of decay resistance in heartwoods. In contrast to heartwood, sapwood, with its low level of extractives, has little resistance to decay (Humphrey, 1916). Scheffer and Cowling (1966) present an excellent and detailed review of wood extractives and their importance in promoting decay resistance. The fact that extractives from heartwood inhibit microbial growth was demonstrated over 60 years ago (Hawley *et al.*, 1924). Nevertheless, most ecological studies have failed to take wood extractives into account wfien comparing species and often seek to explain differences solely on the basis of N content, size, or climate.

Not all wood extractives inhibit microbes, so total extractive content may not predict decay resistance. Of the substances promoting decay in wood, sugars and protein are probably the most important. Smith and Zavarin (1960) measured the mono- and oligosaccharide content of outer bark, inner bark, sapwood, and heartwood in nine California conifers, which contained an average of 0.07, 2.68, 0.32, and 0.05%, respectively, of these substances on a dry weight basis. The protein content of cambium and sapwood was found to be 20–30% in the former, but only 0.83–1.37% in the latter (Allsopp and Misra, 1940). Both studies indicate that inner bark is more nutritious than outer bark or heartwood.

General classes of extractives inhibiting fungal decay include terpenoids, flavonoids, tropolones, and stilbenes (Scheffer and Cowling, 1966). Most decay-inhibiting compounds are polyphenolic in nature (Hillis, 1962, 1977; Scheffer and Cowling, 1966). Extremely small quantities of extractives give heartwood a large degree of decay resistance. For example, thujaplicins, which make *T. plicata* very decay resistant, occur in concentrations of less than 1.2% (Scheffer and Cowling, 1966). The ability of one compound to inhibit decay is often enhanced by the presence of other extractives, e.g., *Libocedrus decurrens*, where a mixture of four terpenoids, carvacrol, *p*-methoxycarvacrol, *p*-methoxythymol, and hydrothymoquinone, imparts greater decay resistance than these compounds acting alone (Scheffer and Cowling, 1966).

Extractive concentrations vary both between and within substrates. The distribution of dihydroquercetin in *P. menziesii* provides an excellent example of these patterns (Gardner and Barton, 1960; Hancock, 1957). The highest concentration of dihydroquercetin occurs in bark (7%), while the lowest generally occurs within the sapwood (<0.5%). Although heartwood contained large quantities of this compound, the concentration is highest near the sapwood–heartwood boundary (1.0–1.5%) and lowest near the pith (<0.6%). In fact, in many trees the heartwood near the pith had less dihydroquercetin than the sapwood. This radial decline in extractive content of heartwood occurs in many species including *Larix occidentalis* (Gardner and Barton, 1960), *T. plicata* (MacLean and Gardner, 1956), *Juglans nigra* (Nelson, 1975), and *Quercus rubra* (Nelson, 1975). Field and laboratory tests indicate heartwood decay resistance increases from the pith outward in genera such as *Quercus* (Scheffer *et al.*, 1949), Robinea (Scheffer and Hopp, 1949), and *Taxodium* (Campbell and Clark, 1960). Decay resistance also varies with height along the tree stem, with the highest degree of resistance in the outer heartwood at the base of the bole (Scheffer and Cowling, 1966).

Decay-resistant species appear to have more variability in decay resistance than nonresistant species. In soil-block tests, Clark (1957) demonstrated that species losing the least weight on mean basis also had the highest variation in weight loss. For example, in one trial lasting 3 months using *Poria monticola* as the decay fungi on *Pinus palustris*, a mean of 47% of the weight was lost. However, losses ranged from 16 to 52%. In contrast, *Abies* sp. lost a mean of 61% of the original weight, but the range was only 57–65%. Other studies indicate decay resistance varies considerably between populations of *Pseudotsuga* (Scheffer and Englerth, 1952), *Quercus* (Scheffer *et al.*, 1949), and *Taxodium* (Campbell and Clark, 1960).

Although extractives initially give the heartwood of many species high decay resistance, this protection gradually decreases with time. The time needed to deactivate extractives introduces a lag time into the decay of species having decay-resistant heartwood. Scheffer and Cowling (1966) listed four mechanisms whereby decay-inhibiting compounds can be inactivated: deactivation by enzymes in the heartwood, self-oxidation of the compounds, microbial degradation, and loss via leaching. Scheffer and Cowling (1966) indicate that microbes responsible for deactivation of extractives are often unable to decay cell walls. Thus, the presence of non-wood-decay fungi, such as molds, may be crucial for the colonization of wood-decay fungi. The role biotic deactivation and leaching play in reducing concentrations of decay-retarding substances is not known. Removal of substances allows factors such as N content to affect decoy processes. This is suggested by the work of Eslyn and Highley (1976), who found that under moderate decay conditions, sapwood of American tree species varied in its resistance to decay in spite of the lack of decay-inhibiting extractives. They hypothesized that species low in N decayed slower than those high in N.

e. Nutrient Elements. The nutrient content of CWD influences decomposition rates, nutrient cycling, and the suitability of CWD as a rooting medium. The elements required by microbes, invertebrates, vertebrates, and plants differ. There are 17 elements essential to higher plants: carbon (C), hydrogen (H), oxygen (O), phosphorus (P), potassium (K), nitrogen (N), sulfur (S), calcium (Ca), iron (Fe), magnesium (Mg), boron (B), manganese (Mn), copper (Cu), zinc (Zn), chlorine (Cl), and molybdenum (Mo). Fungi require all of these except Ca and B, but gallium (Ga), scandium (Sc), and vanadium (V) are possibly essential for some fungi (Lilly, 1965). Nitrogen-fixing prokaryotes require cobalt (Co). Animals require iodine (I), sodium (Na), selenium (Se), chromium (Cr), tin (Sn), vanadium (V), fluorine (F), silicon (Si), and nickel (Ni), in addition to the basic set required by plants. Of these elements, C, H, and O comprise all but a few percent of undecayed CWD. The order of the remaining elements from least to most abundant in undecayed wood is approximately $Ca = N > K \gg Mg = Mn > S = P > Na > Fe = Zn > Cu > B > Mo$ (Table 4).

The concentration of nutrient elements in CWD is generally an order of magnitude less than that of leaves, flowers, or fruits (Likens and Bormann, 1970; Whittaker *et al.*, 1979; Woodwell *et al.*, 1975; Young and Guinn, 1966). The organisms feeding on CWD appear adapted to grow on very low nutrient concentrations (Swift, 1977a). Cowling and Merrill (1966) reviewed adaptations of fungi in terms of N; they may apply to other elements and organisms as well. Physiological adaptations include reduced N content in structural substances so more N is available for enzymes; simplified metabolic pathways so the number of enzymes required is minimized; and increased number of times an enzyme is used so the amount of N needed for a metabolic task is minimized. Wood-destroying fungi also conserve nutrients by autolysis and thus reuse elements otherwise locked up in inactive mycelium. Although fungi grow on wood without an outside nutrient source, elements may be taken up from the soil from throughfall and from litter leachates. Conservation of elements from these outside sources would also enable fungi to decay nutrient-poor CWD.

Invertebrates appear to have adapted to the low nutrient concentrations of CWD (see Section V.B.3). Bark beetles attack the relatively nutrient-rich inner bark and generally avoid wood altogether. The cerambycids only attack the wood after feeding extensively on inner bark. Thus, essential

Table 4 Mean Elemental Content of Bark and Wood for Selected Temperate Species[a]

Element	Mean elemental content (ppm)			
	Gymnosperm		Angiosperm	
	Bark	Wood	Bark	Wood
Nitrogen	2904	704	5129	1189
Phosphorus	589	54	445	110
Potassium	1711	505	2417	1082
Calcium	7232	741	15747	1248
Magnesium	620	157	709	234
Sulfur	540	117	554	176
Manganese	772	149	636	97
Iron	84	10	60	17
Zinc	57	9	83	16
Copper	7	6	8	5
Sodium	90	25	69	29
Boron	11	1	13	1
Molybdenum	4	<1	7	<1
Aluminum	100	—[b]	22	12

[a] From Alban *et al.* (1978), Attiwill (1980), Duvigneand and Denaeyer-DeSmet (1970), Fahey (1983), Hart (1968), Lambert *et al.* (1980), Likens and Bormann (1970), Pastor and Bockheim (1984), Whittaker *et al.* (1979), Woodwell *et al.* (1975), Young and Guinn (1966).
[b] Not available.

elements are gathered from a nutrient-rich region and then moved into a nutrient-poor one. Other insects circumvent low-nutrient concentrations, eating fungi or other insects that have already concentrated essential elements. Yet another adaptation to growing in CWD is an extended life cycle. While most insects complete one to many life cycles in a year, species feeding on wood, such as *Ergates spiculatus* (LeConte) and *Lara avara* (LeConte), take several years to complete their life cycles.

Initial concentrations of many inorganic elements in wood have been reviewed by Ellis (1965); data on other tissues, however, appear to be scant or only available for a few elements. Data from a number of studies have been summarized in Table 4, and although various analysis methods were used, a general indication of the concentrations to be found is represented.

Elemental concentrations vary with species, structural components, and position within trees. Angiosperm wood generally has higher elemental concentrations than gymnosperm wood (Table 4). When bark is considered, angiosperms have higher concentrations of N, K, and Ca than gymnosperms. Both P and Fe appear to be more concentrated in gymnosperm bark than angiosperm bark. Bark contains higher concentrations of elements than wood in both angiosperms and gymnosperms; sapwood and heartwood have similar concentrations of K, Ca, S, Zn, and Cu (Likens and Bormann, 1970; Woodwell *et al.*, 1975). In contrast, P, Mg, Fe, and Na are more concentrated in the sapwood, while Mn is more concentrated in the heartwood. In the few species in which the inner bark has been separated from the outer bark, it appears that N, P, and K concentrations are higher in the inner bark, but concentrations of other elements vary from species to species (Woodwell *et al.*, 1975).

Merrill and Cowling (1966) found N was most concentrated in cambium and least concentrated in heartwood. Within sapwood, N content steadily declined from the youngest annual ring to the sapwood–heartwood boundary. Concentrations were constant throughout heartwood except in the pith region, where they were very high. Certainly, much remains to be learned about the distribution of elements in both undecayed and decayed CWD.

The exact role nutrient content of CWD plays in controlling terrestrial decomposition rates has yet to be explored. Most work has involved additions of various forms of N, and the results of early experiments were summarized by Cowling and Merrill (1966). N content of natural wood is suboptimal for fungi, and addition of inorganic or organic N increases decay rates. In nature, the exact degree to which nutrient content controls decay rates of CWD is less than clear because of the confounding influences of decay-resistant extractives.

Wood decomposition in streams is strongly influenced by the nutrients available in the surrounding water. Lignin and cellulose degradation were stimulated by addition of either organic or inorganic N (Aumen *et al.*, 1983). The relative increases in rates of breakdown caused by N addition were greater for lignin than for cellulose. Furthermore, lignin degradation was sensitive to the form of inorganic N, whereas cellulose degradation was not. Additions of NO_3^- stimulated lignin decay more than additions of NH_4^+. Repression of nitrate metabolism by NH_4^+ may have caused the lesser stimulation of lignin degradation (Aumen *et al.*, 1983).

Addition of either nitrate or phosphate to the water surrounding wood can increase rates of cellulose breakdown, and addition of both speeds degradation even more dramatically (Aumen, 1985). In contrast, lignin decomposition was enhanced only by addition of nitrate and phosphate in combination. The species of wood modifies the potential influence of exogenous nutrients on wood decay. In a laboratory microcosm study, Melillo *et al.* (1985) found that addition of inorganic phosphorus stimulated decay rates of *Alnus* wood shavings, but had no effect on the decay of *Picea* wood shavings. The authors hypothesized that the higher lignin content of *Picea* limited the stimulatory effects of phosphorus.

In studies of nitrate stimulation of lignocellulose degradation, Aumen (1985) observed that organic N content of wood samples doubled and that ammonium accumulated in solution. Subsequent experiments with ^{15}N tracers revealed that the rapid depletion of nitrate was related to breakdown of lignin and cellulose. Introduced nitrate was rapidly immobilized as organic N on the wood substrate. Enrichment of the ammonium in solution with ^{15}N lagged behind the incorporation of $^{15}NO_3$ into the organic fraction, suggesting a subsequent release of ammonium from an organic N source. Rates of N fixation and denitrification were insignificant. ^{15}N appeared in the ammonium fraction in solution within 1 hr of introduction of nitrate, possibly as a result of dissimilatory reduction of nitrate to ammonium. Nitrogen transformations associated with microbial decomposition of CWD are potentially complex and tightly coupled, and our understanding of their role in stream ecosystems requires a thorough integration of laboratory and field studies.

Availability of inorganic P in stream water may increase the rate of CWD decay. Decay rates of *Alnus*, *Betula*, *Populus*, *Abies*, and *Picea* were greater in a first-order than in a ninth-order stream (Melillo *et al.*, 1983). Concentrations of inorganic P in the first-order stream were more than double the P concentrations in the ninth-order stream (0.005 mg liter^{-1} versus 0.002 mg liter^{-1}).

5. *Size Effects*

Larger boles decay slower than twigs, branches, and small boles. Studies of small pieces of wood have demonstrated a negative correlation between size and decay rate. Abbott and Crossley (1982) measured decay of *Q. prinus* branches and found those with diameters of 0–1, 1–3, and 3–5 cm had half-times of 5.5, 6.2, and 7.1 years, respectively. Harris *et al.* (1972) reported that logs averaging 11.4 cm in diameter had half-times of 6.2 years, while those averaging 6.4 cm in diameter had half-times of 2.8 years. Comparisons of studies using different sizes of the same species also strongly suggest a negative correlation between size and decay rate. Fogel and Cromack (1977) found small *P. menziesii* branches in Oregon had half-times of 8 years. In contrast, Graham (1982) reported half-times of 58 years for *P. menziesii* boles in the same region. Boles of *A. balsamea* and *P. rubens* in New Hampshire had half-times of 23 and 22 years, respectively (Foster and Lang, 1982; Lambert *et al.*, 1980), whereas twigs of the same two species 5 mm in diameter had half-times of 6 and 3 years, respectively (Gosz *et al.*, 1973).

Studies of CWD often fail to demonstrate a negative size-to-decay rate correlation. For example, Graham and Cromack (1982) reported *Picea* and *Tsuga* logs over 60 cm in diameter did not decay significantly slower than smaller logs. In the Oregon Cascade Range, smaller *Tsuga* and *Pseudotsuga* snags fragmented faster than large ones, but there was no size effect for logs (Graham, 1982). Decay rate of snags < 20 cm in diameter in the southern Appalachians was not significantly affected by size (Harmon, 1982). Harmon and Cromack (unpublished) failed to find a significant effect of size on decay rates in *A. concolor* logs, although they observed that the lower, larger diameter part of the bole remained sound after the upper stem had decomposed. The lack of correlation between size and decay rate in these studies of boles may be in part methodological because the high degree of variation associated with decay curves indicates a very large number of samples is required to detect size differences.

In spite of the relationship sometimes observed between size and decay rate, the cause remains unexplained. Many studies have invoked the changes in the ratio of surface area to volume (SA/V) to explain the pattern (Abbott and Crossley, 1982; Fogel and Cromack, 1977; Harris *et al.*, 1972; Triska and Cromack, 1980). As the size of an object increases, SA/V declines, which reduces the rate of gas and liquid exchange. However, in absolute values, the greatest change in SA/V occurs for small diameter material. The failure to detect a negative correlation between decay rate and size in boles may be caused by small changes in SA/V for large-diameter pieces.

The negative correlation between size and decay rate may also be caused by changes in substrate quality and/or colonization patterns. The proportion of heartwood increases as diameter increases (Figure 7) (Hillis, 1977); therefore, decay rate should decline with increased size for decay-resistant species. The contribution of the inner bark to total mass also decreases as diameter increases, which reduces the nutrients available for decomposers. Colonization patterns may also cause a negative size-to-decay rate correlation. Decomposers colonize leaves and small branchwood within a period of days to months, but in tree boles, this process takes years. Timber salvage studies after fires and windstorms indicate the rate fungi and insects colonize boles. Decay radially penetrated fire-killed *P. menziesii* boles in Oregon and Washington at rates of 0.8–1.8 cm year^{-1} (Kimmey and Furniss, 1943). In fire-killed trees in California, the proportion of volume colonized by decomposers increased with snag age, but decreased with diameter at breast height (Kimmey, 1955). A similar pattern was observed by Buchanan and Englerth (1940) in windthrown timber in Washington (Figure 7). Colonization patterns are also important in aquatic ecosystems, where microbial activity is often restricted to

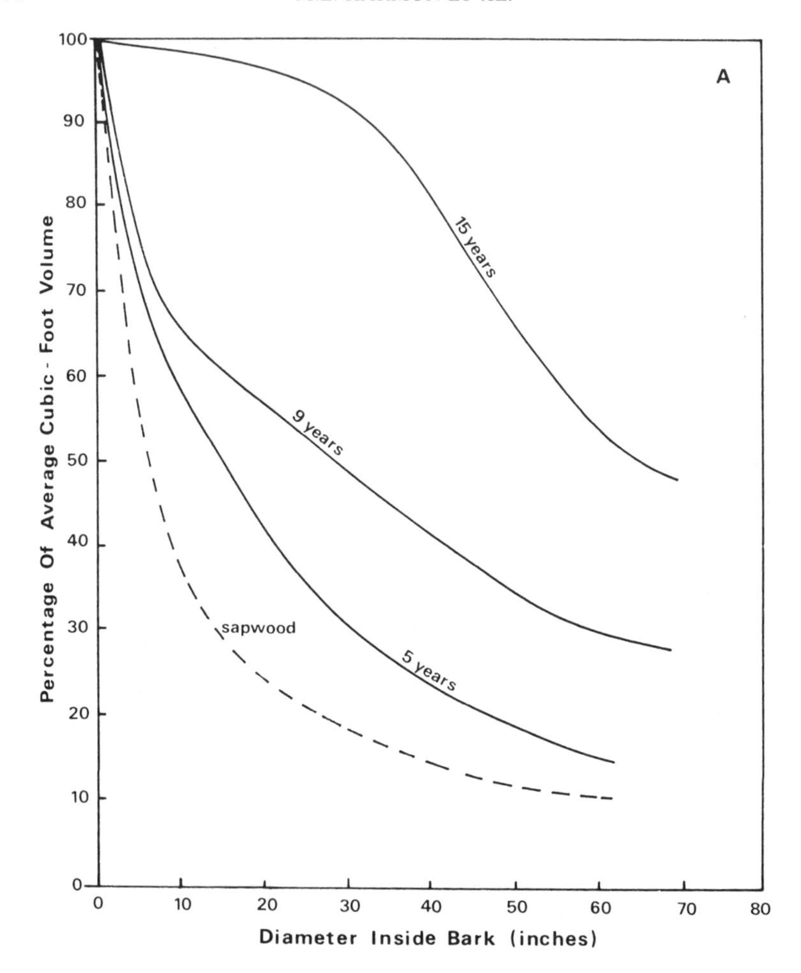

Figure 7 Fungal colonization of windthrown (A) *Picea sitchensis* and (B) *Pseudotsuga menziesii* boles. The dashed line indicates sapwood volume and the solid line indicates the volume of decayed wood 5, 9, and 15 years after the windstorm. Colonization of *P. sitchensis* is considerably faster than that of *P. menziesii*, but decreases with size for both species.

superficial layers. These fungal colonization patterns indicate larger pieces should decay slower than small pieces.

6. Decomposer Organisms

Perhaps the most important factors determining the rate and type of decay are the organisms involved. Moreover, it is through these organisms

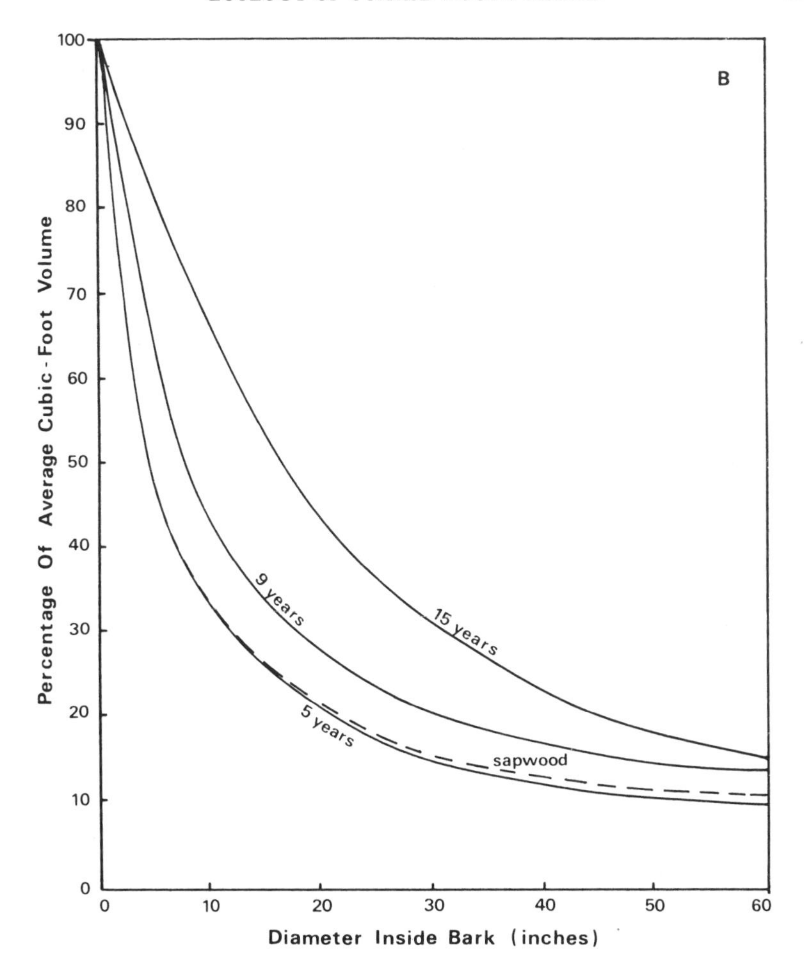

Figure 7 Continued.

that the effects of temperature, moisture, aeration, substrate quality, and size on decay rates are expressed. The following section discusses several taxonomic groups and their importance in terms of CWD decay. It must be borne in mind, however, that most of these taxa act in concert and that they may interact synergistically or antagonistically.

a. Microbes. The microbes that attack CWD can be considered from a taxonomic, synecological, or nutritional and functional perspective. For our purposes, the nutritional and functional perspective is most relevant. However, taxonomic and synecological relationships are also important and have been reviewed by Watling (1982) and Rayner and Todd (1979, 1982), respectively. Functionally, microbes decomposing wood can be divided into those that live on cell contents versus those that degrade cell wall components (Käärik, 1974; Levy, 1982; Swift, 1977a). The former group is frequently divided into molds and staining fungi, while the latter group is divided into bacteria, soft rots, brown rots, and white rots.

The molds are ascomycetes, fungi imperfecti, and phycomycetes that generally feed on cell contents, although some degrade cell walls (Käärik, 1974). Levy (1982) divided molds into primary and secondary types. Primary molds are among the first to colonize CWD and feed on sugars and other simple carbohydrates occurring in parenchyma tissues and sapwood. In contrast, secondary molds are associated with the brown- and white-rot basidiomycetes. The limited amounts of sugars and other simple polysaccharides in wood tissue mean that primary mold fungi do not cause a major loss of biomass from CWD. However, the use of nonstructural carbohydrates by these fungi may inhibit the colonization of wood by basidiomycetes (Hulme and Shields, 1970).

Stain fungi are ascomycetes and fungi imperfecti that have pigmented hyphae and stain-invaded tissues. As with primary molds, stain fungi feed on materials stored in the sapwood and quickly colonize CWD. As in the case of primary molds, the actual loss in weight caused by stain fungi is very small.

Bacteria break down cell walls, but at rates that are much slower than basidiomycetes (Käärik, 1974). These taxa appear to be most important in moist environments. Earlier studies (e.g., Willoughby and Archer, 1973) isolated many fungal species from submerged wood. The biological activity of these species has recently been called into question, however. For example, scanning electron microscopy (SEM) indicates that fungal hyphae are uncommon on the surface of submerged CWD (Aumen *et al.*, 1983). In contrast to fungi, single-celled bacteria and actinomycetes are very common and are probably the major CWD decomposers in aquatic environments (Aumen *et al.*, 1983). Several specific bacteria have been isolated and implicated in soft rot in wood found in aquatic environments—*Bacillus polymixa*, *Bacillus cereus*, and *Bacillus macerans* (Ellwood and Ecklund, 1959; Knuth and McCoy, 1962). Other bacteria found on submerged wood include *Klebsiella pneumoniae*, *Enterobacter agglomerans*, and *Enterobacter* spp. (Aho *et al.*, 1974; Buckley and Triska, 1978; Knuth, 1964). As with molds and stain fungi, the parenchyma tissue is decomposed by bacteria. Other important sites of bacterial activity are pit chambers where bacteria are involved in the breakdown of pit membranes (Sutherland *et al.*, 1979). This in turn increases decay rates by allowing gases and water to move more freely within the wood and also by increasing access to those organisms unable to penetrate cell walls (Levy, 1982). Some bacteria and actinomycetes are able to fix atmospheric nitrogen, which may increase the decomposition rates of other decomposers. *Enterobacter* may play an important role in nutrient cycling because of its ability to fix nitrogen in both terrestrial and aquatic environments (Aho *et al.*, 1974; Buckley and Triska, 1978). Bacteria are also undoubtedly important in decomposing dead fungal hyphae.

Soft rots are ascomycetes and fungi imperfecti that break down cell walls, but primarily utilize cellulose and hemicellulose while modifying lignin to a small extent (Levi, 1965). Soft rots are most important in degrading wood in contact with the soil or those pieces with a high or variable moisture content because they tolerate poor aeration better than basidiomycetes (Duncan, 1961). Thus, their action is probably highest in aquatic environments or in riparian zones. Angiosperm wood is more susceptible to attack by soft rots than gymnosperm wood. The manner in which soft rots attack wood is unique on both a macro- and microscale. On the macroscale, these fungi cause a slow degradation, advancing inward from the surface after destroying outer layers (Käärik, 1974). On the microscale, soft rots form distinctive cavities in the S2 layers of tracheids and fibers (Levi, 1965; Savory, 1954a).

Brown rots are caused by basidiomycetes that decompose cellulose and hemicellulose, but not lignin. Their chemical influence on CWD is therefore quite similar to that of soft rots. Brown-rot fungi do not thin cell walls until late in decomposition (Montgomery, 1982). Along with white rots, brown-rot fungi are responsible for the majority of CWD decomposition in terrestrial ecosystems.

White rots are also basidiomycetes, but, unlike brown rots, are able to decompose lignin as well as cell wall polysaccharides. The action of white rots is confined to near the hyphae (unlike the brown rots), and cell wall material is lost from the cell lumen toward the middle lamella. This sequential loss of material causes furrows to be formed in the cell wall, and in advanced decay, these furrows may join (Montgomery, 1982).

Much is known about how these functional groups modify wood chemically and structurally. Most current knowledge, however, is based on pure cultures, and much remains to be learned about how these organisms interact. Using a scanning electron microscope, Blanchett and Shaw (1978) observed that yeast, bacteria, and basidiomycetes grew in close association in decomposing wood. Furthermore, they demonstrated that mixed cultures decayed wood faster than basidiomycetes alone. In another situation, the presence of mold fungi inhibited the colonization of wood by brown rot (Toole, 1971) and thus inhibited decay.

b. Invertebrates. Invertebrates, chiefly insects, play a significant role in the decomposition of CWD by attacking wood directly or by influencing other organisms. The impact invertebrates have on CWD depends upon several factors: the stage of the tree during attack, the part of the tree utilized, and the other associated organisms. Many factors influence the abundance of invertebrates and hence the rate they decompose CWD. Substrate quality and extractives profoundly influence the extent and rate to which CWD is colonized. For example, the wood-boring beetle fauna is very limited in decay-resistant tree taxa such as *Sequoiadendron* and *Taxus* (Smythe and Carter, 1969). Differential feeding trials of *Reticulitermes flavipes* (Kollar) on wood indicate high survival rates on wood of most tree species, but low survival on *J. nigra* and certain death when fed non-oven-dried *Sequoiadendron* and *Taxodium*.

Specific examples of the way invertebrates influence decay include the following. Carpenter ants reduce CWD to dust and deposit this material outside the branch or log without ingestion. Wood-boring bees may do the same. Many insects eat wood, thereby reducing the particle size and modifying the wood during digestion. Insects introduce microbes into the CWD at the time of attack, thus hastening decay (Ausmus, 1977; Swift, 1977b). Invading invertebrates have an associated fauna of parasites and predators that follow and their feces and decomposing body parts contribute nutrients to the CWD ecosystem. The cavities that are created, combined with the variously modified wood particles, create suitable environments for microorganisms. These same microorganisms, besides decomposing CWD, serve as food for invertebrates such as springtails (Collembola), mites, and dipteran larvae. The presence of galleries also allows nonboring invertebrates such as millipedes, centipedes, and wood lice to invade CWD. Invertebrates attract larger vertebrate predators (skunks, bears, and woodpeckers) that fragment CWD while searching for prey.

Despite these general observations on decay caused by invertebrates, few field studies have measured process rates. Hickin (1963) provides a detailed account of the major wood-boring insects of Britain. He includes information on *Xestobium rufovillosum*, the death-watch beetle, including the relationship between the life cycle of the beetle and the reduction of the wood over time. The degree of wood decay at the time of oviposition determined the length of the life cycle of the insect. The life cycle was as long as 55 months when wood (oak sapwood plus white rot) showed 18% weight loss and <12 months when wood showed 73% weight loss. Swift *et al.* (1976) discuss the decomposition of branch wood in a mixed deciduous woodland in Britain and include estimates of decay rates following branch death, branch fall, animal invasion, and termination. Swift (1977b) reports a net loss of nutrients occurred in branches invaded by animals; in this instance, the dominant organism was a crane fly larva.

Laboratory studies of termites indicate wood consumption rates of 10–90 mg (g termites)$^{-1}$ day^{-1} (dry weight of food/fresh weight of termites) (Wood, 1976). Field estimates of consumption rates by populations of known size are virtually nonexistent (Wood, 1976, 1978). Wood and Sands (1978) calculated a mean weight-specific consumption rate of 30 mg g^{-1} live weight per day for all groups of termites except the Macrotermitinae (chiefly fungus feeders). Smythe and Carter (1969) cite consumption figures for *Reticulitermes flavipes* in feeding trials of wooden blocks of different tree species. Their procedure involved placing blocks of wood (oven-dried and non-oven-dried) on sand or sawdust for 8 weeks. Consumption rates ranged from 0.003 mg termite^{-1} day^{-1} (*Sequoia sempervirens*, non-oven-dried, on sawdust) to 0.088 mg termite^{-1} day^{-1} (*Pinus palustris*, oven-dried, on sand). Wood (1976) cautions on extrapolating from feeding trials where only a single food source is provided. Gentry and Whitford (1982) studied several species of subterranean termites in Georgia by placing pinewood blocks (fresh and preinoculated with fungi) on the ground in different habitats and monitoring feeding activity. Blocks placed in the *P. palustris* habitat had lost $\sim$17% of the original weight in 9 months due to termite feeding compared to a 5–7% consumption loss due to termites in *Q. nigra* and *Liriodendron* habitats.

Savely (1939) estimated the amount of wood eaten by the larvae of the cerambycid *Callidium antennatum* ranged from 1.26 to 3.37 g. Expressed as grams of dry wood per gram of dry larvae, the average was 77.9 g. Similar calculations for the buprestid *Chrysobothris* sp. revealed that 0.39–2.06 g of wood was eaten by larvae (Savely, 1939). Expressed as grams of dry wood eaten per gram of dry larvae, the average was 79.0 g, which is remarkably similar to the value for *Callidium*.

IV. AMOUNT AND DISTRIBUTION OF COARSE WOODY DEBRIS

The amount of CWD in an ecosystem represents the balance between additions from tree mortality and breakage, on one hand, and losses caused by respiration, fragmentation, and transport, on the other. However, the quantity of CWD added to ecosystems varies considerably both spatially and temporally (see Section II.C–D), leading to large fluctuations in CWD mass away from the predicted steady-state level of biomass. Moreover, these pulses of CWD input decay slowly (see Section III.D), and a great deal of time must elapse before these departures from the steady-state level are eliminated. This strong historical influence introduces wide fluctuations in CWD amounts, making it difficult to evaluate the influence that temperature, moisture, and substrate quality have on controlling the amount of CWD.

A. Estimating Biomass

1. Volume Estimation

Volume, total surface area, and projected area can be estimated by recording the length, diameter, decay status, orientation, and species of pieces within a quadrat or unit of stream area (Froehlich, 1973; Swanson *et al.*, 1984). Various plot sizes and shapes have been used in this type of sampling; Warren and Olsen (1964) found long rectangles more efficient than circular plots.

The line-intersect method is used extensively to estimate volume of woody fuels and logging residue (Brown, 1974; Pickford and Hazard, 1978; Van Wagner, 1968; Warren and Olsen, 1964) or fine woody debris in streams (Froehlich, 1973; Swanson *et al.*, 1984). In upland areas, a series of lines of known length are established in random directions; lines for sampling streams are oriented at a 90° angle to the channel and located at regular intervals along streams. The diameter, species, and decay status are recorded for all pieces intersecting a horizontal plane defined by the line. Assuming the pieces are cylindrical, horizontal, and randomly oriented, volume can be calculated using the formula

$$V = \pi^2 \Sigma d^2 / 8L$$

where V is the volume, d is the diameter of a piece, and L is the transect length.

2. Mass Determination

Conversion of volume to mass requires subsampling of logs and snags for density. The measurement of density is discussed in detail in Section III.A.

In the case of aquatic studies, density has rarely been measured, and a density of 0.5 Mg m^{-3} has often been applied. While this value probably does represent the average for undecayed wood of all species, it overestimates the density of undecayed coniferous wood by as much as 20–40% and underestimates the density of undecayed hardwood by the same amount.

In many fire-fuels studies, wood is separated into sound and rotten classes (Brown and See, 1981; Sacket, 1979), with densities of 0.4 and 0.3 Mg m^{-3}, respectively. However, our studies (Table 5) indicate the mean density of wood for coniferous forests ranges between 0.16 and 0.20 Mg m^{-3}, suggesting that use of the former wood densities may have resulted in overestimates of CWD biomass by a factor of 1.6–2.0.

The density of CWD has been estimated by using a decay-class system. Wood-decay classifications commonly place CWD into three to five decay classes that are then sampled to determine the mean density for each class. Definition of decay classes varies between studies. Using the penetration of a pointed 2.5-cm metal rod into a log as an indicator, Lambert *et al.* (1980) defined three classes: (1) slightly decayed, rod penetrates <0.5 cm; (2) moderately decayed, rod penetrates 0.5 cm to half the diameter; and (3) advanced decay, rod can be pushed completely through the log. Lang and Forman (1978) also used the penetration depth of a pointed rod to define three decay classes, but in the moderate decay class the rod penetrated to 2.5 cm.

External characteristics of CWD have also been used to define decay classes (Fogel *et al.*, 1973; MacMillan, 1981; Sollinś, 1982; Triska and Cromack, 1980). These include bark cover; the presence, color, and abundance of attached needles, twigs, and branches; the cover of bryophytes and lichens; species and size of fungal fruiting bodies; the color, crushability, moisture, and structure of the wood; the type of decay present (e.g., brown cubical versus white stringy rot); whether the exposed wood is bleached; whether the log supports itself or has collapsed under its own weight; the cross-sectional profile, ranging from circular to extremely elliptical; the age, size, and density of tree seedlings and saplings growing on the log; the presence and distribution of roots throughout the wood; and the presence of various decay processes such as sapwood sloughing. Although this is a long list, in many cases only the presence or absence of a single characteristic is needed to classify the log or snag. In many studies, the criteria used to define the classes are chosen subjectively by the investigator (Fogel *et al.*, 1973; MacMillan, 1981; Sollins, 1982; Triska and Cromack, 1980). However, more objective classification methods, such as those traditionally used in plant community analysis, are applicable (see Gauch, 1982; Pielou, 1977). Cline *et al.* (1980), for example, successfully used cluster analysis to define five snag decay states.

3. Statistical Considerations

Statistical distributions of biomass parameters need to be considered along with average values when comparing CWD in ecosystems.

Table 5 Volume, Biomass, and Projected Area of Logs and Snags for Various Temperate Forested Ecosystems

Ecosystem	Stand age (years)	Lower limit (cm)	Volume (m^3 ha^{-1})		Biomass (Mg^3 ha^{-1})		Projected area (%)		Refs.[a]
			Logs	Snags	Logs	Snags	Logs	Snags	
Coniferous forests									
Abies amabilis	23	10	—[b]	—	20	60	—	—	7
Abies amabilis	130	10	—	—	75	157	—	—	7
Abies balsamea	12	3	~90	~80	35	27	—	—	14
Abies balsamea	22	3	~107	~52	30	17	—	—	14
Abies balsamea	33	3	~130	~30	32	12	—	—	14
Abies balsamea	46	3	~60	< 10	15	0.25	—	—	14
Abies balsamea	52	3	~50	< 10	13	0.25	—	—	14
Abies concolor	—	15	216	178	49	52	6.8	2.8	11
Larix occidentalis	—	7.5	—	—	40	—	—	—	1
Picea–Abies	200+	7.5	416	—	97	—	12.1	—	10
Picea–Abies	—	7.5	—	—	53	—	—	—	1
Picea–Tsuga	200+	15	—	—	93	26	6.1	0.1	5
Pinus–Abies	—	15	151	130	33	43	4.5	1.9	11
Pinus jeffreyi	—	15	82	2	28	0.7	2.1	0.04	11
Pinus contorta	—	7.5	—	—	32	—	—	—	1
Pinus contorta	105	—	—	—	6	6	—	—	2
Pinus contorta	70	—	—	—	1	2	—	—	2
Pinus contorta	240	—	—	—	1.5	41	—	—	2
Pinus ponderosa	—	7.5	—	—	23	—	—	—	1
Pinus ponderosa	—	7.5	54	—	18	—	—	—	17
Pinus mixed	50+	7.5	~30	—	7	—	1.5	—	10
Pinus mixed	50+	7.5	—	—	11	—	—	—	9
Pinus mixed	—	2.5	—	—	17.8	—	—	—	12

Pseudotsuga/Abies–Picea	—	7.5	125	—	42	—	—	—	17
Pseudotsuga–Tsuga	450	—	500	65	190	24.6	—	—	8
Pseudotsuga–Tsuga	450	15	396	270	81	54	—	—	18
Pseudotsuga–Tsuga	100	15	491	153	108	34	13.7	1.6	3
Pseudotsuga–Tsuga	130	15	309	167	65	42	12.3	2.9	3
Pseudotsuga–Tsuga	250	15	488	339	93	80	20.0	4.5	3
Pseudotsuga–Tsuga	450	15	490	349	82	84	16.4	3.9	3
Pseudotsuga–Tsuga	750	15	554	635	97	105	20.1	5.0	3
Pseudotsuga–Tsuga	1000+	15	574	376	115	81	20.2	3.5	3
Pseudotsuga–Tsuga	3	20	673	—	249	—	—	—	13
Pseudotsuga–Tsuga	19	20	981	—	342	—	—	—	13
Pseudotsuga–Tsuga	110	20	389	—	138	—	—	—	13
Pseudotsuga–Tsuga	181	20	376	—	130	—	—	—	13
Pseudotsuga–Tsuga	515	20	1421	—	490	—	—	—	13
Sequoiadendron–Abies	—	15	722	51	247	22	7.0	1.0	11
Thuja–Tsuga	—	7.5	—	—	66	—	—	—	1
Tsuga heterophylla	130	15	—	—	212	—	—	—	6
Tsuga–Picea	200+	15	—	—	167	40	11.3	0.1	5
			Deciduous forests						
Acer–Betula	20	3	—	—	31.6	6.2	—	—	20
Acer–Fagus	30	3	—	—	12.8	7.9	—	—	20
Acer–Betula	40	3	—	—	7.6	8.4	—	—	20
Acer–Fraxinus	57	3	—	—	4.7	5.1	—	—	20
Acer–Fagus	60	3	—	—	10.9	12.2	—	—	20
Acer–Betula	83	3	—	—	16.8	8.2	—	—	20
Acer–Fagus	~200	3	—	—	38.4	10.9	—	—	20
Fagus–Betula	200+	7.5	82	—	29	—	3.0	—	10

(Continued)

Table 5 Continued

Ecosystem	Stand age (years)	Lower limit (cm)	Volume (m^3 ha^{-1})		Biomass (Mg^3 ha^{-1})		Projected area (%)		Refs.[a]
			Logs	Snags	Logs	Snags	Logs	Snags	
Liriodendron	40	7.5	51	—	14	—	4.0	—	10
Liriodendron	—	2.5	—	—	7.5	—	—	—	12
Populus tremuloides	—	—	—	—	14	11	—	—	4
Prunus pensylvanica	10	3	—	—	31.3	4.0	—	—	20
Quercus–Carya	—	2.5	—	—	15	—	—	—	12
Quercus mixed	330+	10	—	—	21	—	—	—	15
Quercus mixed	220+	5	46	—	16	—	1.6	—	16
Quercus mixed	—	—	—	—	—	—	2.0	—	19
Quercus mixed	200+	7.5	94	—	24	—	1.6	—	10
Quercus prinus	200+	7.5	132	—	21	—	1.7	—	10
Quercus prinus	40+	7.5	—	—	11	—	—	—	9
Quercus prinus	—	2.5	—	—	22	—	—	—	12

[a] (1) Brown and See (1981), (2) Fahey (unpublished), (3) Franklin *et al.* (unpublished), (4) Gosz (1980), (5) Graham and Cromack (1982), (6) Grier (1978), (7) Grier *et al.* (1981), (8) Grier and Logan (1977), (9) Harmon (1980), (10) Harmon (unpublished), (11) Harmon and Cromack (unpublished), (12) Harris et al. (1973), (13) Huff (1984), (14) Lambert *et al.* (1980), (15) Lang and Forman (1978), (16) MacMillan (1981), (17) Sacket (1979), (18) Sollins (1982), (19) Thompson (1980), (20) Tritton (1980). Unpublished data is on file at Forestry Sciences Laboratory, RWU-1251, Corvallis, Oregon.

[b] Not available.

These quantities are usually not normally distributed, but tend to be positively skewed (Figure 8), with the mode lower than the mean or the median. Sample variances associated with CWD biomass are very high, necessitating a large number of samples.

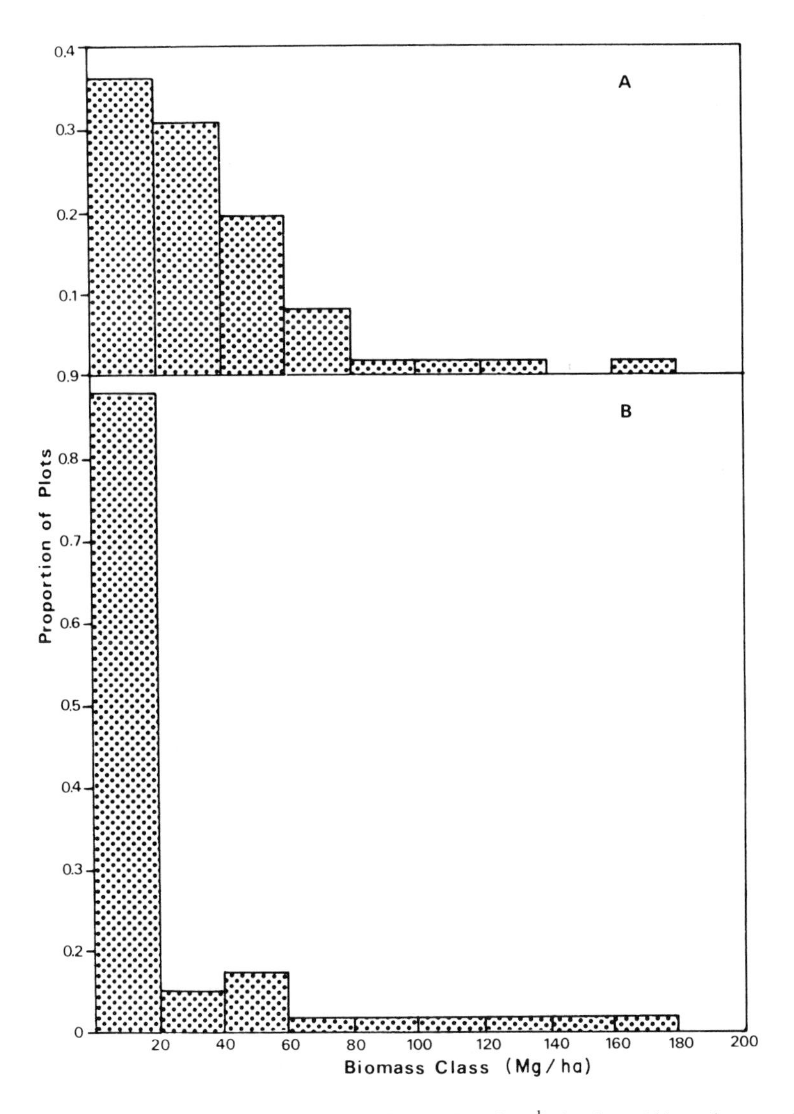

Figure 8 The distribution of biomass classes (Mg ha^{-1}) for logs (A) and snags (B) from a *Sequoiadendron–Abies* forest in Sequoia–Kings Canyon National Park, United States. Seventy-two 625-m^2 plots were sampled.

B. Biomass of Coarse Woody Debris

Surface areas, volumes, and biomass of CWD in temperate forest and associated aquatic systems are presented in this section (see Section V.A). Problems arise when interpreting existing data on CWD biomass. Lack of information on the disturbance history, decay rates, input rates, and site characteristics such as slope, aspect, temperature, and soil moisture makes it difficult to explain differences. Data are not entirely comparable because of variations in the lower size limits of included CWD and assumed values for wood density. Another serious problem is differences in scope of the studies—many include downed logs but not snags, a few are concerned only with snags, and some exclude large branches.

1. Area Covered by Coarse Woody Debris

Although less frequently considered than volume or biomass, the area covered by CWD is important ecologically. Area is most commonly reported as projected area for terrestrial sites, but total surface area is sometimes also calculated, particularly for aquatic ecosystems. Projected area is probably most relevant for soil-forming processes and habitat for animals utilizing the interface between soil and CWD, while total surface area is most pertinent for species inhabiting CWD surfaces. Plants usually colonize the upper half of the surface of CWD; therefore, surface areas somewhere between the total and projected are probably most meaningful for plant habitat considerations (see Section V.A). Projected area is not usually measured in stream studies, in part because it does not relate well to the geomorphic functions of wood. For example, CWD may be buried and stacked and yet may control geomorphic processes. It is more common to consider the area of stream influenced by CWD (see Section V.D).

Values for projected cover of CWD range widely in forest systems (Table 5). The highest projected areas—14–25% of the forest floor—are found in the *Pseudotsuga–Tsuga* forests of the Pacific Northwest. The lowest reported cover of CWD is in *Quercus* forests in eastern North America, with log covers of 1.6–2.0% of the ground surface. Snags were not measured in these studies, but would probably add <0.1% cover to the total CWD cover of *Quercus* forests.

The form of CWD drastically affects the amount of forest floor covered, but has little influence on total surface area. Snags contribute very little to the overall projected area, even though they may comprise a large volume or mass of CWD. In stands with a high proportion of CWD formed by snags, the projected cover will be low.

2. *Volume of Coarse Woody Debris*

Some representative data on volumes of CWD are provided in Table 5. The total volume of CWD in terrestrial ecosystems ranges from a low of ~ 60 m^3 ha^{-1} in an *A. balsamea* forest to 1189 m^3 ha^{-1} in a *Pseudotsuga–Tsuga* forest. Unfortunately, snag volumes have not been reported for deciduous hardwood forests. In terms of logs alone, it appears that many coniferous forests have an order of magnitude more volume than deciduous forests. The volume of CWD in aquatic ecosystems also varies considerably, ranging from 2.5 to 4500 m^3 ha^{-1} (Table 6).

3. *Biomass of Coarse Woody Debris*

A close relationship exists between volume and biomass because of the relatively limited range in wood densities. For example, in *Pseudotsuga–Tsuga* forests, the estimated mean log density ranges from 0.16 to 0.20 Mg m^{-3} for stands between 100 and 1000 years old (Table 5). The mean density of snags in these forests was more variable, ranging from 0.16 to 0.30 Mg m^{-3} over the same age range (Table 5).

Reported values of total CWD mass vary from ~ 6 to more than 269 Mg ha^{-1} for intact temperate forests. Particularly large values are associated with the coniferous forests of western North America, including old-growth *Pseudotsuga* and *Sequoiadendron* forests. In general, deciduous forests have lower log masses than coniferous forests. The range of values for deciduous forests is 11–38 Mg ha^{-1}, whereas the range of values for coniferous forests is 10–511 Mg ha^{-1} (Table 5). The biomass of snags also varies widely, with masses of 1–157 Mg ha^{-1} reported. Values for biomass of CWD in terrestrial systems are difficult to compare because both logs and snags are often not included within a single study. Relative proportions of logs and snags vary widely (see Section IV.D); therefore, total biomass is difficult to estimate from one component alone.

Biomass of CWD in streams shows a similar large variation in mass (Table 6). Reported amounts for channels undisturbed by management activities range from 1 to 1800 Mg ha^{-1}. In addition to forest type, disturbance history, and successional stage, which also affect CWD amounts in upland areas, channel size influences CWD amounts in lotic ecosystems. Biomass measurements in streams are also strongly influenced by geomorphic processes (see Section V.D). Erosion and deposition of sediment in channels may also change the biomass estimated in streams. Burial and exposure of CWD change the number of pieces that can be inventoried. Furthermore, changes in channel width alter the area of inventory, which may also change the biomass estimate.

Table 6 Volume and Biomass of Coarse Woody Debris (> 10 cm diameter) in Streams Flowing through Natural (Unmanaged) Temperate Forests[a]

Stream name	Stand age (year)	Drainage area (ha)	Mean channel width (m)	Coarse woody debris		Biomass (m^3 ha$-^1$)	Refs.[b]
				Reach length sampled (m)	Volume (m^3 ha^{-1})		
Picea sitchensis–Tsuga heterophylla, Prince of Wales Island, Alaska							
Cabbage Creek	300	—[c]	4.1	100	55	22	10
Aha Creek	300	—	4.8	300	230	91	10
Hohngren Creek	300	—	2.1	90	140	57	10
3/10 Mile Creek	300	—	3.0	90	240	94	10
Bonnie Creek	500	—	14.3	200	300	120	10
Picea sitchensis–Tsuga heterophylla, coastal British Columbia, Canada							
Government Creek	200	390	11.0	575	450	180	3
Government Creek	200	680	18.0	696	580	230	3
Government Creek	200	690	16.0	981	580	230	3
Hangover Creek	200	2020	20.0	1852	520	210	3
Reach 5	200	850	—	75	850	340	11
Reach 6	200	850	—	75	550	220	11
Reach 7	200	850	—	75	1700	670	11
Reach 8	200	850	—	75	320	130	11
Reach 4	200	900	—	75	550	220	11
Pseudotsuga menziesii, Klamath Mountains, California							
Horse Flat Tributary	300	7	3.2	74	45	18	5
Rotten Log Creek	300	59	3.8	101	320	130	5

Hidden Creek	300	75	1.6	91	520	210	5
Hidden Creek	300	84	2.6	98	95	38	5
Horse Flat Creek	300	110	3.2	90	250	100	5
Lost (slave Creek	300	124	3.6	90	450	180	5
Horse Flat Creek	300	128	2.8	85	1200	460	5
Horse Flat Creek	300	176	3.9	90	75	30	5
Shinbuster Creek	300	600	7.2	185	120	50	5
Shinbuster Creek	300	640	7.2	183	18	7	5
Coon Creek	300	1360	11.8	193	10	4	5
Sequoiadendron giganda, Sierra Nevada Mountains, California							
Suwanee Creek	1000	160	4.7	210	550	220	7
Crescent Creek	1000	190	5.0	210	1000	420	7
Sequoia semperuirens, California							
Gans West Creek	500	50	4.8	90	1100	450	8
Low Slope Schist Creek	500	70	3.8	90	1000	420	8
Prairie Creek at Hope Creek	500	70	2.3	115	4500	1800	4
Hayes Creek	500	150	4.5	132	3500	1400	4
Little Lost Man Creek	500	350	6.4	253	2800	1100	4
Prairie Creek at Little Creek	500	350	3.9	107	240	96	4
Prairie Creek at Forked Creek	500	660	4.7	200	250	100	4
Prairie Creek at Zig Zag Creek	500	820	6.7	232	420	170	4
Little Lost Man Creek	500	910	9.6	596	980	390	4
Prairie Creek at National Tunnel	500	1120	8.0	269	2200	880	4
Prairie Creek at Brown Creek	500	1670	11.0	335	1700	680	4
Prairie Creek at Campground	500	2720	18.5	395	400	160	4

(*Continued*)

Table 6 Continued

Stream name	Stand age (year)	Drainage area (ha)	Mean channel width (m)	Coarse woody debris		Biomass (m^3 ha^{-1})	Refs.[b]
				Reach length sampled (m)	Volume (m^3 ha^{-1})		
Picea engelmannii, Idaho							
4th of July Creek	200	900	3.7	90	88	35	6
Roaring Creek	200	1030	2.3	90	50	20	6
Pinus, Idaho							
Squaw Creek	200	6670	7.3	90	120	48	6
West Pass Creek	200	6750	4.7	85	2.5	1	6
East Fork Salmon River	200	19600	14.4	270	2.5	1	6
Bruno Tributary	200	450	1.1	60	42	17	6
Picea–Tsuga, White Mountains, New Hampshire							
Bessililul Creek	100	30	1.9	90	80	32	9
Flume Creek	100	190	6.6	180	30	12	9
Pseudotsuga menziesii, Cascade Mountains, Oregon							
Watershed 10S	300	2	0.6	120	1200	470	1
Watershed 10N	300	4	0.9	120	1200	490	1
Watershed 9	400	8	3.5	170	500	200	6
Allen Creek Tributary 1	300	8	0.9	120	780	310	1
Allen Creek Tributary 2	300	9	0.9	120	420	170	1
Watershed 10	300	10	1.8	120	580	230	1
Mortality Creek	85	15	2.0	90	700	280	6

Devils Club Creek	500	20	2.6	85	880	350	6
Poodle Creek	190	28	2.3	120	400	160	1
Quartz Creek Tributary 1	300	34	3.4	120	1400	550	1
Quartz Creek Tributary 3	250	36	2.7	120	680	270	1
Christy Creek Tributary	300	39	2.7	120	400	160	1
Quartz Creek Tributary 2	250	49	2.7	120	620	250	1
Clover Creek	135	52	2.8	90	300	120	6
Blue River Tributary	300	57	3.4	120	45	18	1
Cottonwood Creek	90	62	3.0	90	170	69	6
Cold Creek	500	70	7.0	200	850	340	7
Watershed 2	500	80	5.2	150	750	300	6
Hagan Creek	90	85	2.5	90	480	190	6
Winberry Creek North Fork	300	114	8.5	120	700	280	1
Quartz Creek Tributary 4	300	117	5.8	120	320	130	1
Stoney Creek	250	163	4.9	120	780	310	1
Mack Creek	500	536	9.1	120	650	260	1
Mack Creek	500	600	12.0	350	570	228	6
School Creek	160	645	4.6	120	400	160	1
Lookout Creek	500	1200	15.5	550	340	136	6
Lookout Creek	300	3044	16.8	120	150	61	1
Lookout Creek	500	6000	24.0	240	230	92	6
McKenzie River	100	102000	40.0	800	60	24	6

(*Continued*)

Table 6 Continued

Stream name	Stand age (year)	Drainage area (ha)	Mean channel width (m)	Coarse woody debris		Biomass (m^3 ha^{-1})	Refs.[b]
				Reach length sampled (m)	Volume (m^3 ha^{-1})		
Hardwood, Smoky Mountains, Tennessee							
Minnie Ball Branch	200	44	5.7	90	70	28	2
Pardon Branch	50	52	3.6	90	40	16	2
Ekaneetlee Branch	200	70	5.2	90	160	64	2
Ramsey Prong	200	73	6.5	90	60	24	2
Trillium Creek	200	96	4.6	90	300	120	2
Picea–Abies, Smoky Mountains, Tennessee							
Noland	200	21	6.2	90	140	56	2
Salola	200	70	3.4	90	220	88	2

[a] Specific gravity equals 0.40 for all sites, although many authors used other values.
[b] (1) Froehlich (1973), (2) Gregory and Lienkaemper (unpublished), (3) Hogan (1985), (4) Keller *et al.* (1985), (5) Lehre (unpublished), (6) Lienkaemper (unpublished), (7) Luchessa (unpublished), (8) Swanson (unpublished), (9) Swanson and Sedell (unpublished), (10) Swanson *et al.* (1984), (11) Toews and Moore (1982). Unpublished data is on file at Forestry Sciences Laboratory, RWU-1653, Corvallis, Oregon.
[c] Not available.

Unfortunately, there are no data available concerning CWD biomass in large rivers. The historical record shows that fast, turbulent rivers as well as low-gradient rivers had large amounts of wood influencing their channels. Great drift jams of logs were reported and described for the Amazon, Congo, Orinoco, Ganges, Mississippi, and McKenzie Rivers as well as for rivers in China (Lyell, 1969). For sixth-to eighth-order streams, log jams were a common feature of pristine streams (Michigan Historical Society, 1883; Sedell and Luchessa, 1982; Triska, 1984).

The lower Siuslaw River and lower North Fork Siuslaw River in Oregon were so filled with fallen trees that trappers in 1826 were unable to explore much of these river systems (Ogden, 1961). The Willamette River between Corvallis and Eugene flowed in five separate channels in 1840 (Secretary of War, 1875). The Captain of the Portland district reported that the "obstacles were so great above Corvallis" and that the river banks were heavily timbered for a distance of 1–3 km on either side. In a 10-year period, over 5500 snags and drift trees were pulled from an 80-km reach of river, as the river was confined to one channel by engineering activities. These trees ranged from 1.5 to 2.7 m in diameter and from 27 to 36 m in length (Secretary of War, 1875). Sedell and Frogatt (1984) and Sedell and Luchessa (1982) list rivers in Oregon and Washington that were completely blocked in their lower main channels by driftwood. The Skagit River, Washington, drift jam was 1–2 km long and 0.5 km wide. The Stillaguamish River had six debris jam closures from the mouth to 28 km upstreams. Snags were so numerous, large, and deeply inbedded in the bottom that a stream snag boat was required to operate for 6 months to open a channel 35 m wide on the Stillaguamish (Secretary of War, 1881). Another lower-gradient stream system, the North River, had 11 drift jams along the main river system. Drift jams in high-gradient systems often occurred where the channel gradient decreased abruptly. The Nooksack River is an example (Figure 9).

C. Factors Controlling Biomass

1. Effects of Forests Type on Biomass

Biomass values illustrate the tendency for greater accumulations of CWD in coniferous than hardwood forests (see Section III). A major cause of these differences appears to be faster decay in deciduous forests. Smaller sized material, higher substrate quality, and/or a climate more favorable to decay make CWD decay faster in deciduous than in

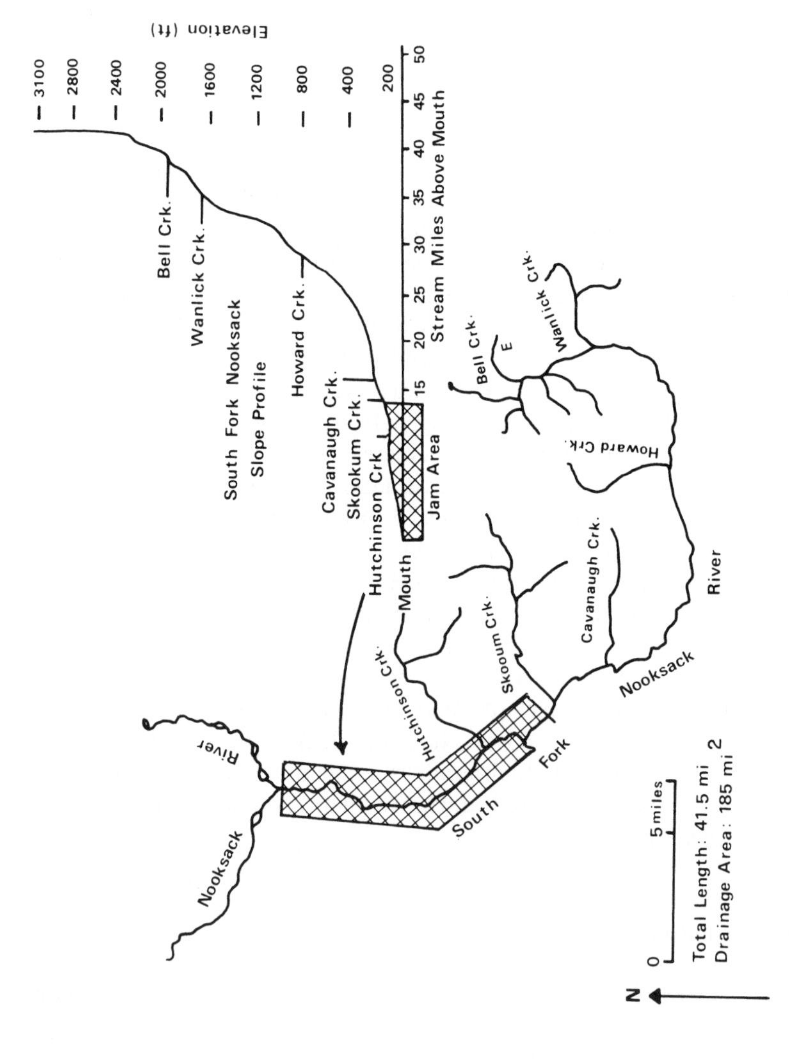

Figure 9 Map and longitudinal profile of the South Fork of the Nooksack River of western Washington indicating the position of an extensive log jam.

conifer ecosystems. However, input rates are also partially responsible for the difference, as illustrated by a comparison of decay and input rates in *Pseudotsuga–Tsuga* and *Quercus* forests. Assuming the input rate of logs is similar for *Pseudotsuga–Tsuga* and *Quercus* forests, the biomass differences would depend upon decay rates alone. Because the reported decay rate constants are 0.03–0.04, and 0.01–0.03 year^{-1} for *Quercus* (MacMillan, 1981) and *Pseudotsuga* (Graham, 1982; Sollins, 1982), respectively, the latter ecosystem should have $\sim$1–3 times the steady-state log biomass of the former. Biomass values of logs show a 4- to 12-fold difference between the two ecosystems, however, so differences in decay rate account for, at most, only a portion of the difference, and input rates must be considered to fully explain the differences. Direct comparisons of input rates are, unfortunately, not possible, but they can be approximated. MacMillan (1981) presents input as logs ha^{-1} year^{-1} instead of volume or mass. Assuming a mean volume of 0.44 m^3 per log (based on MacMillan's Tables 6 and 8) and a wood density of 0.58 Mg m^{-3}, the input of 2.52 logs ha^{-1} year^{-1} converts to 0.64 Mg ha^{-1} year^{-1}. This value is considerably below the 4.54 Mg ha^{-1} year^{-1} rate reported for old-growth *Pseudotsuga–Tsuga* forests (Table 1). Dividing the input rates for both ecosystems by the reported decay rates gives ideal steady-state biomass of 16–21 Mg ha^{-1} for the *Quercus* forest and $>$150–450 Mg ha^{-1} for the *Pseudotsuga–Tsuga* forests. These values reflect the differences much more completely than decay rates alone.

The effect of input rates on CWD biomass seems to have been generally overlooked and helps explain why some coniferous ecosystems with slow decay rates have values comparable to those for hardwood stands where decay is rapid. In *P. contorta* forests of Wyoming (T. J. Fahey, personal communication), CWD mass ranges from 0.6 to 20.8 Mg ha^{-1}, but the decay rate constant for this species is 0.01—comparable to *Pseudotsuga*. The *P. ponderosa* forests of Arizona also have a low CWD mass of 18 Mg ha^{-1} (Sacket, 1979), but these also appear to be caused by very low input rates of 0.25 Mg ha^{-1} year^{-1} (Avery *et al.*, 1976).

Coniferous forests can also have low biomass of CWD due to high decay rates. Harmon and Cromack (unpublished) sampled an *A. concolor* forest in the Sierra Nevada, California, with 49 Mg ha^{-1} of logs decaying with a rate constant of 0.06 year^{-1}. A *Sequoiadendron–Abies* forest had 247 Mg ha^{-1} of logs, which illustrates how adding a single species with very slow decay rates can influence biomass. Removing the contribution of *Sequoiadendron* to the CWD in this stand results in log mass values similar to the *A. concolor* ecosystem.

Amounts of CWD in stream ecosystems are also strongly influenced by forest composition (Table 6). Considering stream reaches draining <1000 ha, *S. semperoirens* stands have more than 10 times the CWD amount

($\bar{x}$ = 660 Mg ha^{-1}, N = 9) observed in *P. menziesii* ($\bar{x}$ = 120 Mg ha^{-1}, N = 10) in northwestern California. Streams of drainage area <1000 ha flowing through old-growth *P. menziesii* forests in the Oregon Cascades ($\bar{x}$ = 300 Mg ha^{-1}, N = 18) and *P. sitchensis–T. heterophylla* forests of British Columbia ($\bar{x}$ = 280 Mg ha^{-1}, N = 8) exhibit intermediate debris amounts. Streams in a very limited sampling of hardwood forests in Tennessee, *Tsuga–Abies* forests in New Hampshire, and *Picea–Tsuga* forests of southeastern Alaska exhibit average amounts of 100 Mg ha^{-1}.

2. *Effects of Disturbance and Succession on Biomass*

Disturbances affect biomass by adding CWD and by initiating a new sere. Some disturbances such as fire, floods, and mass movement of soil also remove CWD. Successional patterns are similar for the various parameters, whether area, volume, or mass is considered. Successional stage affects the size, species, amount, decay class, and distribution (logs versus snags) of CWD.

Effects of disturbance on mass of CWD have been documented for at least four terrestrial ecosystems (Franklin and Waring, 1980; Huff, 1984; Lambert *et al.*, 1980; Tritton, 1980). Changes in CWD following a wave pattern of mortality in *A. balsamea* stands provide one of the clearest examples of disturbance effects (Lambert *et al.*, 1980). Highest overall biomass, ~62 Mg ha^{-1}, occurred 12 years after the onset of mortality, presumably because the entire stand did not die immediately. As the time since disturbance increased, overall biomass and the proportion of snags in the CWD decreased so that after 46 years CWD leveled off at 13–15 Mg ha^{-1}.

Successional trends in CWD mass were also observed by Tritton (1980) in a chronosequence of clear-cut northern hardwood forests. The highest CWD amounts occurred in stands 10 years old and very old stands with 34 and 49 Mg ha^{-1}, respectively. The lowest amount of CWD occurred 40–57 years after clear-cutting, with a total CWD mass of 9.8–16.0 Mg ha^{-1}. The high mass of CWD immediately following clear-cutting was attributable to slash created by logging operations, and 10–20 years after cutting, slash composed 97–98% of the CWD. The decay of slash, coupled with the input of material from the new forest, yield an overall successional curve that is "U"-shaped.

CWD mass appears to be continually high during the later stages of successional development of *P. menziesii–T. heterophylla*-dominated forests in northwestern North America (Franklin and Waring, 1980). The stands examined were 100 to over 1000 years old and developed after catastrophic wildfires. Lowest overall biomass, 107–142 Mg ha^{-1}, occurred at 100–200 years. Older stands had 173–202 Mg ha^{-1} of CWD. CWD biomass would presumably have been much higher earlier in succession, and CWD biomass exceeding 500–1000 Mg ha^{-1} (Grier and Logan, 1977) might occur immediately after a catastrophic fire.

Huff (1984) examined changes in log biomass after catastrophic fires in *Pseudotsuga–Tsuga* forests of the Olympic Peninsula in Washington State. Highest log mass occurred later in succession and lowest values occurred in stands 110–131 years old. This pattern is believed to be the consequence of decay of the fire-killed trees combined with reduced input because as much as 40 years are required for trees to completely colonize the burned sites. Had snags been included, the total CWD biomass would probably have been highest early in succession when the fire-killed stand was contributing significantly to the total.

The recognition that high amounts of CWD can be expected in young, natural stands appears to be relatively recent. High CWD biomass in young stands is a consequence of the fact that most disturbances that kill trees consume relatively small amounts of the wood. Hence, large amounts of CWD as snags and downed logs are the heritage of forests regenerating after wildfire, windstorms, insect epidemics, or other catastrophic disturbances.

CWD in streams and rivers is also strongly affected by disturbances and the successional stage of surrounding forests. The recent occurrence of fire, timber harvest, major floods, and other events drastically affects the rates and size distribution of CWD contributed to the streams and are essential in interpreting the abundance and arrangement of CWD in streams. Swanson and Lienkaemper (1978) examined CWD in small streams flowing through coniferous forests in western Oregon burned by high-intensity wildfires 70–135 years earlier. Material from the prefire stand was found under the smaller diameter pieces of CWD that had fallen in from the postfire stand. In these streams, CWD was continuously present because the residence time of prefire CWD was greater than the time necessary for CWD production by the postfire stand. However, total abundance of CWD in the four streams sampled, was about half that observed in streams bordered by old-growth forests.

Disturbance to forests surrounding streams often changes the species of CWD input and this, in turn, may affect CWD biomass. For example, logging and burning of an old-growth *Tsuga*-dominated watershed in Tennessee resulted in a conversion to *Liriodendron*-dominated forests (Silsbee and Larson, 1983). Mean CWD volume in four old-growth streams was 338 m^3 ha^{-1}, whereas in the logged watershed it was 84 m^3 ha^{-1} (Silsbee and Larson, 1983). Although this decline was due in part to the interruption of CWD input caused by disturbance, the conversion to the much faster decaying *Liriodendron* from *Tsuga* also probably caused a decrease in biomass. Presumably CWD biomass will remain low until large *Tsuga* stems are added to the CWD standing crop.

The interactions between succession and stream size also influence CWD mass. Bilby and Likens (1980) studied streams flowing through forests logged ~60 years earlier in the White Mountains of New Hampshire. After estimating CWD production through succession and for first-, second-, and third-order stream channels, they concluded the larger stream would experience a pronounced decline in CWD because ~200 years would elapse before streamside stands produced pieces large enough to remain in the channel for a significant period of time.

Disturbance regimes affect mass of CWD in streams directly by removing or depositing materials, as during floods, and indirectly by influencing the vegetation in and along the channels. CWD production is severely limited where stream and riparian vegetation is removed or pruned frequently by erosion or transported bedload, CWD, and ice. Vegetation may persist along streams, but may never reach a size or substrate quality sufficient to produce large quantities of CWD.

These studies on successional trends in CVTD mass suggest a simple general model to explain the different successional patterns (Figure 10). We assume, for simplicity, that decay rates are constant and that input rates vary only linearly with succession and are not affected by changes in forest type or stochastic events. In this model, CWD is classified as pre-disturbance, disturbance related, and postdisturbance input. Overall shape of the CWD biomass curve over succession is determined by the amount of original materlal removed by the disturbance, timing of input, and decay patterns of each category. After disturbance, preexisting CWD declines; the greater the amount of preexisting material removed, the less it contributes to the overall curve. Conversely, the greater the amount of freshly created CWD, the higher the pulse following disturbance. Both the predisturbance and disturbance-created components of CWD will decline with time, and the overall biomass curve will also decline unless counterbalanced by new input.

Development of the new stand after disturbance and creation of "new" CWD strongly influence the shape of the mass curve. Severe disturbances or site conditions retard forest reestablishment, and this factor delays the addition of new CWD which, in turn, allows decay to reduce CWD mass below predicted steady-state values. The longer the time needed to create new CWD relative to the residence time of old CWD, the deeper the depression in the total biomass curve. When regeneration is rapid, the overall pattern will be a peak in CWD mass followed by a decline to a steady-state value. In the case of disturbances, such as debris avalanche or timber harvest that remove CWD, the input curves dominate the successional pattern, leading to the classic linear increases presented

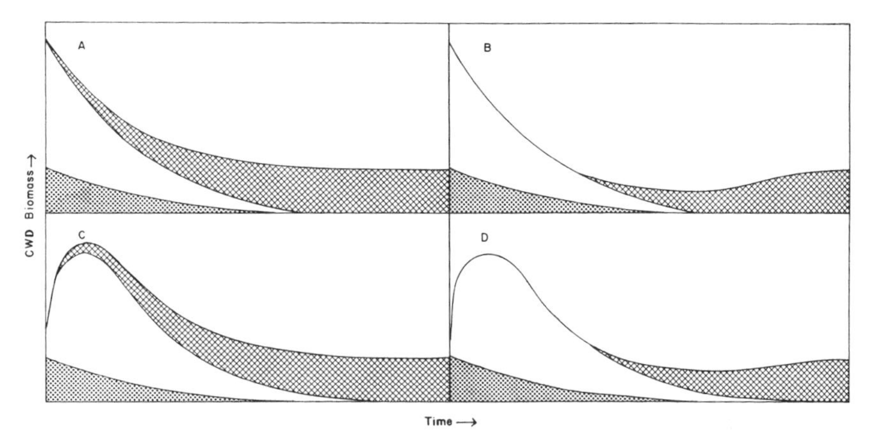

Figure 10 Four hypothetical patterns of CWD mass following disturbance. CWD is-divided into that present before the disturbance (dotted), that created by the disturbance (open), and that added by the stand growing after the disturbance (crosshatched). In each case, we assume the decay rate does not change with time and that postdisturbance input increases linearly to a constant value. (A) In this case, postdisturbance starts immediately after the forest's destruction, which results in an exponential decline to a steady-state value. (B) Same as (A) except that postdisturbance input is delayed. This causes a U-shaped trough to occur during the middle time period, and as postdisturbance inputs offset decay losses, mass increases to a steady-state value. (C) Pattern caused when the initial stand is killed gradually, but the postdisturbance input starts immediately. (D) Same as (C) except that postdisturbance input is delayed.

by Triska and Cromack (1980). Although this model is conceptually very simple, it does accommodate the wide array of responses that might be encountered after disturbances.

3. Effects of Stream Size on Coarse Woody Debris

Amounts of CWD are generally highest in the smallest streams and decrease with increasing stream size, given a particular physiographic and forest setting (Table 6). The general downstream decrease in CWD results from several factors. Larger streams have a greater ability to transport CWD downstream and out of channels onto floodplains. Larger channels also have a more limited source of CWD because forests do not grow in the most active channels, while very small streams can flow through forests without influencing overall stocking amounts. Furthermore, woody vegetation adjacent to rivers may be maintained in an immature stage by frequent disturbance, and the resultant lower biomass of these streamside forests thus limits availability of CWD for channels. However, when large channels migrate laterally and entrain mature forests, a large amount of CWD can be added. Along large floodplain rivers, CWD moves both into and out of the channel during floods, but it is not known whether the floodplain is a net source or sink for CWD.

Comparisons of CWD amounts in and out of a river channel have been made by Wallace and Benke (Wallace and Benke, 1984). They observed less CWD on floodplains than in a fourth-order segment of Black Creek and sixth-order segment of Ogeechee River on the coastal plain of Georgia. They attributed this to transport of CWD into the channel, slower decomposition in the channel, and higher input rates along eroding stream banks. In four small channels in southeast Alaska, Swanson *et al.* (1984) measured no significant differences in CWD amounts between floodplain and channel areas.

D. Distribution of Coarse Woody Debris

The previous section discussed the amount of CWD within various ecosystems, but it is also important to consider how this material is distributed with respect to size, decay state, position (snags versus logs), and spatial arrangement. CWD is not homogeneously spread over the landscape and exhibits patterns that, although not always highly predictable, influence decomposition of CWD and its geomorphic and habitat functions.

1. Position

The proportion of CWD biomass composed of snags and logs is highly variable and exhibits little pattern (Table 5). The lowest percentage of CWD as snags is 2% in a *P. jeffreyi* stand (Harmon and Cromack, unpublished), while the highest was 96% in. a *P. contorta* stand (Fahey, 1983). Even within a single watershed there seems to be a wide variation in the fraction of CWD composed of snags. For example, 6–22% of the CWD mass was snags within a watershed dominated by a 450-year-old *Pseudotsuga–Tsuga* forest (Grier and Logan, 1977). This is a considerably lower proportion than the 38–49% reported for forests of similar ages and composition (Franklin and Waring, 1980; Sollins, 1982). Several controlling factors need to be considered before comparisons can be made, including the cause of mortality, the time since disturbance, and the size and decay rate of the material.

Lambert *et al.*'s (1980) study of fir waves details changes in proportion of CWD as snags after disturbance. Snags constituted 58% of the CWD immediately after canopy death. The snag fraction declined with time so that ~41, 33, 24, and 6% of the CWD were snags at 10, 20, 30, and 40 years after stand death, respectively. In this case, the cause of death led to a large input of standing dead trees. Very similar patterns would be expected for disturbances such as fire and beetle outbreaks that also leave standing dead trees. In disturbances such as windthrow, which increase the proportion of logs, the situation would be reversed.

2. *Size Class Distributions*

Few studies have examined size distributions of CWD. Abundance of pieces declines as diameter increases, which probably is a reflection of the reverse "J"-shaped distribution typical of living tree populations (Harper, 1977). The geometric decline in numbers of logs as diameter increases is exemplified by data from a *Sequoiadendron–Abies* stand (Figure 11). Of the logs, 50% are in the 15- to 35-cm class, and each subsequent diameter class contains half the number of the preceding class. MacMillan (1981) examined the size distribution of logs in a *Quercus*-dominated forest and also found an inverse relationship between log diameter and numbers. Although the larger size classes have fewer individuals, they often compose the majority of the biomass (Figure 11) because volume increases geometrically with size. Another factor that may influence this relationship is the faster decay rate of the smaller material (see Section III.E.5).

In addition to the distribution of biomass in CWD size classes, it is also of interest to know the proportion of above-ground woody debris composed by CWD. A complicating factor in making these calculations is the variable lower limit of CWD but, in spite of this problem, it appears that CWD composes the majority of above-ground woody detritus (Table 7). Brown and See (1981) report that 78–89% of the downed woody biomass was >7.5 cm in six forest types examined in Montana and Idaho. Wood >7.5 cm composed 94% of the downed woody biomass in a North Carolina *Quercus–Carya* forest (Triska and Cromack, 1980), while 79% of the dead and downed wood in a mixed *Quercus* forest in New Jersey was >10 cm (Lang and Forman, 1978). Pieces >7.5 cm composed 70–98% of the downed wood biomass in forest types of Tennessee and North Carolina (Harmon, 1980). Although none of these studies examined the fraction of fine (e.g., attached branches) versus coarse standing material, it is likely that the latter would dominate the standing dead biomass. It appears that CWD dominates the terrestrial woody detritus pool and usually composes >80% of the total biomass. CWD also appears to dominate the biomass in stream systems, although the range in proportions seems larger than in the terrestrial setting. CWD composed only 26% of the woody debris in a *Populus*-dominated

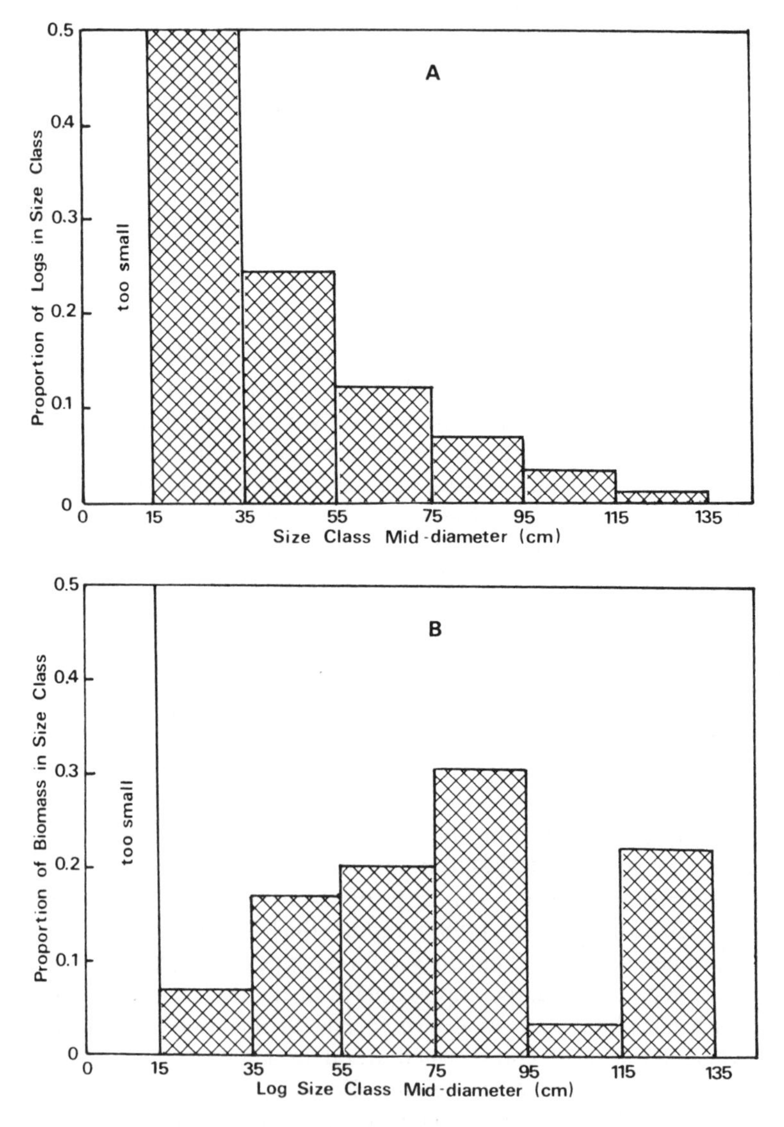

Figure 11 The distribution of (A) numbers of logs and (B) biomass for 20-cm-diameter size classes in a *Sequoiadendron–Abies* forest.

stream in Idaho (Lienkaemper, unpublished). Most other stream systems examined have 40–98% of the wood in the coarse fraction. Swanson *et al.* (1984) summarize data showing coarse debris comprising 93% of wood-debris loading in stream reaches in old-growth *Pseudotsuga* forests of Oregon and 91% in old-growth *Picea–Tsuga* forests in southeastern

Table 7 Proportion of Dead and Downed Woody Detritus Composed of Coarse Woody Debris in Selected Forests and Streams

Ecosystem	Location	Percentage as CWD	Refs.[a]
Forests			
Fagus–Betula	Tennessee	91[b]	3,4
Larix occidentalis	Montana/Idaho	83[b]	1
Liriodendron tulipifera	Tennessee	79[b]	3,4
Picea–Abies	Montana/Idaho	89[b]	1
Picea–Abies	Tennessee	98[b]	3
Pinus contorta	Montana/Idaho	82–86[b]	1
Pinus mixed	Tennessee	70–76[b]	3,4
Pinus ponderosa	Montana/Idaho	78–81[b]	1
Pseudotsuga menziesii	Montana/Idaho	80–82[b]	1
Quercus–Carya	North Carolina	94[b]	10
Quercus mixed	New Jersey	79[c]	5
Quercus mixed	Tennessee	85–92[b]	3,4
Quercus prinus	Tennessee	74–83[b]	3,4
Thuja–Tsuga	Montana/Idaho	88[b]	1
Tsuga canadensis	Tennessee	87[b]	4
Streams			
Hardwoods mixed	Tennessee	40–91[c]	2
Picea–Abies	Tennessee	78–79[c]	2
Picea–engelmannii	Idaho	37–80[c]	6
Picea–sitchensis	Alaska	73–93[c]	8
Picea–Tsuga	New Hampshire	60–72	9
Populus trichocarpa	Idaho	26[c]	6
Pseudotsuga–Tsuga	Oregon	63–94[c]	6
Sequoia sempervirens	California	98[c]	7

[a] (1) Brown and See (1981), (2) Gregory and Lienkaernper (unpublished), (3) Harman (unpublished), (4) Harmon (1980), (5) Lang and Forman (1978), (6) Lienkaemper (unpulished), (7) Swanson (unpublished), (8) Swanson *et al.* (1984), (9) Swanson and Sedell (unpublished), (10) Triska and Cromack (1980). Unpublished data is on file at Forestry Sciences Laboratory, RWU-1251, Corvallis, Oregon.
[b] Lower size limit 7.5 cm.
[c] Lower size limit 10 cm.

Alaska. In *Picea–Tsuga* and hardwood forests of Tennessee, coarse debris in small streams constituted 76% of the total wood loading (Gregory and Lienkaemper, unpublished).

3. *Decay Class Distribution*

In many ecosystems, intermediate decay classes or states tend to compose the largest fraction of CWD biomass, while the most and least decayed comprise the smallest fraction (Franklin *et al.*, unpublished; Graham and Cromack, 1982; Harmon and Cromack, unpublished; Sollins, 1982). Although this may be true, it is helpful to realize that biomass distribution of CWD under steady-state conditions with respect to decay classes (Figure 12) is dependent on the residence times of the classes. If the residence times of the decay classes are equal, i.e., 1–10 years,

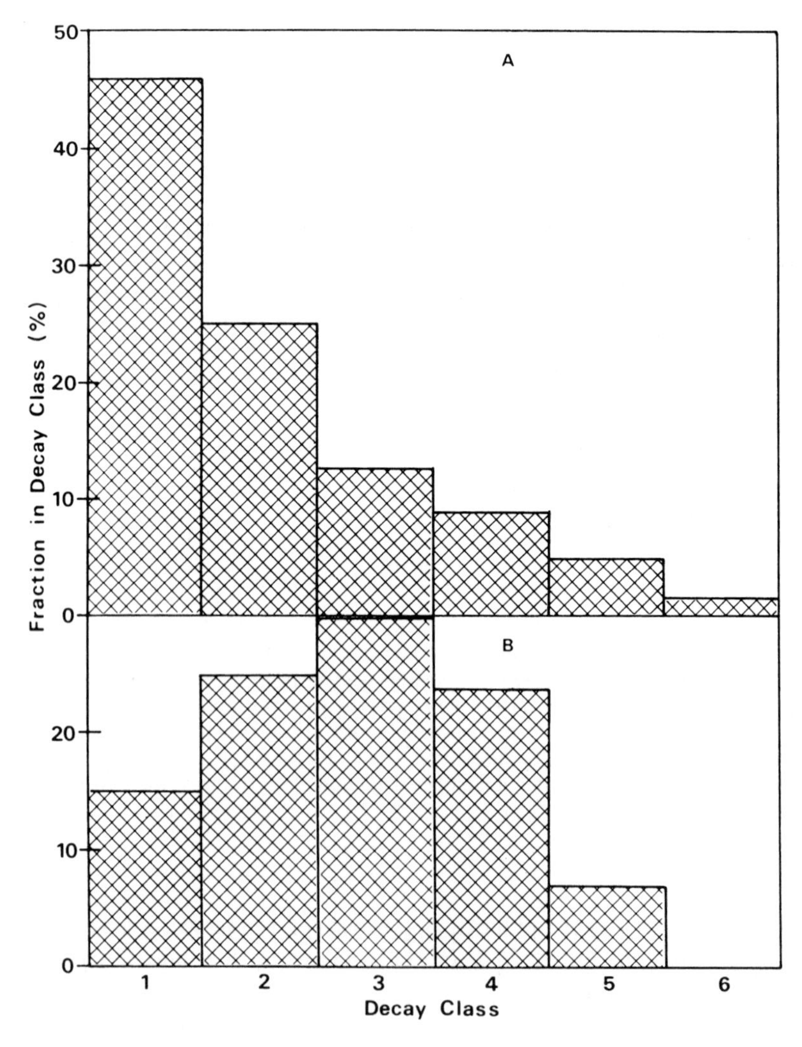

Figure 12 Hypothetical example of how the distribution of CWD mass over decay class is a function of the residence time of the decay class. In both cases, a steady-state system is assumed, but in (A) the residence time in each class is equal, whereas in (B) the residence time increases geometrically with each class.

11–20 years, 21–30 years, and so on, then the youngest decay class will contain the most biomass. If, on the other hand, residence time of decay classes increases geometrically, i.e., 0–1 year, 2–3 years, 4–7 years, and so on, then the intermediate-aged decay classes will have the most biomass.

Several factors, most importantly disturbance, cause deviations from the expected steady-state decay class pattern. After the death of an *A. balsamea* forest, there was a rapid increase in the fraction of biomass composed of slightly decayed boles, and as the time since death increased, this cohort of boles progressed from the moderately decayed to the very decayed classes (Lambert *et al.*, 1980). Ten years after stand death, all the boles were in the slightly decayed class. In stands 20 years old, 75% of the boles were slightly decayed and 25% were moderately decayed. In stands 30 years old, the slightly decayed boles composed the lowest fraction of the biomass (8%), while the very decayed class contained 50%o of the biomass. The effect of disturbance is also evident in a chronosequence of *Pseudotsuga* stands created by catastrophic fires (Franklin *et al.*, unpublished). In a 100-year-old stand, 75% of the CWD is in an advanced state of decay (class 4), probably representing material from the previous stand. At 130 years, much of this biomass appears to have decayed to the very advanced state (class 5). In stands older than 130 years, the majority of CWD biomass is divided between moderate and advanced states of decay (classes 3 and 4).

Other factors influencing the distribution of biomass in decay classes are the size and decay resistance of the material added. An example from a *Sequoiadendron–Abies* stand illustrates the point. *Sequoiadendron* is very decay resistant; when included, the distribution of biomass in the slightly, moderately, advanced, and very advanced decay classes was 74, 15, 7, and 4%, respectively (Harmon and Cromack, unpublished). However, when *Sequoiadendron* is excluded, the distribution for the same classes was 26, 28, 29, and 17%, respectively.

4. *Spatial Patterns*

a. Riparian and Stream Environments. In riparian and stream environments, the arrangement of CWD and architecture of accumulations are influenced by both the mechanisms adding CWD and the geomorphic processes within streams. Several types of spatial arrangements are common in small (i.e., first- and second-order) channels, but randomly distributed CWD appears to be the most widespread. Small streams cannot move CWD, and the spatial pattern of CWD in these streams reflects the spatial pattern of input.

In small- and intermediate-sized streams, CWD forms large accumulations with a very open structure when windthrow and earthflows are the major agents adding CWD to channels. Addition of CWD to channels by rapid soil mass movements and snow avalanches results in formation of large accumulations with a tight, interlocking fabric. Similarly, rapid mass movements of colluvium, alluvium, and CWD down channels result in tightly meshed accumulations. In these cases, CWD is entrained by and pushed in front of a debris flow moving down a channel. As channel gradient decreases, the debris flow eventually stops, leaving a wedge of sediment up to 10 m thick and several hundred meters long trapped behind a CWD accumulation that has been impregnated with sediment.

CWD has a moderately clumped distribution in intermediate-sized streams where few pieces are large enough to be stable during floods (see Section V.D). Pieces wider than the active channel can remain in place for up to several centuries (Keller and Tally, 1979; Swanson and Lienkaemper, 1978; Swanson *et al.*, 1976) and often trap smaller pieces of CWD as they float downstream. Therefore, stream-transported CWD accumulate above large pieces of CWD as well as above boulders and other stable structures.

In large channels, a high proportion of CWD pieces can be moved during floods. CWD accumulates at sites such as heads of islands, mouths of secondary channels, heads of point bars, and outside of meander bends (Keller and Swanson, 1979; Singer and Swanson, 1983; Swanson and Lienkaemper, 1982; Wallace and Benke, 1984). This leads to a highly clumped distribution of CWD in large channels. The importance of these sites are controlled both by trapping efficiency and proximity to a CWD source.

Published information on the arrangement of CWD in individual stream reaches is limited to Hogan (1985). He measured orientation and clustering of CWD in 11- to 45-m-wide channels in clear-cut and mature *P. sitchensis–T. heterophylla* stands in British Columbia. There was some tendency (not examined statistically) for debris to be oriented diagonally across the channel in unlogged channels and parallel to the main axis of logged channels. CWD pieces formed accumulations with an alongstream spacing of three to four channel widths.

b. Terrestrial Environments. Spatial aggregation occurs in the terrestrial environment, but patterns are not as clear as in streams and are more dependent on patterns of input than on transport mechanisms. Spatial aggregation can result from aggregated mortality such as blowdown patches, very localized insect attacks, and the spread of pathogens such as *Pheilinus weirii* or *Fomes annosus* by root contact (see Section II.C).

Yet another cause of CWD aggregations might be spatial discontinuities in the trees that are the source of CWD.

Although there are many potential methods to examine spatial patterns of terrestrial CWD (see Pielou, 1977, p. 113), little quantitative work has been published. By using the techniques employed in analyzing the spatial distribution of live trees, Cline *et al.* (1980) found that snags in unmanaged *Pseudotsuga* forests of western Oregon tended to be random in distribution, although in 25% of the stands the snags were aggregated into patches of 5–10 trees that apparently had died simultaneously. As far as we can tell, no one has examined spatial patterns in logs, although their compass orientation has been explored. Clearly, much remains to be learned about spatial patterns of CWD and the causes of these patterns.

5. *Orientation*

The arrangement of logs on hillslopes is important because it influences sampling procedures (see Section IV.D), use of CWD as habitat (Maser and Trappe, 1984), and the geomorphic role of CWD. Direction of tree fall should be influenced by prevailing storm winds, slope steepness and aspect, and type of input (e.g., snags versus live trees). Most current work centers on influence of storm winds, but the other factors may be equally important. Topography influences tree form through processes such as soil creep and snow pressure, and biases direction of tree fall. Steep slopes also allow pieces to roll or slide from their original orientation. The combination of slope steepness, topographic configuration, and slope direction can also modify the direction and strength of storm winds (e.g., Gratkowski, 1956).

The role of wind in orientation of logs has been extensively studied. Prevailing storm winds appear to be very important on gentle topography. In their reconstruction of a hardwood forest in New Hampshire, Henry and Swan (1974) found evidence that four storms had blown over 76% of the trees in a southwestern direction. In Poland, Falinski (1978) found that 63% of the *Picea abies* were uprooted in a southeastern direction, reflecting autumn storm winds. Gratkowski (1956) observed that 90% of the windthrows in the Cascade Range of Oregon pointed to the northeast, reflecting the southwestern direction of storm winds. Orientation patterns do vary within a forest type. An example is presented in Figure 13, which depicts log orientation in two *Tsuga-Picea* stands from coastal Oregon and Washington. Vegetation is similar, but windthrows at Neskowin Crest, Oregon, have a strong northerly orientation, whereas those at Quinault, Washington, have a

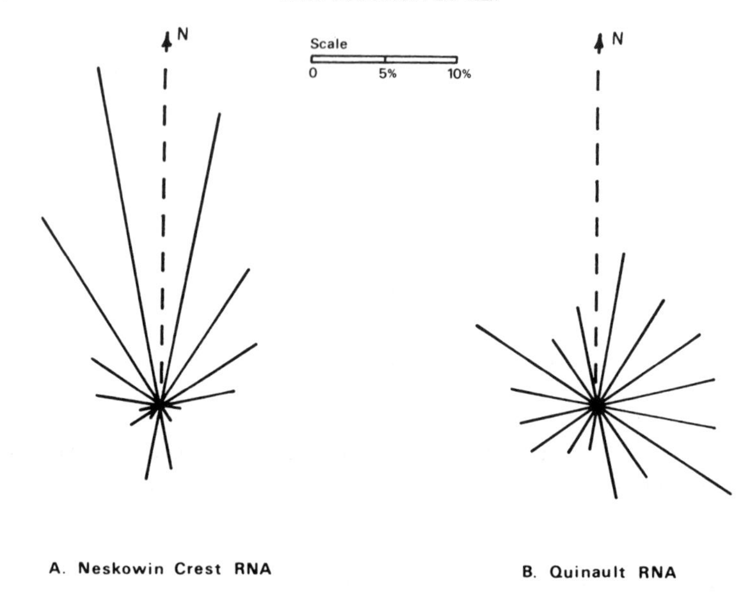

Figure 13 Orientation of logs in two *Picea sitchensis–Tsuga heterophylla* forests from the northwestern United States. The length of the fines indicates the percentage of logs pointing in a given direction. (A) Neskowin Crest Research Natural Area, Oregon. (B) Quinault Research Natural Area, Washington. In spite of similar canopy composition, the logs at Neskowin Crest have a strong northerly orientation, white those at Quinault have a fairly random orientation pattern.

weak easterly orientation. At Neskowin, log orientation reflects consistent, strong southwesterly winter winds at a site only 1 km from the Pacific Ocean. In contrast, the Quinault stand is located 60 km inland and is presumably exposed to less consistent wind directions.

V. ROLES OF COARSE WOODY DEBRIS

The previous sections have discussed the creation, decay, and amount of CWD found in various ecosystems. In this section, we explore the numerous functions that CWD performs in these ecosystems. Knowledge of these roles is very important to effectively manage ecosystems because removal of CWD may lead to an "unexpected" alteration of ecosystems unless these roles are appreciated. For many microbes, invertebrates, vertebrates, and plants, CWD and its associated microenvironment represent a habitat and/or food source. Of these taxa, perhaps microbes are the most closely associated with CWD. We will not treat this group specifically in this discussion, however, as it has been reviewed by Frankland *et al.* (1982). CWD is also important as a pool of energy, carbon, and nutrients in ecosystems. The relationship of these CWD pools and nutrient cycling in most ecosystems has yet to be explored. In both terrestrial and aquatic environments, CWD is intimately related to certain geomorphic processes.

A. Coarse Woody Debris as Plant Habitat

The importance of CWD as a substrate for autotrophs has been recognized for some time despite the frequent emphasis on decomposer organisms. The autotrophic taxa associated with CWD are quite varied and include green algae, diatoms, blue-green algae, lichens, liverworts, mosses, clubmosses, horsetails, ferns, gymnosperms, and angiosperms.

1. Use of Coarse Woody Debris by Plants

Autotrophs vary in their use of CWD. Many species are superficially attached to the surface of CWD and, thus, are epiphytes. Vascular plants may send their roots into rotting wood and bark to extract water and nutrients. These roots may originate from plants that established themselves on CWD or from those initially established on soil (Lemon, 1945). Other vascular plants root in the mat of decaying fine litter that often accumulates on the surface of CWD. These three types of exploitation are not mutually exclusive; an individual may first be epiphytic and then send roots into the surface organic mat, then into CWD, and, finally, into the underlying mineral soil.

Rotting wood has lower concentrations of the nutrients required for plant growth than most mineral and organic soils; therefore, growth can be slower on CWD than on the rest of the forest floor. Minore (1972), for example, found tree seedlings grew faster when rooted in needle litter than in rotten wood. CWD develops greater concentrations of nutrients and a higher water-holding capacity and therefore becomes a more favorable rooting medium as decay proceeds. Despite these changes, rotten wood remains a nutritionally poor substrate when compared to mineral soil. The nutrition of plants rooted in CWD appears to be intertwined with mycorrhiza; this association may make it possible for higher plants to extract sufficient nutrients from woody debris. Although some species such as *T. heterophylla* may survive on logs for up to 1 year without mycorrhizal inoculation, their survival and growth is improved when inoculation occurs (Christy *et al.*, 1982). Several species of mycorrhiza-forming fungi have been isolated from rotten wood (Kropp, 1982).

Plants that are not rooted in or on CWD may also benefit from its presence. This is probably most important on severely disturbed sites where CWD may ameliorate environmental extremes and provide shaded microsites. Woody debris may also protect seedlings from being buried by material moving downslope.

2. *Autotrophs and Decomposition*

Autotrophs influence CWD decomposition and, conversely, decomposition processes influence the composition and structure of autotrophs using CWD. Superficial growths of plants add organic matter either as their own detrital remains or by trapping fine litter that otherwise would be lost from the surface. These organic matter accumulations serve as a rooting medium for larger, more demanding species (Harmon, 1985; Minore, 1972) and are a potential source of nutrients for organisms decomposing CWD. Plants rooted directly in CWD also influence decay by adding nutrients and more labile carbon by either root turnover or secretions.

Plants growing on CWD influence fragmentation rates of the debris in many ways, both positively and negatively. Root systems may bind the decayed material into a coherent structure (Triska and Cromack, 1980), and cover by bryophytes may reduce the erosive effects of rainfall. On the other hand, plants falling from CWD increase fragmentation rates when their root systems tear off pieces of bark and wood. The expansion of the root systems also contributes to fragmentation of CWD. The occurrence of browse plants on CWD increases fragmentation rates by attracting large mammals that pull plants from the substrate. Sloughing of bark and wood influences the autotrophic community by removing individuals. The larger and more deeply rooted the individual, the less likely it will be disturbed by fragmentation.

3. *The Importance of Coarse Woody Debris as Habitat*

White CWD is generally known to serve as plant habitat (Falinski, 1978; Lemon, 1945; Thompson, 1980), there is little information on the proportion of plants associated with CWD versus other habitats. Similarly, little information exists on facultative versus obligatory use of CWD.

A few lists of vascular herbs growing on CWD are available (Dennis and Batson, 1974; Lemon, 1945; McCullough, 1948; Sharpe, 1956). However, even a listing of the genera found by these few studies lies beyond the scope of this review. Herbs rarely appear restricted to CWD. Thompson (1980), for example, found that none of the 31 species examined was confined to logs. Dennis and Batson's (1974) study was an exception; they found 11 flood-sensitive species that were restricted to floating logs and stumps in a North Carolina swamp.

Numerous tree species can apparently grow on wood. In the southeastern United States, *Acer rubrum*, *Pinus caribaea*, *P. palustris*, *Pinus rigida*, *Pinus rigida* var. *serotina*, and *Q. nigra* grew on rotten wood, although only *P. caribaea* reached maturity on this substrate (Lemon, 1945). *Acer rubrum*, *Populus heterophylla*, *Populus deltoides*, *Quercus lyrata*, and *Taxodium distichum* seedlings grew on floating logs and stumps in a North Carolina swamp (Dennis and Batson, 1974). None of these species appeared to reach maturity on CWD in this setting, however. *Picea rubens* grew on stumps and logs in the southern Appalachian Mountains (Korstian, 1937). Stupka (1964) also reported that *Betula lutea* and *Betula lenta* frequently grew on stumps and logs in this area also. *Abies balsamea*, *Picea glauca*, and *P. rubens* seedlings grew on logs in New England and New York (Westveld, 1931). In the virgin forests of the Adirondack Mountains of New York, Knechtel (1903) observed *Pinus strobus*, *Picea rubens*, and *Tsuga canadensis* growing on CWD. *Betula alleghaniensis* rooted on stumps in northern Wisconsin (Kozlowski and Cooley, 1961). Rotten wood was a good seedbed for establishment of *Picea mariana* seedlings (LeBarron, 1950). *Picea engelmannii* and *Abies lasiocarpa* were observed growing on logs in Colorado (McCullough, 1948) and in British Columbia (Griffith, 1931; Smith, 1955; Smith and Clark, 1960). Lowdermilk (1925) reported *P. engelmannii* was often rooted on rotting logs throughout the northern Rocky Mountain region.

In northwestern North America, trees frequently grow on logs and stumps in the *P. sitchensis*, *T. heterophylla*, and *Abies amabilis* zones (Franklin and Dyrness, 1973). *Picea sitchensis* and *T. heterophylla* are frequently found growing on CWD in the coastal (*P. sitchensis*) zone (Hines, 1971; Kirk, 1966; McKee *et al.*, 1982; Minore, 1972). In addition to these species, *Alnus rubra*, *P. menziesii*, and *T. plicata* grow on CWD in this zone (Harmon, 1985). In the Cascade Mountains, *T. heterophylla* commonly grows on CWD (Christy and Mack, 1984; Fogel *et al.*, 1973; Franklin *et al.*, 1981; Maser and Trappe, 1984; Thornburgh, 1969; Triska and Cromack, 1980). Taylor (1935) observed *P. sitchensis* and *T. heterophylla* seedlings growing on rotten wood in southeastern Alaska.

Although all of these references are from North America, trees grow on CWD in other areas as well. Baldwin (1927a,b), Jones (1945), and Lachaussée (1947) indicate that *Picea* often establishes on rotten wood in European forests. *Betula pubescens* was noted growing on stumps in a heather moor in England (Dimbley, 1953).

Few studies note the importance of CWD in terms of tree recruitment, and it is therefore difficult to judge if trees rooted on CWD are little more than a curiosity. Because CWD covers a small fraction of the forest floor in most ecosystems, one might assume CWD plays a minor role. While this is often true, there are ecosystems in which CWD is an important seedbed. The *P. sitchensis–T. heterophylla* forests of the north Pacific Coast appear to be an outstanding example. McKee *et al.* (1982) found that 94–98% of the tree seedlings in a forest of this type were growing on CWD, and yet only 6–11% of the forest floor was covered by CWD (Graham and Cromack, 1982).

Logs are important seedbeds for trees in other forests witfin the Pacific Northwest as well. Christy and Mack (1984) found that 98% of *T. heterophylla* seedlings in an old-growth *Pseudotsuga–Tsuga* forest were rooted on rotten wood that covered 6% of the forest floor.

Subalpine *Picea–Abies* forests in British Columbia provide another example of the importance of dead wood as a seedbed for seedlings. Smith (1955) observed 75% of the tree seedlings grew on rotten wood that covered only 9% of the forest floor. Griffith (1931) found that in another *Abies–Picea* stand, 65% of the *P. engelmannii* and 48% of the *A. lasiocarpa* seedlings grew on rotten wood.

Qualitative observations indicate that CWD can be an important seedbed in eastern North American forests. For example, Knechtel (1903) studied regeneration in virgin forests in the Adirondack Mountains and noted *P. strobus*, *Picea rubens*, and *T. canadensis* "were reproducing almost entirely on the old decaying tree trunks lying in the forest."

The proportion of seedlings growing on CWD can be misleading in terms of overall reproductive success because CWD is not a stable habitat and fragmentation markedly reduces overall survival rates. Thus, many seedlings may initially establish on CWD, but the chances of survival to maturity may be higher on the forest floor.

4. Factors Controlling the Importance of Coarse Woody Debris as Habitat

The importance of CWD as plant habitat is controlled by moisture, species interactions, decay state of the substrate, and presence of a snowpack. Moisture conditions influence both the type and abundance of plants on CWD. Logs emerging from water may be one of the only sites on which mesophytic species can grow in swamps and bogs (Dennis and Batson, 1974; Hall and Penfound, 1943; Lemon, 1945). Within drier terrestrial environments, the combined effects of moisture and decay state determine the species using CWD. As moisture decreases, CWD becomes less favorable as a habitat for larger, complex life forms. This tendency is offset by decay, which increases the water-holding capacity of CWD (see Section III.E). The general pattern is illustrated by McCullough's (1948) successional study of logs in Colorado. Herbs, shrubs, and trees colonized logs in a less advanced state of decay in the mesic or bog environment than they did in the xeric environment. In the xeric environment, lichens and bryophytes dominated most of the successional sequence, and shrubs and trees colonized logs only in the most advanced stages of decomposition. There is some indication that rotten wood retains water better than humus. Place (1955) observed that rotten wood remained moist under a forest canopy, while the adjacent humus dried out. Since this would increase seedling survival, rotten wood might become an important seedbed in xeric environments. In northern Rocky Mountain forests, the higher moisture content of rotting wood resulted in a higher level of ectomycorrhizal activity during dry periods than was observed in the surrounding soils (Harvey *et al.*, 1976, 1979).

The state of decay affects growth of root systems and this in turn may influence plant survival. Thornburgh (1969) found that the roots of *T. heterophylla* growing on very rotten logs tended to ramify throughout the wood and did not penetrate to the mineral soil. In contrast, trees growing on partially decayed logs sent their roots around the solid inner core and into the underlying mineral soil. *Tsuga* growing under the latter situation had a more stable rooting medium and were less apt to uproot. Root growth patterns are no doubt modified by the size of the CWD. When large, sound pieces of CWD are involved, a tree might not be able to get its roots into the soil before it "outgrows" the support offered by the surface mantle of humus, bark, and rotten sapwood.

Deep snowpacks may increase the importance of CWD. Litter accumulating on the snowpack surface in a subalpine *A. amabilis–T. heterophylla* forest in the Cascade Range of northwestern North America tended to smother *Tsuga* seedlings germinating on the forest floor (Thornburgh, 1969). Because the log surfaces were raised above the surface of the forest floor, litter accumulations and seedling burial were reduced. Positioning high on logs could also lengthen the growing season. Christy and Mack (1984) extended the titter-burial hypothesis to lower elevation forests where snowpack is unimportant. They hypothesize that logs in *Pseudotsuga–Tsuga* forests of the Cascade Range shed much of the litter that falls on them so that litter accumulations on logs are less apt to bury newly germinated seedlings. In the wetter *P. sitchensis–T. heterophylla* forests of this region, log surfaces that retain litter best also retain seeds best and have the highest rates of seedling recruitment (Harmon, 1985).

Competitive and amensalistic interactions can restrict tree seedlings to logs in areas with heavy bryophyte and herb cover. In the *P. sitchensis–T.*

heterophylla forests, competition with bryophytes and herbs dramatically reduced tree-seedling survival on the forest floor and on very old, stable log surfaces (Harmon, 1985). In addition to competitive interactions, deep bryophyte layers may reduce survival by preventing seedling root systems from reaching the mineral soil before drying and nutrient deficiencies occur (Harmon, 1985).

CWD may also provide refuges for plants that are prone to herbivory when growing on the forest floor proper because plants on large logs and stumps are more difficult for animals to reach. Concentrations of logs can also form natural exclosures and allow patches of ungrazed vegetation to develop (Franklin and Dyrness, 1973; Sharpe, 1956).

5. *Successional Patterns on Logs*

A complex plant succession is initiated as soon as a bole falls to the forest floor. Succession on CWD is complex because a number of processes, including colonization, decomposition, fragmentation, and species interaction, are involved and because boles are added to the forest floor in a continuum of states ranging from sound to very decayed. Moreover, the control these processes exert on community structure varies over the sere.

During the earliest stages of log succession, the community is dominated by the epiphytes that inhabited the living tree. For some of these species the microclimatic changes caused by tree fall may lead to death. For example, the lichen *Letharia vulpina* (L.) Hue usually dies after trees or snags fall in the Sierra Nevada Mountains of California because it cannot tolerate burial under a snowpack (Harmon and Cromack, unpublished). For other species, the change in microclimate may lead to a temporary increase. Thus, the liverwort *Ptilidum californicum*, which spreads rapidly over surfaces of newly fallen logs on the Olympic Peninsula, eventually is replaced by larger mosses such as *Hylocomium splendens* (Sharpe, 1956).

The ability of the log surface to retain both seeds and needle litter increases as the newly created CWD surfaces become colonized with lichens and/or bryophytes (Harmon, 1985). Soon an organic soil accumulates, which allows forest floor species to invade log surfaces. In mesic environments, these processes allow succession to proceed independent of wood decay. For example, Harmon (1985) found that *Picea* and *Tsuga* seedlings could grow on undecayed logs as long as superficial humus deposits were present. However, in xeric environments, wood decay may have to proceed substantially before shrubs and trees can invade.

As colonization proceeds, intra- and interspecific competition increases. There is a tendency for more complex and larger life forms to displace simpler, smaller forms. Sharpe (1956), for example, stated that larger feather mosses such as *Hylocomium splendens and Rhytidiadelphus loreus* replaced smaller species such as *Mnium punctatum* and *Dicranum fuscesens* during succession in the Olympic rain forests. However, there are many exceptions to this pattern, and larger plants are often added as succession proceeds, without eliminating previous layers (McCullough, 1948; Sharpe, 1956).

Smaller species may exclude larger species by competing with them during critical life stages. Thus, deep carpets of *Hylocomium* and *Rhytidiadelphus* mosses can prevent log colonization by tree seedlings (Harmon, 1985). In mesic environments, a high density of tree seedlings and shrubs can accumulate on the surface of logs (McKeè *et al.*, 1982). This led Sharpe (1956) to conclude that severe competition at this point in succession causes a rapid loss of individuals. Larger individuals may send their roots into underlying soil before smaller individuals do. This greatly increases the amount of nutrients available to the former trees and gives them a strong competitive advantage over the surrounding smaller trees.

Fragmentation tends to offset successional trends and reinitiate the colonization process. Fragmentation may result from sloughing of bark or wood, toppling of individual plants from the log, impacts of falling trees or snags, or animal impacts. Although fragmentation removes plants, it can also accelerate succession. First, it may reduce competition and allow survivors to grow faster. Second, fragmentation may expose uncolonized areas on logs covered with deep bryophyte carpets and allow establishment of tree and shrub seedlings.

B. Coarse Woody Debris as Animal Habitat

1. Terrestrial Vertebrate Habitat

CWD provides habitat for many terrestrial vertebrates, including amphibians, reptiles, birds, and mammals. Elton (1966) recognized this function and noted, "When one walks through the rather dull and tidy woodlands—say in the managed portions of the New Forest in Hampshire [England]—that result from modem forestry practices, it is difficult to believe that dying and dead wood provides one of the two or three greatest resources for animal species in a natural forest, and that if fallen timber and slightly decayed trees are removed the whole system is gravely impoverished of perhaps more than a fifth of its fauna." This situation is by no means restricted to European forests. Thomas (1979) identified 179 vertebrate species using CWD in the Blue Mountains of Oregon and Washington, which is 57% of the species breeding in that region.

An extensive literature exists that describes the relationship between CWD and animals. For example, a bibliography on cavity-nesting birds alone contained 1713 references (Fischer and McClelland, 1983)! A thorough synthesis of vertebrate–CWD interactions is beyond our scope, and we have selected only a few articles to illustrate important points.

a. Factors Affecting Vertebrate Use of Coarse Woody Debris. Factors influencing the type and extent of animal use include physical orientation (vertical or horizontal), size (diameter and length), decay state, species of CWD, and overall abundance of CWD. Whether CWD is standing or down is a major factor influencing vertebrate use. Birds and bats use snags, for example, whereas mammals other than bats, amphibians, and reptiles typically use logs. Relatively few species use both logs and snags. For example, in the Blue Mountains of Oregon, only 20% of the CWD-using species use both snags and logs (Thomas, 1979).

Initial size of CWD is an important variable and influences the type and duration of use. Cavity-nesting birds (CNB) select trees with larger than average diameters for nesting (Carey, 1983; Mannan *et al.*, 1980; McClelland, 1977; Raphael and White, 1984). Species size dictates the minimum snag diameter for nests. The contrast between two species of woodpecker, *Dryocopus pileatus*, which is ~38 cm long, and *D. villosus*, which is ~19 cm long, illustrates this point. Snags with minimum diameters of ~50 and 25 cm are required by each species, respectively (Thomas, 1979). The influence of log size on wildlife is unknown, but Maser *et al.* (1979) suggested that larger logs are more useful, as they provide more cover than smaller logs. Snag and log size also determine the duration of use because larger CWD generally lasts longer than smaller CWD (Cline *et al.*, 1980; Maser and Trappe, 1984; see also Section III.D–E).

Decay state strongly affects vertebrate use of snags and incipient-to-advanced decay state is needed by most CNB (Connor *et al.*, 1976; Miller and Miller, 1980). Zones of rotten wood are probably selected to reduce the energy required by birds to excavate nests, and heart rots may allow CNB to use freshly created snags (Harris, 1983). Some CWD-using species can only excavate snags in advanced stages of decay (Thomas, 1979).

Log use by vertebrates also differs with decay state. Thomas (1979) describes changes in species and utilization patterns during log decay. Initial use is external (e.g., as perches or cover for runways), reflecting the hard condition of the log. As decay begins, utilization becomes internal. Loose bark, for example, provides spaces for hiding and thermal cover. Very decayed logs are soft enough to be excavated by the burrowing of small mammals, and this activity, in turn, allows amphibians and reptiles access to the log. As logs decay, the types of food, such as invertebrates and fungal fruiting bodies, available to vertebrates change. Feeding by vertebrates probably peaks toward middle to late stages of decay when logs are softer and many prey species are most abundant.

Vertebrate use of CWD is strongly influenced by spatial distribution of CWD as well as by the abundance of CWD. Aggregations of snags in small patches may enhance nesting habitat for some CNB, such as woodpeckers (Bull, 1975; Davis *et al.*, 1983; Jackman, 1974). Similarly, patterns and levels of log use may vary dramatically between isolated logs and those that form a continuous network. Continuity of CWD may allow some species to move through an otherwise hostile environment such as a clearcut or recently burned area.

b. Patterns of Vertebrate Use of Coarse Woody Debris. Terrestrial vertebrates use CWD for many functions. Thomas (1979) recognized cover, feeding, and reproduction as major uses; resting, preening, bedding, lookout, drumming, sunning, bridge, roosting, and hibernating were considered minor uses. A given species may use CWD for all, several, or only one function; therefore, the dependence of species on CWD varies. Some, such as salamanders and CNB, are probably obligatory, as opposed to facultative or opportunistic users of CWD. For many species, however, the level of dependence on CWD is unknown.

The largest and best recognized use of snags for shelter is by cavity-dwelling species. Primary cavity species create cavities in snags, while secondary cavity species use and/or enlarge preexisting cavities. Thomas (1979) recognized 39 bird and 23 mammal species using cavities in snags in the Blue Mountains of Oregon. At least 42 species of CNB are commonly found in temperate forests in North America (Table 8). In addition to cavities, protected sites associated with loose bark are important for bat roosting.

The permanent and winter residents of forest avifauna are generally the species using CWD. Most CNB are nonmigratory (Von Haartman, 1957), while migratory or transient birds generally do not use CWD (Snyder, 1950; Williams, 1936). Across five successional stages of *P. menziesii* forests, CNB accounted for 60% of the species of the winter avifaunas (Mannan, 1977). In an old-growth *Fagus–Acer* forest, 89% of the bird species that were permanent residents and fall and winter visitors used CWD (Williams, 1936).

Table 8 Cavity-Nesting Birds Commonly Censused in Temperate Forest Ecosystems of North America

Common name	Scientific name[a]	Type of cavity use[b]
Common flicker	*Colaptes auratus*	P(L)[c]
Pileated woodpecker	*Dryocopus pileatus*	P(L)
Rod-bellied woodpecker	*Centurus carolinus*	P
Gila woodpecker	*Centurus uropygialis*	P
Red-headed woodpecker	*Melanerpes erythrocephalus*	P
Acorn woodpecker	*Melanerpes formicivorus*	P(L)
Lewis' woodpecker	*Asyndesmus lewis*	P(L)
Yellow-bellied sapsucker	*Sphyrapicus varius*	P
Williamson's sapsucker	*Sphyrapicus throideus*	P
Hairy woodpecker	*Dendrocopos villosus*	P(L)
Downy woodpecker	*Dendrocopos pubescens*	P
Ladder-backed woodpecker	*Dendrocopos scalaris*	P
Nuttall's woodpecker	*Dendrocopos nuttallii*	P
Arizona woodpecker	*Dendrocopos arizonae*	P
White-headed woodpecker	*Dendrocopos albolarvatus*	P(L)
Black-backed three-toed woodpecker	*Picoides arcticus*	P(L)
Northern three-toed woodpecker	*Picoides tridactylus*	P(L)
Wied's crested flycatcher	*Myriarchus tyrannulus*	S
Ash-throated flycatcher	*Myriarchus cinerascens*	S(L)
Olivaceous flycatcher	*Myriarchus tuberculifer.*	S
Western flycatcher	*Empidonax difficilis*	S(L)
Violet-green swallow	*Tachycineta thalassina*	S
Tree swallow	*Iridoprocne bicolor*	S
Black-capped chickadee	*Parus atricapillus*	P(L)
Carolina chickadee	*Parus carolinensis*	P
Mountain chickadee	*Parus gambeli*	S(L)
Boreal chickadee	*Parus hudsonicus*	P
Chestnut-backed chickadee	*Parus rufescens*	P
Tutted titmouse	*Parus bicolor*	S
Plain titmouse	*Parus inornatus*	S
Bridled titmouse	*Parus wollweberi*	S
White-breasted nuthatch	*Sitta carolinensis*	S(L)
Red-breasted nuthatch	*Sitta canadensis*	S(L)
Pygmy nuthatch	*Sitta pygmaea*	S (L)
Brown creeper	*Certhia familiaris*	S(L)
House wren	*Troglodytes aedon*	S(L)
Winter wren	*Troglodytes troglodytes*	S(L)
Bewick's wren	*Thryomanes bewickii*	S(L)
Carolina wren	*Thryomanes ludovicianus*	S(L)
Eastern bluebird	*Sialia sialia*	S(L)
Western bluebird	*Sialia mexicana*	S(L)
Mountain bluebird	*Sialia currucoides*	S

[a] After Peterson (1961).

[b] P, Primary excavator; S, secondary nonexcavator; after McClelland (1977), Raphael and White (1984).

[c] (L), Also uses logs; after Thomas (1979).

Logs are used as shelter by many animal species, including a wide range of small mammals (Table 9). Logs provide protective cover immediately after their creation (Thomas, 1979). Loose bark provides thermal protection for salamanders and other temperature-sensitive species in addition to other protective functions (Maser and Trappe, 1984). Burrowing into decayed logs by some species provides opportunities for denning, feeding, and reproduction for other species. The mammals *Peromyscus maniculatus*, *Tamiasciurus hudsonicus*, and *Pituophus melanoleucus* utilize log burrows as reproduction sites in the Blue Mountains of Oregon (Thomas, 1979). Three species of predatory salamanders, *Batrachoseps wrighti*, *Ensatina eschscholtzi*, and *Aneides ferreus*, deposit eggs within logs of the coastal coniferous forests of western Oregon (Maser and Trappe, 1984). Hollow logs are used as cover and dens by larger animal species, including bears, and by rodents to store food (Thomas, 1979).

The plants, fungi, and animals inhabitating and decomposing CWD are a major food for many vertebrates. Snags are heavily utilized as feeding sites by insectivorous bird species, such as those of the genera *Centurus*, *Drycopus*, *Dendrocopos*, and *Picoides*. The *P. menziesii*–*T. heterophylla* forests of northwestern North America provide examples of log use for vertebrate food resources (Maser and Trappe, 1984). The vole *Clethrionomys californicus* extensively uses logs for shelter and food. This species eats mostly fungi and prefers truffles (Maser *et al.*, 1978; Ure and Maser, 1982), some species of which fruit mostly in rotten wood, where their mycelium forms a mycorrhizal association with conifers. Vole feeding disperses fungal spores to other suitable habitats. Several salamander species, *B. wrighti*, *A. ferreus*, and *E. eschscholtzi*, feed in logs, eating most invertebrate species found in logs. Ants, beetles, isopods, and common earwigs are important foods for *A. ferreus*, for example. The shrew *Sorex trowbridgei* and shrew-mole *Neurotrichus gibbsi* are important mammalian predators in rotting logs. The shrew-mole is a particularly efficient burrowing predator, feeding heavily on earthworms, centipedes, and flies (Maser *et al.*, 1981).

CWD provides spatial and temporal continuity of habitat that may be important to the survival and migration of animals. After catastrophic disturbances, logs continue to serve as shelter and fulfill other functions even though other habitat features are drastically altered. Similarly, logs provide a corridor that allows log-related species to migrate.

c. Importance of Coarse Woody Debris as Habitat. The dependence of CNB on the presence of snags makes them ideal species to illustrate the importance of CWD as vertebrate habitat. The contribution CNB

Table 9 Small Mammals Using Coarse Woody Debris in Temperate Forest Ecosystems of North America and Europe[a]

Scientific name[b]	Common name[b]	Type of log use[c]
Order Inseetivora		
Family Soricidae		
Blarina brevicauda	Short-tailed shrew	P
Sorex cinereus	Masked shrew	P
Sorex fumeus	Smoky shrew	P
Sores obscurus	Dusky shrew	P
Sorex trowbridgii	Trowbridge shrew	P
Sorex vagranis	Vagrant shrew	P
Order Rodentia		
Family Sciuridea		
Eutamias amoenus	Yellow-pine chipmunk	P(C)[d]
Glaucomys sabrinus	Northern flying squirrel	S(C)
Spermophilus beecheyi	California ground squirrel	S
Tamias striatus	Eastern chipmunk	S
Tamiasciurus hudsonicus	Red squirrel	S(C)
Family Cricetidae		
Subfamily Cricetinae		
Neotoma cinerea	Bushy-tailed wood rat	S(C)
Peromyscus leucopus	White-footed mouse	P
Peromyscus maniculatus	Deer mouse	P(C)
Subfamily Microtinae		
Clethrionomys gapperi	Southern red-backed vole	P
Clethrionomys glareolus	Common red-backed vole[e]	P[f](C)
Family Muridae		
Subfamily Murinae		
Apodemus flavicollis	Yellow-necked field mouse[e]	P[f](C)
Family Zapodidae,		
Zapus hudsonicus	Meadow jumping mouse	S
Zapus princeps	Western jumping mouse	S
Zapus trinotatus	Pacific jumping mouse	S
Order Carnivora		
Family Mustelidae		
Mustela erminea	Ermine	P(C)

[a] Based on references cited in text regarding species richness and abundance of small mammals.
[b] From Golley *et al.* (1975), pp. 361–370.
[c] P, Primary, use logs to fulfill the three major life-history functions: reproduction, feeding, and cover, S, secondary, use logs to fulfill only one or two of the major life-history functions.
[d] (C), Also use snag or tree cavities or nest boxes; North American species, Thomas (1979); European, Truszkowski (1974).
[e] European species, Hansson (1971), Grodzinski (1971), Grodzinski *et al.* (1970).
[f] Based on similar ecology, Corbet and Southern (1977).

make to the avifauna, however, varies markedly between deciduous and coniferous forests and over the course of succession. CNB account for 9–39% and 8–62% of the total bird species in deciduous and coniferous forests, respectively. Species richness and abundance of CNB appear to be greater in coniferous than in deciduous forests, hence the avifauna of the coniferous ecosystems might be more responsive to changes in the amount and quality of CWD present than those of deciduous ecosystems. In deciduous forests, the contribution CNB make to the total number of bird species increases from < 10% (one species) in early seral stages to 30% (seven species) in old-growth forests (Anderson, 1972; Holmes and Sturges, 1975; Holt, 1974; Johnston and Odum, 1956; Kendeigh, 1948; Martin, 1960; Odum, 1949, 1950; Salt, 1957; Shugart and James, 1973; Shugart *et al.*, 1978; Stewart and Aldrich, 1949; Stiles, 1980). In terms of the total number of individuals, CNB increase from ~1–18% of the total bird population between the same two seral stages. In coniferous forests, CNB species comprise 30% of the species during early seral stages, but this proportion declines to 20% as the new forest develops. As succession proceeds, the proportion of the avifauna comprised of CNB increases again to 30%. The mean contribution CNB make to the total number of individuals increases from 20% (70 individuals 40 ha^{-1}) to 30% (80 individuals 40 ha^{-1}) between early seral stages and old-growth coniferous forest, respectively (Bock and Lynch, 1970; Haapanen, 1965; Hager, 1960; Mannan, 1977; Mannan *et al.*, 1980; Manuwal and Zarnowitz, 1981; Marcot *et al.*, unpublished; Martin, 1960; Raphael, 1980; Raphael *et al.*, 1982; Salt, 1957; Scott *et al.*, 1982; Szaro and Balda, 1979; Wiens and Nussbaum, 1975).

The relative species richness and abundance of secondary CNB (SCNB) and primary CNB (PCNB) have different patterns during deciduous versus coniferous forest succession. The ratio of SCNB species to PCNB species (SCNB:PCNB) does not consistently exceed 1 until late (mature and old-growth) seral stages of deciduous and coniferous forests: The SCNB:PCNB ratio was consistently > 1 in middle (sapling/pole timber and small saw timber) to late seral stages of deciduous forests; in contrast, SCNB:PCNB abundance did not consistently exceed 1 until late seral stages of coniferous forests. The relationship between species richness and abundance of SCNB and PCNB is important because SCNB depend partly upon cavity abundance (Brush, 1981; Von Haartman, 1957), and the abandoned cavities of PCNB are one source of nest cavities for SCNB. Another major source of nest sites of SCNB, however, is natural cavities in living and dead trees formed by processes independent of PCNB activity. The major source of nest sites for SCNB is abandoned PCNB cavities in coniferous forests and naturally formed cavities in deciduous forests (Carey, 1983; Raphael, 1980). This difference in the relative importance of natural versus abandoned cavities between forest types may partly explain the tendency for SCNB:PCNB abundance to exceed 1 earlier in deciduous than in coniferous forest succession. SCNB in deciduous forests are apparently able to colonize forests irrespective of PCNB abundance because SCNB depend more upon the abundance of natural cavities than on cavities abandoned by PCNB.

Small mammal communities of forests are, as noted earlier, comprised of many species using CWD. Species using CWD average 70–90% (4–7 species) of the total number of small mammal species richness in both deciduous and coniferous forests over a wide range of seral stages (Ahlgren, 1966; Aldrich, 1943; Gashwiler, 1970; Gunther *et al.*, 1983; Hirth, 1959; Hooven and Black, 1976; Kirkland, 1977; Manville, 1949; Martell and Radvanyi, 1977; Morris, 1955; Odum, 1949; Pearson, 1959; Raphael, 1983; Raphael *et al.*, 1982; Ryszkowski, 1969; Scott *et al.*, 1982; Storer *et al.*, 1944; Wetzel, 1958). Averaged over forest type and successional stage, small mammal species using CWD comprise 75–99% of the total number of individuals. Abundance of mammals using CWD generally follows the abundance of CWD and appears to be highest during the earliest and latest stages of forest succession and lowest during middle successional stages.

The herpetofaunal communities of forests have been studied relatively little qualitatively and even less quantitatively. The available literature shows, however, that CWD-using species may predominate within the herpetofauna. For example, the reptiles and amphibians using CWD comprise 93% (mean of 8 species) and 99% of the individuals (54 individuals ha^{-1}) ranging from the early seral stages to old-growth *Pseudotsuga* (Raphael, 1983; Raphael *et al.*, 1982).

2. *Influences on Fish Populations*

The perspective of stream ecologists concerning the relationship between CWD and fish populations has changed dramatically during the past two decades. Before 1970, wood generally was considered a hindrance to fish migration and a cause of oxygen depletion in streams. In contrast, recent investigations have emphasized the beneficial role CWD plays in the formation and stabilization of fish habitat.

a. Blockage to Migration. Logging often increases the amounts of CWD in streams above natural levels and causes massive log jams that can potentially block the upstream migration of anadromous salmonids

(Elliot, 1978; Meehan *et al.*, 1969; Narver, 1971). After 1936, fishery management agencies in the Pacific Northwest removed log jams from streams to increase fish access to spawning and nursery areas (Hall and Baker, 1982; Sedell and Luchessa, 1982). Early logging practices doubtlessly added debris and increased the potential for blockages, but debris removal operations continued for many years with little evaluation of their need. Narver (1971) felt complete barriers were relatively rare and that migration was only hindered at certain flows.

Although some log jams block fish passage, they rarely remove a major fraction of the potential spawning or rearing habitat in a drainage basin. For example, log jams prevented fish migration to 12% of the length of potential fish-producing streams in the Coquille River basin in Oregon in the 1940s and early 1950s (Sedell and Luchessa, 1982). Only 5.5% of the length of potential fish-bearing streams in the Siuslaw National Forest in western Oregon were blocked by log jams in the late 1970s (Sedell and Luchessa, 1982).

Removal of log jams was also thought to decrease bank cutting and streambed instability because CWD deflects currents into streambanks (International Pacific Salmon Fisheries Commission, 1966; Pfankuch, 1978). The resulting siltation and shifting sediments caused by bank cutting would then smother and scour salmonid eggs and reduce the abundance of invertebrates (Gammon, 1970). The concept that CWD contributes to channel instability is inconsistent with most recent geomorphic studies (see Section V.D). In fact, removal of log jams may actually increase the adverse effects siltation and sediment instability have on fish and invertebrate populations (Beschta, 1979). Abundance of sea-run *Salvelinus malma* decreased after removal of debris dams in an Alaskan stream, and 2 years after removal, numbers of trout were only 20% of the preremoval population (Elliot, 1978). Log-jam removal in seven stream reaches in Oregon released stored sediments and destroyed fish habitat (Baker, 1979). However, fish populations did not decline in these streams after CWD removal.

b. Water Quality. CWD contains organic compounds that are potentially lethal to aquatic organisms. Leachates of whole *Tsuga logs*, with and without bark, were not toxic to fry of *Salmo gairdneri* and *Oncorhynchus tshawytscha* in 96 hours of exposure (Atkinson, 1971). Leachates of *Pseudotsuga* needles, *Tsuga* needles, and *Alnus* leaves were toxic to *Poecilia reticulta* and *S. gairdneri*, but at concentrations so high that O_2 depletion would become a threat long before toxic effects could be expressed (Ponce, 1974). Foliage terpenes and heartwood tropolenes were more toxic to *Oncorhynchus kisutch* and aquatic insects than bark extractives and heartwood lignins (Peters *et al.*, 1976). Analysis of water from several natural streams and logging-influenced streams revealed that the adverse effects of these leachates would be restricted to freshly logged areas with large amounts of *Thuja* slash or swampy areas with naturally high accumulation of *Thuja* debris (Peters *et al.*, 1976). Extracts of *P. sitchensis* and *T. heterophylla* bark were toxic to *Oncorhynchus gorbuscha* fry (Buchanan *et al.*, 1976). Salmon fry were more sensitive to extracts of *Tsuga* than *Picea* bark extracts; 50% of the fry were killed at a concentration of 56 mg liter^{-1} of *Tsuga* bark extract and 100–124 mg liter^{-1} of *Picea* bark extract (96-hour LC_{50}). As with *Thuja*, these concentrations are sufficiently high that the toxic effects of leachates would be limited to log-holding facilities or recent clear-cuts with heavy slash deposits.

Although toxic effects are rarely encountered, depletion of dissolved O_2 by microbial respiration and chemical oxidation of wood often affects fish populations adversely. Dissolved O_2 concentrations in stream water and interstitial water in a coastal stream in Oregon decreased to potentially lethal amounts after logging (Hall and Lantz, 1969). Dissolved O_2 concentrations increased after debris removal, but were still significantly lower than prelogging concentrations. Egg-to-fry mortality of *Oncorhynchus nerka* increased significantly when bark debris composed >4% of the volume of spawning gravel (Servizi *et al.*, 1970). Fry emergence was retarded when bark exceeded 1% of the volume. Oxygen in water can be consumed by either microbial respiration or by abiotic oxidation of the substrate. Abiotic oxidation rates (chemical oxygen demand) may exceed biological respiration rates (biological oxygen demand). Ponce (1974), for example, found that the chemical oxygen demand of *Pseudotsuga* wood exceeded biological oxygen demand by an order of magnitude. Thus, addition of CWD to streams may significantly reduce dissolved O_2 even when biological activity is low. Mortality or exclusion of fish by O_2 depletion caused by wood debris is not a major concern under natural conditions and is only likely to occur where unusually large accumulations of fresh CWD occur. Moreover, aeration in moderate- to high-gradient streams will probably offset reductions in oxygen concentrations caused by the presence of CWD.

c. Habitat. CWD plays a major role in stream channel geomorphology (see Section V.D); therefore, fish habitat is intricately linked to CWD dynamics. CWD potentially provides cover, creates important hydrologic features such as pools and backwaters, and stores inorganic sediments. The importance of CWD to fish populations has been recognized in a number of recent review articles (Franklin *et al.*, 1981; Maser and Trappe, 1984; Meehan *et al.*, 1977; Sedell and Swanson, 1982; Swanson *et al.*, 1982b; Triska *et al.*, 1982).

Early investigations of fish habitat in streams identified CWD as a major cover (Boussu 1954; Hunt, 1969; Tarzwell, 1936). Hartman (1965) observed that *O. kisutch* and *S. gairdneri* were associated with debris dams and that these salmonids decreased in winter where log cover was absent. Subsequent investigations have documented the use of wood habitat by fish in streams (Everest and Meehan, 1981; June, 1981; Lister and Genoe, 1979; Osborn, 1981).

In the Pacific Northwest, winter is a period of high flow, low stream temperature, and low light intensity. Stable winter habitat and refuges during high flow are critical for fish survival, and CWD is an important source of cover and a major agent in channel stabilization. Fry and juveniles of *O. kisutch* and juvenile anadromous *S. gairdneri* (age 1+) used logs and upturned tree roots as their major source of winter cover in several streams on Vancouver Island, British Columbia (Bustard and Narver, 1975a). Juvenile *O. kisutch* in Carnation Creek, British Columbia, inhabited deep pools, log jams, and undercut banks with tree roots and debris in winter (Tschaplinski and Hartman, 1983). During winter, stream reaches with these habitat types retained higher populations of juvenile salmon than those without these habitats; furthermore, fewer fish were lost after freshets in reaches with abundant CWD. Logging did not result in a change in numbers of salmon that migrated out of Carnation Creek in the autumn or into the stream in the spring.

In addition to large, stable accumulations of CWD, lateral habitats outside the main channel (e.g., backwaters, sloughs, and side channels) are critical refuges for fish during floods and serve as rearing areas for juveniles. CWD, boulders, and living trees were the major structural features responsible for the creation and maintenance of backwaters and side channels in third-order streams in the Cascade Mountains (Moore, unpublished). Bustard and Narver (1975a) and Bustard and Narvar (1975b) observed that juvenile *O. kisutch* moved into sidepools and small lateral tributaries during winter floods. Off-channel ponds in rivers of the Olympic Peninsula supported the majority of salmonid production in the drainage (Peterson, 1980). Side channels and terrace tributaries contained the highest biomass of juvenile *O. kisutch* in the Hoh River of the Olympic Peninsula (Sedell *et al.*, 1982). Bisson *et al.* (1982) found coho salmon fry predominantly in backwater pools in 19 stream reaches in Washington. In reaches above anadromous zones, backwater pools were the preferred habitat of the trout *Salmo clarkii*. Fry of *S. clarkii* in streams in the Cascade Mountains of Oregon occupied backwater habitats and were not found in the main channel until early fall; even then they remained in main channel habitats close to backwaters (Moore, unpublished).

The abundance of fish populations in streams and rivers is strongly related to the abundance of CWD. Wood debris was a major component of off-channel habitats in rivers of the Olympic Peninsula, and side channels with CWD supported eight times more juvenile *O. kisutch* than side channels without CWD (Sedell *et al.*, 1982). Densities of *Salmo trutta* in a Danish stream declined after removal of small wood debris (Mortensen, 1977). Lestelle (1978) observed numbers and biomass of resident *S. clarkii* declined in winter after removal of 85% of the wood volume in a stream in Washington. Import of wood from upstream reaches was associated with increases of trout numbers and biomass to the original amounts. Yearling anadromous *S. gairdneri* and *S. clarkii* of all ages preferred habitats with abundant wood debris in streams in Washington (Bisson *et al.*, 1982). Densities of juvenile *O. kisutch* declined after removal of wood debris in two Alaskan streams (Bryant, 1982). In southeast Alaska, streams in clear-cut reaches supported higher biomass of young-of-the-year salmonids than streams with buffer strips or old-growth forests (Murphy *et al.*, 1985). However, streams with buffer strips contained significantly more yearling salmonids than streams flowing through clear-cuts or old-growth forests. Evidently, blowdown trees in the buffer strips provided an important source of cover that increased overwintering survival.

Spatial distribution of wood also influences the quality of fish habitat. Fish occupy three-dimensional space in the water column, and therefore the architectural arrangement of wood accumulations affects the potential use of that habitat. Little research has been completed on this aspect of fish habitat, but a recent study in streams in British Columbia found that fish abundance around wood debris increased as complexity of the accumulation increased (Forward, 1984). Intricate networks of logs, branches, roots, and small wood debris create a more complex, diverse array of cover and hydrologic features that may benefit fish populations.

3. *Terrestrial Invertebrate Habitat*

Vast numbers of terrestrial invertebrates use CWD for food, shelter, and as a site for breeding. Some rely entirely on the resources of a single tree species, while others are able to use many. Other invertebrate taxa found in CWD represent groups more commonly found in forest litter (Graham, 1925). This section reviews the major taxa that use CWD, how they use it, and the successional development of CWD in terms of invertebrate taxa.

a. Invertebrate Use of Coarse Woody Debris. Many invertebrates use wood in one form or another as food. Some invertebrates are only attracted to dying or very recently dead trees, while others require decayed wood. The part of a tree that can be eaten also varies between species. Some ingest the nutrient-rich inner bark, while others utilize the less nutritious wood (Parkin, 1940), and still other invertebrates eat the fungi decaying the wood rather than the wood itself. Within a dead tree, the nutrients and energy originating in the wood pass through many trophic levels (Savely, 1939).

Terrestrial invertebrates may also use CWD as protection from environmental extremes. For example, Lloyd (1963) found that slugs, snails, terrestrial isopods, centipedes, and earthworms migrated into branches during warmer weather, but lived in the litter layer during colder weather. Earthworms, slugs, snails, and centipedes have also been found in decayed *Pinus* and *Quercus* logs (Savely, 1939). Other invertebrates use wood as a hibernation site. The carabid beetle *Feronia oblongopunctata*, normally a litter-inhabiting species, hibernates in winter as an adult in cells excavated in the wood and under the bark of logs, and aestivates in these same sites during the summer (Penney, 1967).

Some invertebrates, mainly insects, use wood as a nesting site. These include carpenter ants (*Camponotus*), termites (Isoptera), carpenter bees (Xylocopidae), and domestic honey bees (superfamily Apoidea). Some of the social wasps construct their "paper" nests from masticated wood fibers gathered from CWD.

Organisms spending much of their life cycle in CWD, such as bark beetles, wood-boring beetles, and some mites and collembolans, breed and reproduce there as well. Some organisms only use wood for a small portion of their life cycle. For example, the egg stage and instars I–III of the millipede *Cylindroiulus punctatus* are found under bark, while the later instars and adults are found in the litter layer (Banerjee, 1967).

b. Successional Relationships. The fauna associated with a tree changes as the tree's condition goes from living, to dying, to dead and decaying. During succession there is a shift from hostplant specificity to habitat specificity, so that by the end of succession, the decay state of CWD is more important than the tree species contributing the CWD (Howden and Vogt, 1951). Interest in the succession of insects in dead and dying trees extends back for some time. Some of the earliest workers in this field include Townsend (1886), Packard (1890), Felt (1906), Adams (1915), Blackman and Stage (1918, 1924), Graham (1925), and Savely (1939). More recent work has been conducted by Howden and Vogt (1951) on *Pinus* snags, Fager (1968) on *Quercus* logs, and Deyrup (1975, 1976, 1981) on the insect fauna of dead and dying *Pseudotsuga*. We will cover only the common points because an extensive review would be needed to cover succession of each tree species.

Bark beetles (Scolytidae) are among the first insects to occur in CWD, attacking weakened or recently killed trees. Bark beetles are often host specific and are usually limited in their occurrence to specific areas of the tree (Furniss and Carolin, 1977). An associated and often extensive guild of parasites and predators rapidly follows bark beetles into CWD. For example, Miller and Keen (1960) cited 16 insect species that preyed upon the western pine beetle (*Dendroctonus brevicomis*). Four species of parasitic

Hymenoptera are associated with this beetle and 10 species of beetles (representing 6 families), 1 species of ant, and 1 species of snakefly (Raphidiidae) were reported as predators. The list of parasites and predators would be more extensive if mites had been included. Mites are often important egg predators, but no specific instances were reported for the western pine beetle. However, the authors do report larval predation by mites, and this appears to be the case for many other bark beetle species as well (Rust, 1933).

Other wood-boring organisms follow shortly after the attack by bark beetles. The ambrosia beetles (Scolytidae), round-headed wood borers (Cerambycidae), flat-headed wood borers (Buprestidae), together with horn-tailed wasps (Siricidae) usually attack freshly killed trees. Some species are cambium or phloem feeders (some bark beetles and flat-headed borers), while others may start in the phloem and then tunnel into the heartwood (some round-headed borers, flat-headed borers, ambrosia beetles, carpenter worms, and horn-tailed wasps, among others). These species open up the wood to other decay organisms, either brought in with the insect or entering after an opening is created.

Termites and carpenter ants often enter decaying wood, but species have their own special requirements so far as moisture (e.g., dry- and damp-wood termites) and decay state are concerned. Termites and ants also have large groups of associated organisms that follow them into CWD, called termitophiles and myrmecophiles, respectively.

As the wood decays further, organisms unable to penetrate sound wood appear. A number of beetles are found in very decayed wood, including representatives of such families as Scarabaeidae, Lucanidae, and Passalidae. The larval stages of a number of families of flies are frequently found in very decayed wood, including the Tipulidae and Mycetophilidae. Many collembolans and mites also appear at this stage.

c. Major Taxa Using Coarse Woody Debris as Habitat.

i. Bark beetles. Bark beetles, or Scolytidae, play an important role in the creation and early stages of CWD decay. Stark (1982) recognized three stages in the life cycle of bark beetles: production (mating, gallery construction, oviposition, and brood development), dispersal (flight and host selection), and colonization (aggregation and overcoming host resistance). Only in the dispersal stage is the beetle away from CWD.

Another type of scolytid is the ambrosia beetle. In North America, these beetles attack dead or dying trees (Stark, 1982) and have a symbiotic relationship with fungi. The beetles introduce the fungi into boles and their breeding galleries, where the fungi provide the insects food. Ambrosia beetles differ in their gallery construction from true bark beetles in that their galleries go deep into wood. The Platypodidae, a small family of beetles closely related to the Scolytidae, also are often called ambrosia beetles. They have habits similar to the scolytid ambrosia beetles, but confine most of their boring activities to heartwood (Furniss and Carolin, 1977).

ii. Wood borers. A number of other beetle taxa bore into wood besides bark and ambrosia beetles. Two major families of wood-boring beetles are flat-headed wood borers (Buprestidae) and round-headed wood borers (Cerambycidae). The large beetle family Curculionidae (weevils) also contains some wood-boring species.

The family Buprestidae contains a number of wood-boring species and most enter dying or dead trees (Furniss and Carolin, 1977). Many species are hostplant specific or at least confine their activities to closely related host species. Female buprestids lay eggs in cracks and crevices in the bark. The larvae bore into the cambium region first and then usually into the wood (Miller and Keen, 1960), forming flattened burrows that are packed with boring dust. Life cycles may last from one to many years.

The Cerambycidae function much like the Buprestidae by attacking weakened, dying, or recently dead trees. The eggs are laid in the cracks and crevices of bark or in holes created by the female beetle. The hatching larvae bore into the cambium layer and sometimes into the wood itself. Some species are quite large (adults 40–70 mm long) and create large burrows deep in the wood.

iii. Termites. Termites (order Isoptera) are an important group of wood-eating social insects occurring in many parts of the world. Although most abundant and diverse in tropical and subtropical regions, selected genera and species occur in cooler regions (Weesner, 1960, 1970). There are five living families of termites and more than 2000 species (Weesner, 1960). The more primitive families live in and eat wood, while the more advanced taxa make nests on the ground or in trees and may eat grass, dead leaves, and fungi. The primitive groups have symbiotic protozoa in their hindgut, enabling them to digest cellulose, whereas the more highly evolved groups derive their cellulases from bacteria in the gut or produce their own enzymes.

Termite excavations in wood are normally shielded from light and are usually longitudinal cavities. The cavities are frequently characterized by the presence of cylindrical pellets of excrement, although dry-wood termites may actually move these pellets outside the nest.

Damp-wood termites normally live in damp, generally rotten wood. *Zootermopsis angusticollis* (Hagen) is a common species in CWD along the Pacific Coast of North America (Furniss and Carolin, 1977). Dry-wood termites, as their name implies, enter and live in dry wood (Weesner, 1970). The subterranean termites often establish their colonies in the ground, but feeding frequently extends to wood above ground (Weesner, 1970).

These termites build tubes to food sources, maintaining contact with the colony and preventing desiccation and exposure to light. Some of the species of subterranean termites show a preference for springwood, leaving the harder summerwood generally untouched (Furniss and Carolin, 1977).

iv. Carpenter ants. Carpenter ants, *Camponotus* Mayr, are a conspicuous group of wood-dwelling insects in most habitats containing CWD. Some members of the subgenus *Camponotus* nest in the soil and thus are not considered here. Carpenter ants are common in most northern boreal forests, and yet there is remarkably little information about most species, and even the taxonomy of the genus *Camponotus* is unsettled (Creighton, 1950).

Snags, logs, and stumps are used by carpenter ants as nesting sites, and although they chew the wood to excavate galleries, it is not ingested (Coulson and Witter, 1984; Furniss and Carolin, 1977). Most carpenter ants feed on honeydew produced by homopterans (Fowler and Roberts, 1980; Gotwald, 1968; Sanders, 1972; Tilles and Wood, 1982), but some are known to be predaceous (Ayre, 1963; Fowler and Roberts, 1980; Green and Sullivan, 1950; Myers and Campbell, 1976).

Life-history information on species of *Camponotus* in North America is limited and primarily based on the work of Pricer (1908) in Illinois and Sanders (1964) in New Brunswick, Canada. Articles by Eidmann (1929), Hölldobler (1944), and Marikovskii (1956) provide a similar basis for work on the Old World *Camponotus* species, *C. herculeanus*.

A colony of carpenter ants is normally established by one delated, fertilized queen (Mintzer, 1979). The colonizing queen is capable of boring into wood, but entry is commonly made through an existing opening such as that left by an emerging insect. Initially, colony formation often follows the burrows of wood-boring insect larvae (Breznak, 1982; Parkin, 1940). First-year colonies of *Camponotus* in Illinois contained a single queen and an average of 8.68–9.76 workers and 16.71–18.21 larvae (Pricer, 1908). It takes from 3 to 6 years before a colony produces winged females, at which time there would be ~2000 workers. Some colonies may be quite large. Sanders (1970) recorded a colony of *C. herculeanus* from Ontario that contained 12,240 workers, 1,059 females, 77 males, and 10,280 larvae. As colonies become older, they become decadent, producing large numbers of males but no winged females. Sanders (1970) suggests that a figure of 500 workers is characteristic of decadent colonies.

v. Other Hymenoptera. In addition to the Hymenoptera that are parasitic on other wood-inhabiting insects, two families are frequently found in CWD—the Siricidae, or horntail wasps, and the Xylocopidae, or carpenter bees. Most horntail wasps attack coniferous trees. Many species are polyphagous in the larval stage (Furniss and Carolin, 1977). Symbiotic fungi are reported to be associated with some species, and some larvae have been reported to feed on the fungi (Morgan, 1968). The Xylocopidae, or carpenter bees, often burrow into wood where they rear their young. The bee larvae develop in the cells constructed in the burrow.

vi. Lepidoptera. An enormous number of Lepidoptera are associated with trees; only a few taxa have wood-boring larvae. These include some Hepialidae, Cossidae, and Sesiidae. All of these have larvae that bore into living wood—most often into the phloem layer first—and then into the sapwood, and eventually into the heartwood (Furniss and Carolin, 1977).

vii. Diptera. The Diptera, or true flies, is a large group of insects, but comparatively few terrestrial species are associated with CWD and are chiefly in the immature stages. Teskey (1976) reported representatives of 45 dipteran families that had taxa associated with dead and dying trees. The larvae of some species of crane flies (Tipulidae) and fungus gnats (Xylophagidae) bore into rotten wood. Larvae of fungus gnats are often found beneath the loose bark of stumps and fallen trees where they feed on fungi. The larvae of some Asilidae are found in decaying wood where they prey on round-headed beetle larvae and other arthropods. Some species of Syrphidae have larvae that mine in the cambium layer, particularly of conifers (Furniss and Carolin, 1977). Larvae of the genus *Medetera* (Dolichopodidae) are important predators of bark beetle larvae and adults and some other wood-boring beetles.

viii. Mites. Wallwork (1976) studied the mite fauna of decaying twigs and branches of *B. lutea* and *Tsuga*. Three of the four wood-boring mite species occurred in both tree species. Two of the three species showed a preference for different parts of the branch. For example, *Steganacarus magnus* occurred most often in the heartwood of *B. lutea* and in the bark of *Tsuga*. A variety of fecal-feeding and predaceous mites also were found in both tree species. *Betula lutea* had a larger fauna (12 species) than *Tsuga* (6 species), but fewer total individuals per branch.

4. *Influences on Aquatic Invertebrates*

The distribution and abundance of aquatic invertebrates in streams are intricately linked to CWD. In addition to using wood directly as habitat or food, aquatic invertebrates are strongly influenced by channel structures created by CWD and the storage of inorganic sediments and organic matter (Anderson and Sedell, 1979; Meehan *et al.*, 1977; Sedell and Swanson, 1982; Triska and Cromack, 1980; Ward *et al.*, 1982). In this section, we examine the habitat relationships, feeding dynamics, and life-history strategies of aquatic invertebrates associated with CWD in streams. When CWD is abundant, a specialized fauna has evolved that is closely associated with wood debris. Dudley and Anderson (1982) listed over 50 taxa, representing five orders, as "closely associated" with wood and twice as many as facultative users. Almost all of these taxa were most common in headwater streams.

a. Habitat. Many invertebrates in streams and rivers use CWD surfaces opportunistically as a refuge. Other taxa bore, mine, and ingest decayed wood and associated microbes, feed on periphyton attached to the wood surface, use wood as an oviposition site or entry route into water, or use wood as an attachment site for filter feeding.

It is difficult to determine exactly where "aquatic" habitat ends. We have included floodplains subjected to infrequent flooding as well as the active channel. Species richness (especially of burrowers) is often greatest at the land–water interface, which is submerged during high water. Capillary movement of water allows streamside wood to remain fully saturated throughout most of the year.

Substrate quality is an important factor in aquatic invertebrate colonization of CWD. The species of wood, degree of waterlogging, and decay class are all important. The extent of microbial invasion has a considerable influence on its utilization by insects. Wood with surface decay is exploited by gougers (e.g., the beetle *Lara avara*), shallow tunnelers (primarily chironomids), and surface scrapers (e.g., the mayfly *Cinygma*), as well as taxa using it as an attachment site (e.g., black flies and net-spinning caddisflies). All of these species also occur on CWD with decay throughout. The many grooves, crevices, and cracks in the well-decayed CWD serve as refuges from predation and the abiotic environment. Other uses (oviposition, pupation, case-making, and emergence) also are greater on decayed CWD than on the firm, submerged pieces.

The absence of gallery formation and deep tunneling in submerged wood is a unique attribute of invertebrate–wood associations in aquatic environments (Cummins *et al.*, 1983). In terrestrial ecosystems, the abundance, diversity, and degree of social organization of insect taxa associated with wood are much greater than those found in freshwater ecosystems. In marine environments, insects are largely absent, but Annelida, Mollusca, and Crustacea are major inhabitants and decomposers of wood (Cummins *et al.*, 1983).

Wood is used as a feeding platform or attachment surface by invertebrates in streams or rivers with a shifting sand bed (Benke *et al.*, 1984; Cudney and Wallace, 1980; Dudley and Anderson, 1982; Nilsen and Larimore, 1973). Where CWD constitutes most of the stable substrate, it may be the setting for a significant amount of secondary production

(Benke *et al.*, 1984). Cudney and Wallace (1980) found that submerged wood was the only substrate suitable for net-spinning caddisflies, in the Savannah River of Georgia. Snags in the Satilla River of Georgia were highly productive, not only for net-spinning caddisflies, but also for filter-feeding Diptera and other typical "benthic" insects (Benke *et al.*, 1984). Of the 100 taxa identified from snag, sand, and mud habitats, 63 occurred on snags and 29 of these were "very common." Biomass of insects on snags was 20–50 times higher than in sandy habitats and 5–10 times greater than in mud habitats. Production estimates for the snag habitat are among the highest yet reported for stream ecosystems. The authors hypothesized insect production was limited by the availability of substrate rather than food.

The Elmidae, or riffle beetles, are commonly found on CWD in streams. *Lara avara* occurs in streams west of the Rocky Mountains and is the one obligate xylophagous elmid that has been studied (Anderson *et al.*, 1978, 1984; Steedman, 1983). Other genera of elmid beetles may be xylophagous, such as *Macronychus* and *Ancyronyx*, which are usually found on wood (White, 1982). *Macronychus glabratus* is reported to be a wood feeder (LeSage and Harper, 1976b). In Coast Range streams of western Oregon, the density of *Lara* larvae was 71 per m^2 of wood surface, with densities on coniferous wood about half that on deciduous wood (Steedman, 1983). Larvae were found in similar abundance on large and small sticks of wood in various states of decay (Steedman, 1983).

Wood is used for case construction by Trichoptera in several families and genera, but especially by the Limnephilidae. Of the 92 genera of case-making caddisflies discussed by Wiggins (1977), about one-quarter use bark or wood, at least occasionally, in case construction. Although most of these species add bark or wood chunks to their cases, *Amphicosmoecus* and *Heteroplectron* bore cavities in twigs and wood chips.

Pupation in moist or saturated wood is a common behavior for caddisflies and Diptera. Burrowing in CWD rather than attaching to stones may decrease mortality caused by predation, desiccation, or exposure to lethal temperatures when water level decreases. Burrowing into wood prior to pupation often occurs at or above the water line where oxygen concentration is high, but where the wood is still moist.

Wood is used as an oviposition site both above and below the water line. Many limnephilid caddisflies deposit egg masses on damp wood. Wisseman and Anderson (1984) found that in Coast Range watersheds, oviposition by *Ecclisocosmoecus scylla*, *Hydatophylax hesperus*, and some other species was concentrated on a few large logs overhanging the upper reaches of streams. Submerged branches are often used for oviposition by hydropsychid caddisflies. Eggs of the surface bugs *Gerris* and *Microvelia* are often glued to wood at the water's edge (Anderson, unpublished). On the same sticks, eggs of the free-living caddis, *Rhyacophila*, and the false cranefly, *Ptychoptera townesi* Alexander, were observed. Females of the xylophagous cranefly, *Lipsothrix nigrilinea* Doane, and of Chironomidae have also been observed ovipositing at or below the water line.

b. Wood Consumption. Xylophages consistently occur on or in woody debris and ingest wood particles. Diverse modes of feeding behavior and life-history strategies occur in this group, and all the major nonpredatory aquatic orders of aquatic insects have xylophagous representatives. Xylophages are less well represented in the hemimetabolous orders of insects than in the Holometabola, but some Plecoptera and Ephemeroptera ingest wood.

Plecoptera larvae can remove and ingest the soft, decayed surface of submerged wood by shredding or scraping, but few species have evolved as wood-feeding specialists. Gut-content analysis indicates the nemourids *Zapada* and *Visoka* and the peltoperlid *Yoraperla* are xylophages (Pereira *et al.*, 1982). In New Zealand, the austroperlid, *Austroperla cyrene* (Newman), was shown to be a wood feeder (Anderson, 1982).

Among the mayflies associated with wood, the tropical species *Povilla adusta* sometimes causes economic damage by burrowing into bridge pilings (Bidwell, 1979). This species is also abundant on submerged trees in African reservoirs (McLachlan, 1970; Petr, 1970). *Povilla adusta* burrows into wood or uses the galleries formed by terrestrial wood-boring beetles, but it feeds primarily on the periphyton attached to wood surfaces (Petr, 1970). The heptageneid mayfly, *Cinygma integrum* Eaton, is closely associated with wood debris in Pacific Northwest streams (Anderson *et al.*, 1984). Pereira (1980) reared *Cinygma* larvae using stream-collected wood as food. This species scrapes epiphytic autotrophs (algae) and heterotrophs (fungi and bacteria) from wood surfaces, and fungal mycelia are their primary food source (Pereira and Anderson, 1982).

The families of aquatic Coleoptera that consume wood include borers (Oedemeridae), scraper-collectors (Elmidae, Elminae; Psephenidae, *Acneus*; and Helodidae, *Cyphon*, and *Metacyphon*), and gougers (Elmidae, Larinae; and Ptilodactilidae, *Anchytarsus*, and *Anchyteis*). Scraper-collectors usually do not ingest significant amounts of wood fragments (Anderson, 1982; Pereira *et al.*, 1982). *Anchytarsus* is reported to be entirely xylophagous, but is rare throughout its range (LeSage and Harper, 1976a). Other ptilodactylids that may be wood feeders are also uncommon.

The oedemerid wharf-borer, *Nacerda melanura*, has been reported to be injurious to timbers of wharves along the California coast, and *Copidita* 4-*maculata* are known to bore into wet bridge and mine timbers (Essig, 1942). In Oregon and Washington, Dudley and Anderson (1982) found large numbers of larvae and adults of *Ditylus quadricollis* in a few submerged logs; where they occurred, the wood was riddled by their tunnels.

The Elmidae, or riffle beetles, are the most common xylophagous beetles in stream habitats. *Lara avara* larvae feeding on *Alnus*, *Tsuga*, and *Pseudotsuga* wood that had been in a stream for 5 years produced feces at a mean rate of 9% of body weight per day (range 0–41%) (Steedman, 1983). Steedman calculated that the field population produced 1.1–2.5 $g m^{-2} year^{-1}$ of feces and removed 0.2–0.8% of the available CWD per year.

An investigation of feeding habits of Trichoptera in Oregon streams found that $\sim$20 species ingested wood to some degree (Pereira *et al.*, 1982) and that many of the leaf shredders also feed on wood and associated microbes (Anderson *et al.*, 1978). The Calamoceratidae, Lepidostomatidae, and several genera of Limnephilidae are closely associated with wood in feeding. Some caddisflies listed as grazers by Wiggins and Mackay (1978), such as *Neophylax* and *Ecclisocosmoecus*, were also shown to fragment wood via their feeding activities. A New Zealand leptocerid caddisfly, *Triplectides obsoleta*, and a North American calamoceratid caddisfly, *Heteroplectron californicum*, both construct cases by hollowing out a twig and commonly tunnel into moist wood for pupation (Anderson, 1982). Though *Heteroplectron* consumes large quantities of wood, it cannot complete development with wood as its total diet (Anderson *et al.*, 1984).

The greatest diversity of aquatic and semiaquatic xylophages occurs in the Diptera. Most species are burrowers and collectively are probably the major wood decomposers in aquatic habitats. Dudley and Anderson (1982) listed 10 dipteran families that are closely associated with wood. Chironomidae (10 genera), Tipulidae (4 genera), and fungus gnats of the families Mycetophilidae and Sciaridae are the most common xylophages encountered. Only some genera of Chironomidae are fully aquatic, whereas the other taxa occur in moist wood at or above the water line.

The discovery that chironomid larvae live in decaying wood is fairly recent (Teskey, 1976), but several recent studies indicate wood-boring midges are widespread (Anderson *et al.*, 1984; Borkent, 1984; Cranston, 1982; Kaufman, 1983). Borkent (1984) reviewed the systematics and phylogeny of *Stenochironomus* and related genera and showed that the 65 species available for study were all highly modified for a mining mode of life, with a dorsoventrally compressed head capsule, expanded thoracic segments, and a long, flaccid abdomen. Most species of this genus were wood borers. Wood-mining *Stenochironomus* live in firmly anchored wood in lentic and lotic habitats. The larvae generally only mine angiosperm wood that has a clean surface and only occur in firm wood. The larval chambers are parallel to the wood surface under a layer of firm wood or thin bark. Other important wood-burrowing midges include four genera of Orthocladiinae midges: *Chaetocladius*, *Orthocladius*, *Symposiocladius*, and *Limnophyes* (Anderson *et al.*, 1984).

A large xylophagous midge, *Xylotopus par*, burrows into soft, well-decayed wood (Kaufman, 1983). Growth rates of this midge were greatest in *Tilia* logs that were terrestrially decayed and then submerged for only 2 weeks. In submerged logs of *Fraxinus* and *Populus*, the density of *Xylotopus* exceeded 5000 larvae m^{-2}, and biomass increased exponentially from 70 mg m^{-2} in June to 5000 mg m^{-2} in August.

The large size of their larvae makes the Tipulidae the most conspicuous dipteran wood borers of semiaquatic habitats. Alexander (1931) listed 19 genera and 48 species of xylophagous tipulids on a worldwide basis, and Teskey (1976) recorded 30 Nearctic species associated with dead trees, but noted that his list was probably very incomplete. A succession of tipulid genera occurs in wood (Teskey, 1976). *Gnophomyia* larvae occur in fermenting sap beneath the bark before significant decay has commenced. *Ctenophora* and *Epiphragma* penetrate into relatively hard wood, while *Lipsothrix* larvae burrow into wood only where a portion of a log or branch is continuously immersed in a stream. Partitioning along the moisture gradient was observed in an Oregon stream in which *Austrolimnophila badia* occurred primarily above the water line, whereas *Lipsothrix* spp. occurred at and slightly below the water line (Anderson *et al*., 1984; Dudley, 1982).

Life cycles and habitat preferences of the craneflies *Lipsothrix nigrilinea* and *L. fenderi* differ in streams in western Oregon (Dudley, 1982). The nonadult stages are spent within soft logs in constant contact with water. *Lipsothrix fenderi* larvae are found in a wider variety of wood types, including solid wood, coniferous as well as deciduous species, and in marginal (drier) habitats.

c. Life-History Strategies. Despite the number of examples of wood feeders discussed above, only a small fraction of aquatic insect taxa exploit wood debris as a food source. Life-history and feeding strategies of aquatic xylophages include some combination of the following: a long life cycle to compensate for low growth rates, high consumption rates, a symbiotic gut microflora to aid digestion and to furnish essential nutrients, and the ability to switch to high-quality food in later instars when rapid growth and lipid accumulation are required.

A long life cycle is characteristic of most aquatic insects consuming wood. For example, the beetle *Lara* lives for more than 4 years, the caddisfly *Heteroplectron* has a generation time of 2 years or more, and the cranefly *Lipsothrix* requires 2 years or more to complete its life cycle. Xylophagous midges are apparently univoltine, whereas many of the collector-gatherers and algal feeding midges are multivoltine.

A symbiotic gut flora does not occur in the wood gougers *Lara* or *Heteroplectron* (Cummins and Klug, 1979; Steedman, 1983), which have a simple, straight gut without diverticula or a fermentation chamber to accommodate symbionts. The residence time of ingested particles in the gut of *Lara* is relatively short (~8 hrs); therefore, these insects consume large amounts of wood.

Increased consumption may compensate for low food quality (Cummins and Klug, 1979), but insects also may exhibit high rates of ingestion on optimum diets. High consumption rates are characteristic of the midge *Xylotopus* (Kaufman, 1983), wood-boring tipulids, and the caddisfly *Heteroplectron* (Anderson *et al.*, 1984).

Attached bacteria in the hindgut of wood-boring tipulids, and in some other aquatic insects are assumed to be symbionts (Cummins and Klug, 1979). Kaufman (1983) described a thickening in the midgut region of *Xylotopus* that contained a dense arrangement of attached rods and filaments of a sporulating bacterium. In contrast to the more common situation where the bacteria are associated with the hindgut, this band in *Xylotopus* occurs on the posterior midgut wall, outside of the peritrophic membrane (Kaufman, 1983).

Concentration of N and other nutrients into a smaller volume by microbes is important to most xylophages. Steedman (1983) hypothesized that *Lara* passively absorbed molecules liberated by microbial enzymes and also digested contents of fungal, bacterial, and animal cells mechanically disrupted by feeding. Gougers and tunnelers as well as surface scrapers and shredders exploit the surficial layer of CWD enriched by microbes.

C. Importance of Coarse Woody Debris in Terrestrial Nutrient Cycles and Carbon Budgets

Coarse woody debris represents a substantial, yet little studied accumulation of energy, carbon, and nutrient elements in many forest ecosystems. Compared with other more commonly measured fluxes such as litterfall, the organic matter transfers into and out of the CWD compartment tend to be large. In this section, we compare the amount of nutrients stored in CWD relative to other ecosystem components and the flux of nutrients added to the forest floor in woody debris and leaf litter. A discussion of the mechanisms controlling the accumulation and loss of nutrients in CWD during decay then follows. Using simulation, we then examine the role of CWD in nutrient cycling across the sequence of stand development and consider its impact on forest productivity.

1. Organic-Matter and Nutrient Storage

Despite over 20 years of ecosystem research, we were able to locate only eight terrestrial sites where amounts of organic matter and nutrients in

CWD and other ecosystem components could be compared (Table 10). Overall, logs and snags accounted for 1–45% of total aboveground organic-matter storage, the two lowest values coming from mixed deciduous forests and the highest value coming from a young *A. amabilis* stand in which aboveground living biomass had not yet accumulated in large amounts. A second-growth tropical stand contained 23% of its aboveground mass in CWD, presumably left over from the previous stand. In contrast, two mature tropical forests contained 2–4% of their aboveground mass in CWD. CWD accounted overall for from <1 to nearly 20% of total ecosystem organic-matter storage, depending, in large part, on the amount of soil organic matter. For example, the 121-year-old *T. heterophylla* ecosystem from coastal Oregon has almost as much CWD as the old-growth *Pseudotsuga–Tsuga* ecosystem from the Oregon Cascades, and yet CWD comprises ~10% of the ecosystem total in the former case and ~17% in the latter case. The difference is primarily caused by the fact that soils in the coastal ecosystem accumulated ~6 times more organic matter than soils in the Cascade ecosystem.

Nitrogen and phosphorus in CWD account for 1–21% of the total aboveground storage (Table 10). In general, the proportion of ecosystem N stored in CWD is smaller than the proportion of organic matter stored, indicating that N concentration in CWD is lower than in other aboveground components. Two sites are exceptions to this pattern: the Virelles mixed-oak stand, where CWD accounted for a larger proportion of the N and P than of the aboveground biomass, and the old-growth conifer stand at H. J. Andrews Experimental Forest in Oregon, where CWD accounted for the same proportion of aboveground biomass and N.

Nutrient pools were measured across an age sequence of seven Pseudotsuga-dominated conifer stands in the Oregon and Washington Cascade Range (Table 11). The amount of N and P stored in CWD ranged from 100 to ~244 kg ha^{-1} and from 5 to 13 kg ha^{-1}, respectively, which is within the range reported for other ecosystems (Greenland and Kowal, 1960; Grier, 1976; Sollins *et al.*, 1980). Values were remarkably variable, however, and total storage in CWD showed little pattern with stand age. On average, the more advanced decay classes accounted for most of the N, P, and Ca stored in CWD, although two of the seven stands contained a significant proportion of the total N and P in decay class III (Figure 14). The amount of K stored in CWD tended to peak earlier in the decay sequence (class III or IV), which is to be expected given how quickly K leaches from CWD.

The significance of nutrient storage of CWD to the overall forest economy is difficult to judge. Compared with N stored below ground, the amounts stored in CWD are often relatively small indeed (Table 10). In *T. heterophylla* ecosystems, the soil contains ~200 times as much N

Table 10 Biomass, Nitrogen, and Phosphorus Storage in Fallen Logs, Snags, and Other Ecosystem Components at Forest Sites throughout the World

Stand type	Item[a]	Fallen logs	Snags	Vegetation		Litter	Soil	System total	CWD		References
				Above ground	Below ground				(Percentage of aboveground)	(Percentage of system total)	
Tropical second growth	DW	72	—[b]	236	54	2.27	44	408	23.4	17.8	Greenland and Kowal (1960)
	N	230	—	1,469	316	35	4,605	6,650	13.3	3.5	
	P	19	—	95	22	1	13	150	16.5	12.6	
Mixed deciduous	DW	1.6	—	121	35	3.2	120	280.8	1.27	0.57	Duvigneand and Denaeyer-DeSmet (1970)
	N	11	—	406	127	33	4,480	5,060	2.44	0.22	
	P	0.5	—	32	12	1.4		45.9	1.47	1.09	
Mixed deciduous	DW	3.9	—	122	34	25.6		185	2.57	2.10	Henderson and Harris (1975)
	N	9	—	388	104	265	5,080	5,850	1.36	0.15	Henderson *et al.* (1978)
Coastal *Tsuga heterophylla* (26 years old)	DW	90	—	193	38	22	770	1,113	29.5	8.09	Grier (1976)
	N	95	—	335	32	171	33,800	34,400	15.8	0.28	
	P	11	—	54	3	42	14	124	10.3	8.87	

(Continued)

Table 10 Continued

Stand type	Item[a]	Fallen logs	Snags	Vegetation		Litter	Soil	System total	CWD		References
				Above ground	Below ground				(Percentage of aboveground)	(Percentage of system total)	
Coastal *Tsuga heterophylla* (121 years old)	DW	212	—	920	187	34	776	2,130	18.2	9.96	Grier (1976)
	N	180	—	780	157	265	34,900	36,300	14.7	0.50	
	P	33	—	188	13	65	13	312	11.5	10.6	
Tropical montane rain forest	DW	11	—	492	63	10	330	906	2.14	1.21	Edwards and Grubb (1977)
Primary tropical forest	DW	18	—	406	67	6	248	745	4.19	2.42	Klinge (cited by Edwards and Grubb, 1977)
Old-growth *Pseudotsuga menziesii*	DW	190	25	718	152	51	133	1,270	21.9	16.9	Sollins *et al.* (1980)
	N	190	25	539	197	260	3,720	4,930	21.2	4.36	
	P	6.7	0.8	89.1	22.6	50.2	7.6	177	5.11	4.24	
Abies amabilis (23 years old)	DW	20	63	53	25	48	220	429	45.1	19.3	Grier *et al.* (1981)
Abies amabilis (130 years old)	DW	75	165	447	138	149	273	1,247	28.7	19.2	Grier *et al.* (1981)

[a] DW, dry weight (Mg ha^{-1}); N, nitrogen (kg ha^{-1}); P, phosphorus (kg ha^{-1}).
[b] Not available.

Table 11 Nutrient Storage in Coarse Woody Debris across a Chronosequence of *Pseudotsuga*-Dominated Sites in the Oregon and Washington Cascade Range[a]

Location	Age (year)	DW[b] ($Mg\ ha^{-1}$)	N[c]	P	K	Ca	Mg	Mn	Na
Bagby	250	115	244	12.4	21.9	223	28.8	9.5	5.5
H. J. Andrews 2	450+	79	143	7.5	15.3	153	17.8	7.8	3.4
H.J. Andrews 3	450+	115	227	12.3	24.0	237	28.9	12.7	5.3
Wind River	550	81	144	7.6	17.4	157	17.8	8.8	3.5
Squaw Creek	750	98	213	11.9	20.3	215	26.5	6.8	4.3
Mount Rainier 2	1000+	140	238	13.2	30.5	258	31.0	19.6	6.0
Mount Rainier 3	1000+	90	101	5.3	9.0	94	11.6	2.9	2.0

[a] From Sollins *et al.* (unpublished).
[b] DW, Dry weight.
[c] All elements are measured in kilograms per hectare.

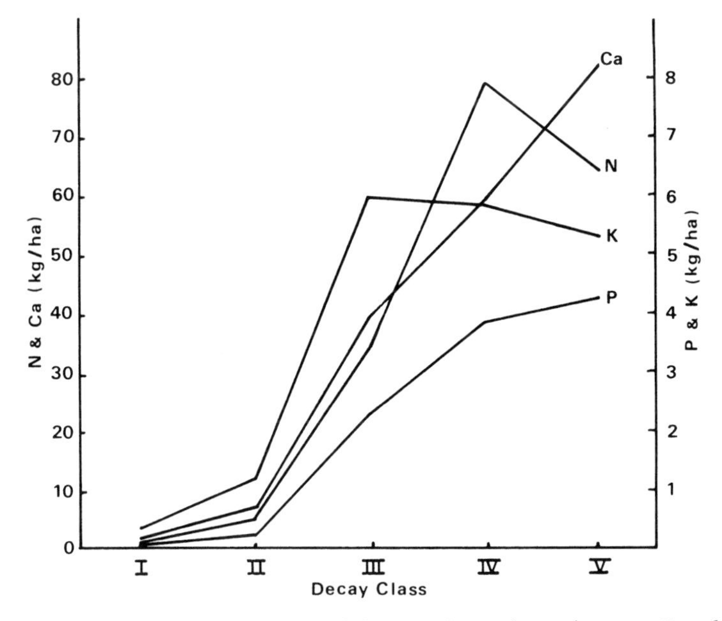

Figure 14 Distribution of N, P, K, and Ca over decay classes in seven *Pseudotsuga menziesii* dominated stands from the Cascade Range of Oregon and Washington.

as the logs. The soil of an old-growth *Pseudotsuga* ecosystem at H. J. Andrews was much less fertile, but the soil–log N ratio was still ~20. The amount of P stored in logs and soil is much more similar, especially if one includes only the readily extractable soil P in the comparison. Ratio of readily extractable soil P to P in logs ranged from 0.4 to 1.3 at the four sites where both were measured. The comparison requires caution, however, because there is much disagreement as to how to measure "readily extractable" soil P and because concentrations in the logs are near detection limits. At present, however, it appears that we lack the data to even roughly gauge the importance of P stored in CWD.

CWD and associated microbial communities greatly influence patterns of nutrient cycling in stream ecosystems. In a first-order stream in an oldgrowth coniferous forest in Oregon, the majority of the N was stored in coarse and fine woody debris (Triska *et al.*, 1984). CWD constituted 32% of the total N pool, while fine woody debris stored 18%. Fine particulate organic matter represented 40% of the total N stored, but a large fraction of this material may have been derived from CWD. Storage of a major fraction of nutrient capital in geomorphically stable, slowly decomposing CWD results in a persistent, stable nutrient supply for streams.

2. *Input of Organic Matter and Nutrients to the Forest Floor*

Input via tree death accounted for a substantial proportion of the total organic matter returned to the forest floor at the few sites where it was measured (Table 12). However, because leaf fall has higher nutrient concentrations than CWD, a larger proportion of nutrients return to the forest floor in foliage than in CWD.

Organic-matter transfer to the forest floor in fine and coarse wood are of similar magnitude. Together they account for 24–39% of total organic matter returning to the forest floor at five of the forest types, but for 60–74% of dry-matter return in the *Pseudotsuga* stands (Table 12). Why the latter ecosystems behave so differently is not known. Input of nutrients via fine wood has been measured in few ecosystems; at H. J. Andrews and Hubbard Brook, the two sites for which data could be obtained conveniently, fine woody litterfall accounted for 16 and 19%, respectively, of the total N returned in litterfall. In contrast, CWD accounted for 16 and 3%, respectively, of the total N returned to the forest floor. These two studies indicate that nutrient return in CWD can be ignored in some ecosystems, whereas return in fine woody debris definitely cannot. However, in forests where the CWD input rate is high, a large amount of nutrient transfer to the forest floor will be overlooked if CWD is not measured.

3. *Nutrient Accumulation and Loss in Coarse Woody Debris*

a. Input to Logs via Throughfall and Litterfall. Logs occupy a large portion of the land area—from 6 to 25% in conifer stands of the Pacific Northwest

Table 12 Annual Input of Organic Matter, Nitrogen, and Phosphorus in Falling Stems, Branches, and Foliage

Stand type	Stems and large branches			Fine wood			Foliage and miscellaneous			References
	DW[a] (Mg ha^{-1})	N (kg ha^{-1})	P (kg ha^{-1})	DW (Mg ha^{-1})	N (kg ha^{-1})	P (kg ha^{-1})	DW (Mg ha^{-1})	N (kg ha^{-1})	P (kg ha^{-1})	
Mixed northern hardwood	1	1.4	0	1.2	10.4	0.9	3.5	42.4	3.1	Gosz *et al.* (1972)
Liriodendron tulipifera	0.91	—[b]	—	0.12	—	—	3.3	—	—	Edwards (1972), Sollins *et al.* (1973)
Mixed deciduous/pine	0.6–1.9	3.6[c]	—	0.5–0.7	—	—	3.4–4.1	34.0	—	Harris *et al.* (1973), Henderson and Harris (1975)
Pseudotsuga menziesii (20–90 yeas old)	2.61	1.32	0.19	0.84	—	—	2.38	21.3[c]	3.1[c]	Gessel and Turner (1976) Sollins (1982, unpublished)
Pseudotsuga menziesii (450 + years old)	7	3.9	0.6	1.1	3.8	0.5	2.9	16.0	4.0	Grier and Logan (1977), Sollins *et al.* (1980)
Populus tremuloides	0.45	—	—	0.85	—	—	3.1	—	—	Gosz (1980)
Abies amabilis (26 years old)	0.3	—	—	0.47	—	—	1.36	—	—	Grier *et al.* (1981)

[a] DW, Dry weight.
[b] Not available.
[c] Includes fine woody litterfall.

(Table 5). Consequently, they can potentially intercept a large proportion of incoming litterfall and throughfall. Nutrients normally added to the forest floor via these two pathways, may, therefore, accumulate within CWD. Grier (1978) found that Na accumulated in fallen logs near the Oregon Coast, whereas amounts of K decreased rapidly. Precipitation at this coastal site contained ~3 times as much Na as K, suggesting that rainfall provided the Na that accumulated during log decay. At other sites where precipitation does not contain unusually large amounts of Na, both Na and K concentrations decline roughly in parallel through time (Foster and Lang, 1982; Sollins and Cline, unpublished).

b. Input to Logs via N Fixation. The existence of asymbiotic N fixers in decayed wood was first postulated by Cowling and Merrill (1966); then demonstrated by Sharp and Milbank (1973) and Cornaby and Waide (1973). Estimates of annual N input average nearly 1 kg ha^{-1} $year^{-1}$ across a wide range of forest types (Table 13). The lack of variation is deceptive, however, because lower rates at some sites were offset by larger biomass of CWD.

The organisms responsible for N fixation in decaying wood have received considerable study. Aho *et al.* (1974) isolated the obligate anaerobes *Klebsiella* and *Enterobacter* with nitrogenase activity from heart rot of living *A. concolor* trees. Spano *et al.* (1982) reported similar findings for fallen *Pseudotsuga* logs infected with *Fomitopsis pinicola*—obligate anaerobes were again responsible for the nitrogenase activity. Silvester *et al.* (1982) found, however, that the diazotrophic bacteria inhabiting fallen *Pseudotsuga* logs were microaerophiles, showing maximum activity at O_2 concentrations of ~5%. They explained the difference between their results and those of previous workers, noting that it was difficult to achieve complete anaerobiosis within the wood samples unless the samples were flushed with N_2 for several hours. With long flushing times, however, nitrogenase activity was zero, indicating that the organisms were not anaerobes.

The accuracy of current methods for measuring low-level N fixation by asymbiotes is subject to serious question. Only acetylene reduction was measured, not ^{15}N fixation, in five of the seven studies (Aho *et al.*, 1974; Cornaby and Waide, 1973; Larsen *et al.*, 1978; Sharp and Milbank, 1973; Spano *et al.*, 1982). Roskoski (1981) measured both processes and reported ratios of acetylene reduction to N_2 fixation ranging from zero to infinity, with a mean of 5.9, rather different from the theoretically expected ratio of 3.0. Excessive incubation time can lead to a substantial error in the ratio. Silvester *et al.* (1982) reported a mean ratio of 3.5 when incubations were kept <6 hrs, whereas ratios ranged from 4.55 to 6.01 when incubations were continued for 6–24 hrs. This confirmed work by David and Fay (1977) indicating that because acetylene inhibits N_2 fixation, amino N amounts in

Table 13 Acetylene Reduction and Estimated Nitrogen-Fixation Rates by Asymbiotic Organisms Inhabiting Coarse Woody Debris

Forest type	Acetylene reduction (nmol g^{-1} day^{-1})	CWD biomass (Mg ha^{-1})	Nitrogen fixation (kg ha^{-1} $year^{-1}$)	References
Mixed deciduous	25–37	11.8	0.9	Cornaby and Waide (1973)
Rocky Mountain mixed conifer	1–11	113	0.7	Larsen *et al.* (1978)
New England hardwoods	0–210	5–38	0.2–2.1	Roskoski (1980, 1981)
Intermountain *Thuja–Tsuga*	3–30	50	0.3	Larsen *et al.* (1982)
Rocky Mountain mixed conifer	36–230	—[a]	—	Spano *et al.* (1982)
Pacific Northwest *Pseudotsuga–Tsuga*	0–17	100	1.4	Silvester *et al.* (1982)

[a] Not available.

the diazotrophs will be depleted, in turn stimulating nitrogenase activity or synthesis.

Despite these methodological problems, it seems unlikely that N fixation in CWD amounts to more than a few kilograms per hectare per year. Such an input rate is much smaller than return in leaf fall and throughfall and is probably also smaller than return in root death. The N fixed in CWD, however, is a net input to the ecosystem; therefore it is more revealing to compare these rates to input by bulk precipitation and from N fixed elsewhere in the forest ecosystem. In an old-growth *Pseudotsuga* ecosystem, total input from external sources was estimated at 5 kg ha^{-1} $year^{-1}$ (Sollins *et al.*, 1980), and an addition of even 1 kg ha^{-1} $year^{-1}$ via fixation in CWD could be important. Input via precipitation is much greater at Hubbard Brook (22 kg ha^{-1} $year^{-1}$) (Likens *et al.*, 1977) and at the Oregon coastal sites studied by Grier (1978). At these two sites, the impact of a 1 kg ha^{-1} $year^{-1}$ input via N fixation in CWD would be correspondingly less.

c. Input to Logs via Fungal and Root Colonization. Fungal rhizomorphs and vascular plant roots abound in rotten logs in all but the earliest stages of decay (see Section V.A). These interconnections with the "outside world" are likely the most important pathways by which N and other nutrients enter logs (Ausmus, 1977). There have been no quantitative studies of nutrient transfer into logs by fungi, although the ability of many fungi to translocate nutrients is well established (e.g., Bowen, 1973). Vascular plant roots may act similarly because as they grow into a matrix largely devoid of available nutrients, nutrients can presumably be mobilized from other portions of the plants and translocated into the roots that colonize the wood. Such roots might then die in place, adding nutrients to the wood.

d. Output from Logs via Fragmentation and Leaching. Although fragmentation is perhaps the dominant transfer of nutrients out of logs and snags, it remains the least studied. Lambert *et al.* (1980) and Sollins (1982) concluded that on a dry weight basis, fragmentation accounted for about half of the total loss from logs. The importance of fragmentation as a nutrient loss from logs and snags depends upon the behavior of the nutrient. For example, N concentrations tend to be highest in the most decayed material, which is most likely to fragment; consequently, the fragmented material is likely to be of higher N content than the more solid material that remains behind. Potassium concentrations, however, tend to be lowest in the most decayed material; fragmentation would therefore tend to be a much more important transfer of N than of K.

Leaching is in many ways the complement to fragmentation—where one is high, the other is low. Leaching accounts for major losses of K and Na from CWD, but is less important for N, P, and Ca losses. This is evident from the fact that K and Na concentrations decline much more rapidly during decay than do concentrations of the other elements (Foster and Lang, 1982; Graham and Cromack, 1982; Grier, 1978; Yavitt and Fahey, 1982). Unfortunately, no one has measured the amount of nutrients transferred from CWD via leaching or fragmentation.

e. Overall Pattern of Nutrient Accumulation and Loss. The above processes interact to control amounts of nutrients in CWD, and their occurrence is well documented even if their quantitative significance is not. One way to view the overall pattern of nutrient loss or accumulation in CWD is to plot proportion of the original nutrient and dry weight remaining through time. Typically, N and Ca remain above the dry weight curve, indicating that there is net transfer into the logs (Figures 15–17). Potassium and P typically track below the dry weight line, indicating net transfer out of logs. These patterns appear to hold true regardless of wood species or ecosystem type (Foster and Lang, 1982; Graham and Cromack, 1982; Grier, 1978; Harris, 1978; Lambert *et al.*, 1980; Miller, 1983), except where precipitation contains unusually large concentrations of a particular element.

Nutrients probably also accumulate in decaying logs within aquatic ecosystems if patterns for fine wood are similar to CWD. In streams, N concentrations in twigs, chips, and bark of *Pseudotsuga* increased during the first 225 days of incubation (Triska and Cromack, 1980). Rates of increase in N content were greater in twigs and chips than in bark. Twigs and chips also decayed faster than bark. Nitrogen content of *Picea*, *Abies*, *Populus*, and *Alnus* wood chips increased with decay in streams of Quebec (Melillo *et al.*, 1983). The percentage of the original N content remaining generally leveled off after 3–12 months, depending on species. The maximum amounts of N immobilized for *Alnus*, *Betula*, *Populus*, *Picea*, and *Abies* were estimated to be 4.06, 6.32, 5.43, 5.21, and 4.67 mg N g^{-1} of initial tissue. Nitrogen only accumulates on the surface of submerged CWD. For example, Anderson *et al.* (unpublished) found that the outer rind of microbially stained *Alnus* wood had ~ 5 times the N content of unstained wood. Abrasion of this outer N-rich layer by flowing water is probably a major loss of N from logs in streams; however, the magnitude of this loss has not been estimated.

The overall nutrient accumulation and loss process can also be summarized as a "k" value—annual input of the element divided by amount in the CWD on the forest floor. An old-growth *Pseudotsuga* stand, where records of mortality span 30 years and ~ 40 ha, provides an example (Sollins, 1982). The k value is highest for Na (0.046 $year^{-1}$), and lowest for N (0.013 $year^{-1}$). The k values of the elements rank $N < Ca < Mg < P < K < Na$. Dry weight decay constant, k, in this ecosystem averaged 0.030 $year^{-1}$, which means that P, K, and Na were lost faster than dry weight; N, Ca, and Mg increased relative to dry weight.

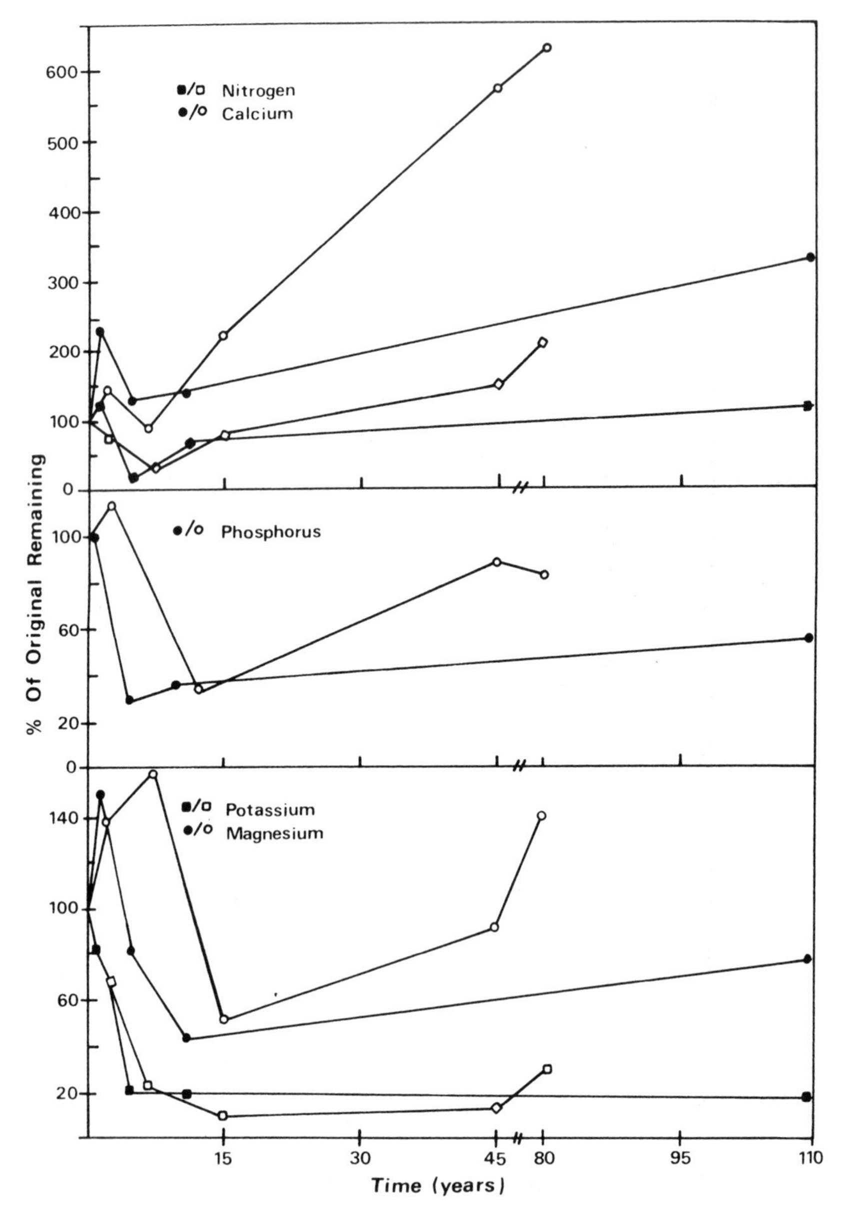

Figure 15 Changes in nutrient content of root wood (2.6–5.0 cm diameter) with decay. Open versus closed symbols indicate different sites. After Yavitt and Fahey (1982).

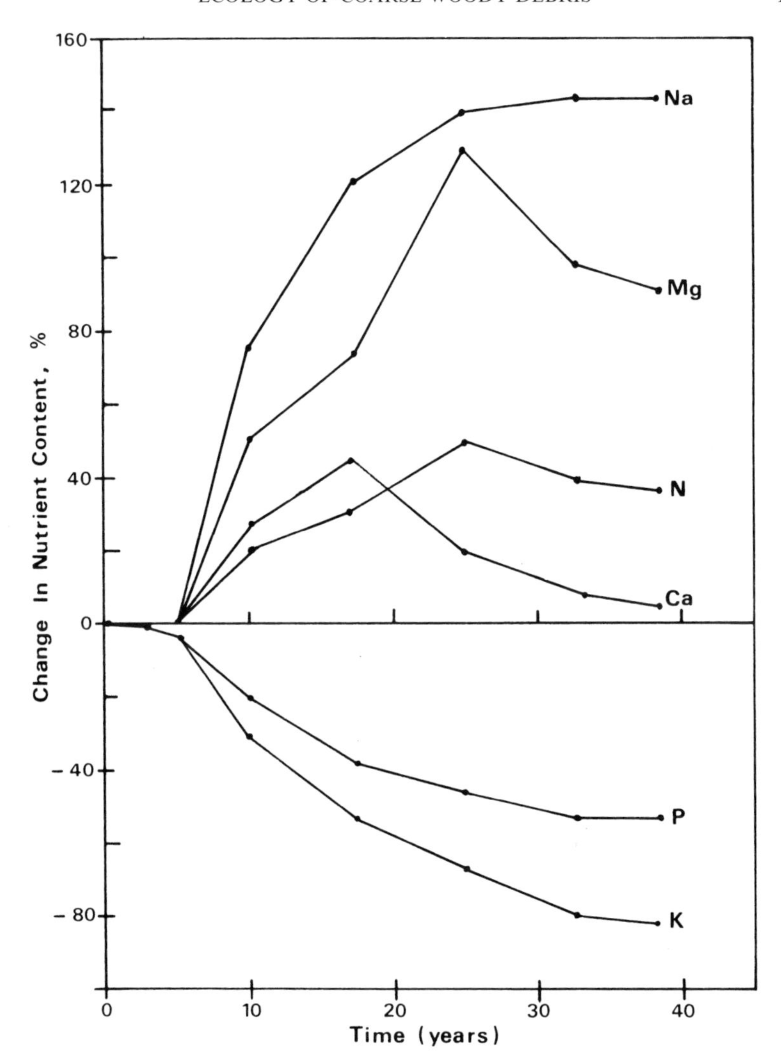

Figure 16 Percentage accumulation or loss of nutrients in fallen logs as related to time since tree death. After Grier (1978).

4. Coarse Woody Debris as a Factor in Site Productivity

Effects of CWD on site productivity can be studied by simulating stand growth over several rotations with and without the initial accumulation of CWD provided by the old-growth stand. Here we use a computer model developed by Kimmins and Scoullar (1979, 1981) and modified to model growth of *T. heterophylla* (Sachs and Sollins, unpublished).

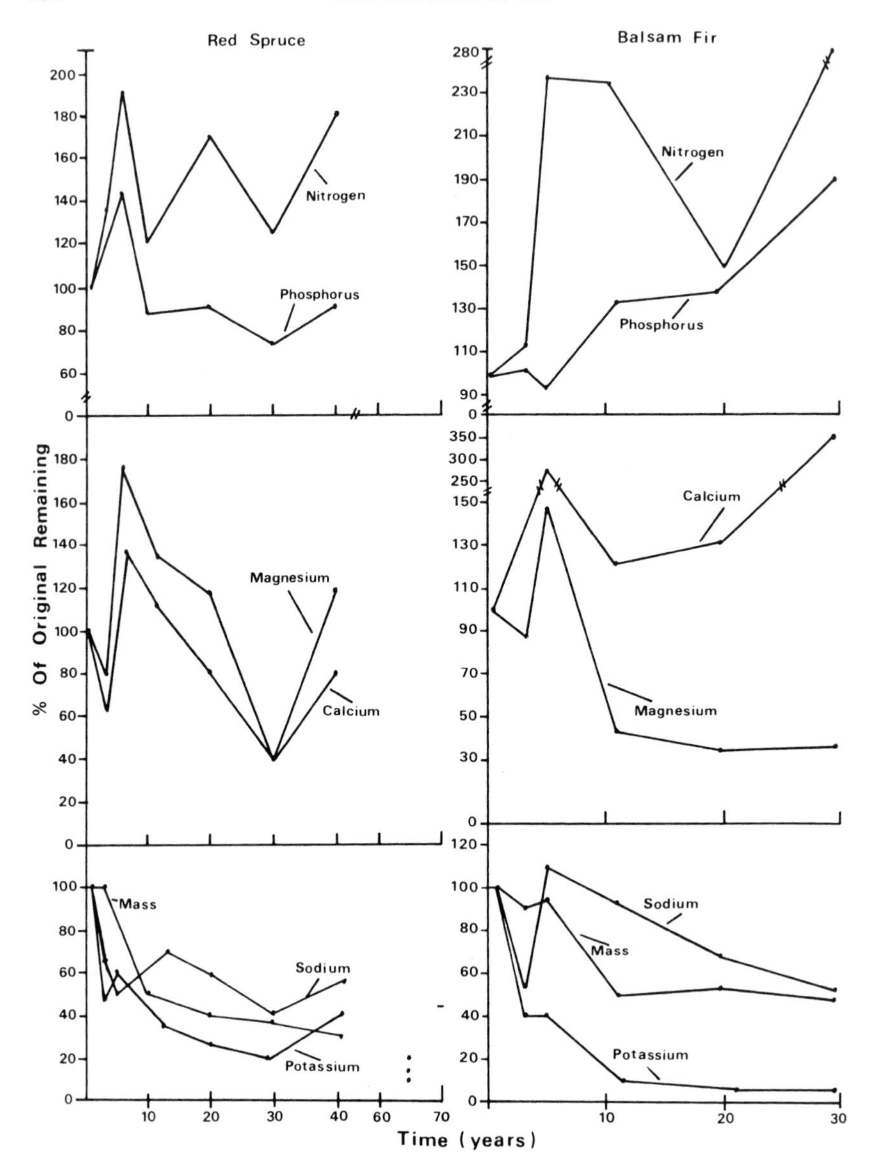

Figure 17 Change in mass and nutrient content with decay of wood from 10–15-cm-diameter boles. The age 64 *Picea* value, from >25-cm boles, is shown to indicate potential long-term trends. After Foster and Lang (1982).

The model, FORCYTE 10, considers N to be growth-limiting and decreases growth when available inorganic N is insufficient to permit stand growth at rates predicted from yield tables for unmanaged stands. The example site, on fertile land near the Oregon coast, had been cut once in the 1920s and the old-growth *P. sitchensis* removed; it was not burned, however, and large amounts of CWD survived into the current *T. heterophylla* stand.

The simulation begins by clear-cutting the existing stand and considers the impact of leaving or removing the old-growth CWD left over from the previous stand. Productivity was modeled over 17 30-year rotations without thinning, approximating the local management system for pulpwood production, or over six 90-year rotations, which approximates the least intensive, economically practical management regime at this site.

Merchantable yield decreased slowly under all scenarios; but more quickly if the old-growth CWD was removed initially from the stand (Figure 18). Yield declined more under 30- than under 90-year rotation. Removal of the old-growth CWD caused short-term increases in yield relative to the control (CWD not removed). This occurred because CWD immobilizes N as it decomposes and makes the N unavailable for tree growth. CWD is thus a short-term N sink but a long-term N source.

Predicted differences in merchantable yield with and without old-growth CWD were perhaps small (about 5%), but the economic impact of even these differences in merchantable yield could be substantial. Moreover, the site modeled here is one of the most N rich in the world (see coastal *Tsuga*, Table 10), and differences in yield after removal of CWD might be even greater at less fertile sites.

D. Geomorphic Functions of Coarse Woody Debris

The geomorphic roles of CWD can be grouped into effects on landforms and on transport and storage of soil and sediment. The importance of these roles differs between hillslopes and stream channels because the types and rates of soil and sediment transport as well as the mobility of CWD differ markedly between these environments. We will consider hillslopes and stream channels separately to highlight differences, but they intergrade in most landscapes.

1. Hillslopes

Geomorphic functions of CWD are rather poorly described and quantified for forested hillslopes. Effects of tree uprooting have received the greatest attention, while the effects of logs received the least.

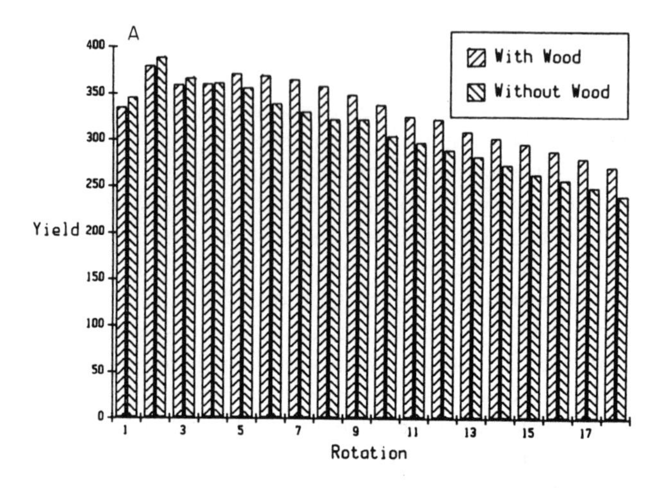

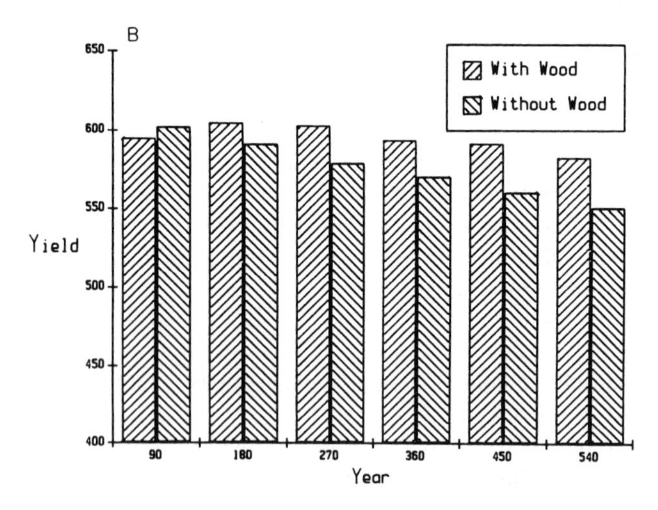

Figure 18 Predicted effect of removal of CWD on subsequent yield in an intensively managed *Tsuga heterophylla* stand from coastal Oregon. Rotation lengths are 30 years (A) and 90 years (B) and predictions are based on the FORCYTE 10 model.

a. Roots and Root-Throw. Tree uprooting creates microtopographic features and mixes soil. Microrelief of up to several meters is created on forested slopes by uprooted trees (Denny and Goodlett, 1956; Lyford and MacLean, 1966; Stephens, 1956). Stone's (1975) review found from 14 to 48% of the landscape was "visibly disturbed" by root-throw. The abundance of this landscape feature reflects the balance between processes creating and erasing this distinctive microtopography. Factors influencing root-throw potential of a site, such as topography, soils, and species wind firmness, as well as the size of the root systems upturned, strongly determine the rate microtopography is created. Surface erosion and soil creep obliterate pit-and-mound topography with rates dependent upon slope gradient and soil erodibility. Denny and Goodlett (1956) and Stephens (1956) estimate that pits and mounds can be distinguished from the surrounding forest floor for 300–500 years after their creation.

Root-throw and soil creep move soil at comparable rates. Denny and Goodlett (1956) calculate a soil transport rate of 1.5 mm $year^{-1}$ in the central Appalachian Mountains. Dietrich *et al.* (1982) estimate a rate of 2 mm $year^{-1}$ for a study area in the Tatra Mountains, Poland, using data from Kotarba (1970). Reid (1981) reports a soil movement rate of 1.8 mm $year^{-1}$ at a study site in the western Olympic Mountains, Washington. Swanson (unpublished) calculated a rate of 0.1 mm $year^{-1}$ for an old-growth *Pseudotsuga* forest in the western Cascade Mountains, Oregon.

The overall importance of root-throw to sediment production of a watershed has been examined using a sediment budget (Swanson *et al.*, 1982c). Reid (1981) estimated that 11% of sediment yield was derived from root-throw in a 10-km^2 basin on the west side of the Olympic Peninsula, Washington. In a 10-ha watershed forested with old-growth *Pseudotsuga* in western Oregon, Swanson *et al.* (1982a) calculated that soil transfer by root-throw accounted for $\sim$2% of long-term sediment production in a landscape dominated by landslide erosion.

Root-throw can be an important process in soil genesis and patterning (Armson and Fessenden, 1973; Denny and Goodlett, 1956; Stephens, 1956; Stone, 1975). Uprooting and subsequent soil and litter transport and storage cause substantial soil mixing and heterogeneity. Although the annual turnover rate of surface soil is commonly slow—Swanson (unpublished) estimates a rate of 0.4% $year^{-1}$ for an old-growth *Pseudotsuga* forest—the imprint of root-throw on soils is widespread because pits and mounds often persist for centuries.

b. Logs. Logs lying on the soil surface control the downslope movement of water, soil, and litter across the surface of the ground. In many forested environments, overland flow of water is rare, but downslope surface transport of particulate matter is well documented (Imeson and Von Zon, 1979; Swanson *et al.*, 1982a). The amount of sediment and organic matter stored upslope of logs in undisturbed forests, however, has not been studied. One would expect that the amount of storage would increase as slope increased and as the residence time of logs on the forest floor increased.

The influence of logs and other woody debris on surface erosion is well demonstrated in the landscape devastated by the May 18, 1980 eruption of Mount St. Helens, Washington. This eruption created a 50,000-ha zone of downed forest vegetation. CWD on and within the tephra deposits interrupted the surface and subsurface flow of water, and erosion rates on areas of downed forest were lower than areas clear-cut before the eruption (Collins *et al.*, 1983; Swanson *et al.*, 1983). Smith (1985) found that CWD stored significant amounts of tephra emplaced during the eruption, but stored little of the material transported by sheet and ril erosion.

2. *Streams and Rivers*

Recent interest in CWD in streams and rivers has stimulated consideration of its importance in controlling aquatic habitat and the movement of sediment and water. Controversy has often fueled this interest. On one hand, wood has been considered a hazard to navigation, life, and property during floods. CWD has also been cleared from rivers and streams to remove blockages to navigation and fish migration (Sedell *et al.*, 1984). This contrasts with the increasingly common practice of introducing and retaining CWD in streams for fish habitat improvement or channel stabilization. These conflicts in management objectives have triggered research on the behavior and functions of CWD in streams and rivers.

a. Channel Morphology. Steps in the channel longitudinal profile are created where a large log or accumulations of CWD form a dam that traps a wedge of sediment (Heede, 1972a,b; Keller and Swanson, 1979; Pearce and Watson, 1983). The tread of such a step is primarily composed of the stored sediment; the CWD accumulation forms the riser. Steps created by CWD reportedly vary widely in their importance to the channel profile. Marston (1982) evaluated effects of log steps in 13 watersheds of up to fifth order in the Oregon Coast Range. Steps created by CWD controlled 6% of total fall in the channels, but he argued that geologic factors, and not CWD, controlled the overall shape of the longitudinal profile. Stream reaches in other areas have much greater proportions of the channel fall occurring over CWD: 50–100% in the Rocky Mountains of Colorado and White Mountains of Arizona (Heede, 1972a,b, 1977); 30–60% in watersheds <5 km^2 in area and $<20\%$ in larger basins in Redwood Creek, California (Keller and Tally, 1979); 10–52% in the White Mountains, New Hampshire (Bilby, 1981); and 30–80% in the western Cascades, Oregon (Keller and Swanson, 1979; Swanson *et al.*, 1976). The degree to which CWD controls stream profile is related to abundance and size of CWD and the ability of channels to bypass obstructions. The importance of CWD in altering longitudinal channel profiles decreases with strewn order. Bilby (1981), for example, observed that the percentage of channel drop formed by CWD decreased from 52 to 42 to 10% from first- to third-order channels in the Hubbard Brook Experimental Forest, New Hampshire.

CWD triggers abrupt changes in channel pattern and position by blocking flow in main channels and chronic changes by deflecting flow against banks, thus accelerating the lateral migration of streams. Entire sections of major rivers can be changed drastically, as was the case of the Red River in northern Louisiana, where an accumulation of CWD grew to a length of more than 300 km over ~200 years (Lobeck, 1939). This jam blocked the main river and the mouths of tributary streams, forming lakes. Less dramatic changes are more typical, such as when CWD accumulates in channels and triggers oxbow cutoff of meander bends (Keller and Swanson, 1979).

Effects of CWD on channel geometry have been quantified over stream reaches and at sites of individual CWD accumulations. CWD partially or completely crossing a channel commonly deflects streamflow laterally or causes it to diverge. Sediment deposits upstream from the CWD as well as downstream in the case of low-gradient channels widen and decrease the depth of the channel (Keller and Swanson, 1979). Zimmerman *et al.* (1967) contrasted channel geometry, particularly channel width, between sod and forested stream banks in the Sleepers River basin, New Hampshire. Average channel width and variation in width was greater along forested reaches in channels draining watersheds $<\sim 6$ km^2. This was attributed to CWD in the channel and tree root mats in the stream banks. In this study, the influence of CWD on variation in width of larger channels was minimal. In another comparison of geometry between channels with sod and forested banks, Murgatroyd and Ternan (1983) observed channels shallower and up to three times wider in the forested reach. The thick turf and root mat were the primary cause of the narrower channel through sod-lined reaches. Channels near accumulations of CWD along five streams flowing through coastal *Sequoia* forests of northern California were 27–124% wider than overall average channel widths on streams with drainage basins 1.1–19.8 km^2 (Keller and Tally, 1979). Hogan (1985) observed a similar, though less pronounced pattern in seven stream reaches in British Columbia.

CWD can be a dominant control on the abundance and geometry of pools and riffles, which has important implications for fish habitat (see Section V.B). Pools are formed or their geometry is modified by scour and deposition associated with stream flow over, under, and around CWD. The architecture and position of wood accumulations subtly alter the functions of wood in stream ecosystems. R. Beschta (personal communication) examined pool formation near model logs in an artificial channel and determined that the position of logs in the water column profoundly affected pool size. Logs located at the water surface at maximum discharge formed the largest pools, and logs resting on the streambed created the smallest pools.

In meandering streams lacking CWD, pools typically form at the outside of meander bends with a spacing of from 5 to 7 bankfull-channel widths (Leopold *et al.*, 1964). CWD can increase pool frequency and variability in pool depths. Lisle and Kelsey (1982) hypothesized that CWD can increase pool frequency and that the location of the deepest portion of the channel is related to large roughness elements such as CWD. Scour at these roughness elements forms pools. In Jacoby Creek, a 30-km^2 basin in northern California, a pool spacing of 4.6 bankfull-channel widths was observed and 86% of the pools were associated with large roughness elements, of which CWD made up a substantial proportion (Lisle and Kelsey, 1982). In an analysis of stream reaches with high and low CWD amounts in British Columbia, Hogan (1985) documented greatest variability in depths of pools associated with CWD. Keller *et al.* (1985) examined 10 stream segments with drainage areas up to 27.2 km^2 in the Redwood Creek basin, northern California, and observed pool spacing of <5 bankfull-channel widths in six of them. CWD significantly influenced the morphology of 50 to 100% of the pools in these study reaches. In six reaches (drainage areas 1.6–9.8 km^2 disturbed by logging, they observed pool spacing of 1.3–4.1 bankfull-channel widths and 43–100% of the pools influenced by CWD.

Several experiments document changes in pool abundance before and after CWD removal. MacDonald and Keller (1983) observed that pools increased from 5 to 8 in a 100-m reach during the first year after removal of CWD. Pool spacing decreased from 2.5 to 1.6 channel widths. Bilby (1984) also reported reduction in number and area of pools after cleanup of CWD in an 11.5-m-wide segment of Salmon Creek, Coast Range, Washington. Even after both cleanup operations, however, CWD exposed by erosion of sediment influenced several of the pools.

The entrainment of live vegetation and transport of CWD by floods and mass-movement events can dramatically change channel geometry, especially in small, steep, forested streams. A debris torrent down such a stream converts a complex channel to a smooth, bedrock- or cobble-lined chute (Keller and Swanson, 1979; Swanson and Lienkaemper, 1978). Accumulations of CWD emplaced by mass movements and floods affect channel location, pattern, and geometry. Subsequent streamflow typically bypasses these obstructions if the valley floor is wide enough to accommodate changes in channel position (Hack and Goodlett, 1960). Moreover, steps in the longitudinal profile of the channel form where debris accumulations collect in narrow valleys (Pearce and Watson, 1983; and others).

b. Sediment and Organic-Matter Storage. The large size and relative stability of CWD result in temporary storage of inorganic sediment and organic matter in stream channels. Research by geomorphologists has emphasized storage of inorganic material, principally potential bedload (Mosley, 1981), while ecologists have concentrated on CWD as stored organic material and as structures for storing finer organic materials. It is difficult to clearly separate these roles because organic and inorganic materials are commonly thoroughly mixed and work together to enhance the stability of CWD-controlled accumulations. CWD provides the main structural elements of an accumulation, while inorganic sediment fills interstitial areas and increases the bulk density of the entire accumulation. The relative importance of organic and inorganic features in stabilizing sediment deposits depends on their relative sizes. In channels with large boulders and bedrock outcrops and small CWD, inorganic material controls retention of CWD. CWD controls storage of material in systems where CWD is much larger than other types of particulate matter.

CWD accounted for 35% of the obstructions and 49% of the stored sediment volume in seven watersheds in Idaho ranging in area from 0.26 to 2.02 km^2 (Megahan, 1982; Megahan and Nowlin, 1976). Total sediment stored in these watersheds was $\sim$15 times the average annual sediment yield, which amounted to 0.23 m^3 of stored sediment per meter of channel length for all types of obstructions. In an old-growth *Pseudotsuga* forest in western Oregon containing much more CWD than in the Idaho channels, 1.92 m^3 m^{-1} of sediment was stored by CWD in a 120-m-long segment of a third-order stream (Swanson and Lienkaemper, 1978).

The importance of CWD in storing sediment can also be assessed by removing accumulations from channels. Bilby (1981) compared sediment yields from a 175-m section of stream in a second-order watershed before and after removal of CWD in New Hampshire. In the first year after removal, export of fine and coarse particulate matter increased 500% over the value expected for the untreated condition. Effects of CWD removal on stored sediment have also been evaluated by repeated surveys of channel cross sections before and after treatment. Using this technique, Beschta (1979) observed that removal of several large CWD jams released 5250 m^3 of stored sediment from a 250-m section of a third-order stream in the Coast Range of Oregon. Baker (1979) examined the effects of removal of major CWD accumulations on sediment storage in seven streams in the Cascade and Coast Ranges of Oregon. Quantities of stored sediment ranged from 800 to 4500 m^3 and lengths of accumulations and associated sediment from $\sim$40 to 100 m. Three of the accumulations were completely removed, releasing 29–97% of the stored sediment within 2 years. A single channel was opened through the other four jams and only 8–20% of stored sediment left these sites during the same period. MacDonald and Keller (1983) observed that 60% of the 90 m^3 of CWD-stored sediment left a 100-m segment of Larry Damm Creek in north Coastal California in the first winter following removal of CWD.

CWD also greatly influences the ability of stream reaches to retain organic matter. Removal of small debris dams in a second-order stream in New Hampshire resulted in a 500% increase in export of fine and coarse particulate organic matter and accentuated transport during high-discharge events (Bilby, 1981). Export of dissolved organic matter in this study also increased with dam removal, but only by 6%. Removal of CWD in a second-order stream in Oregon resulted in a 4-fold decrease in storage of organic matter, and average travel distance of introduced leaves tripled (Gregory, unpublished). Speaker *et al.* (1984) observed that stream reaches with CWD dams retained leaves ~10 times more efficiently than reaches without major debris accumulations. In the southern Appalachians, logged watersheds had less CWD in stream channels than old-growth watersheds and had higher concentrations of particulate matter in the water (Silsbee and Larson, 1983). These higher concentrations were thought to be caused in part by lower organic-matter retention in the logged watersheds.

The sequential trapping of progressively finer sizes of organic matter determines the structure and retention characteristics of CWD accumulations. CWD traps branches and sticks that are in turn the primary trapping sites for leaves (Speaker *et al.*, 1984). Complex structures controlled by CWD are major trapping and storage sites for fine particulate organic matter. In summary, the retention characteristics of streams and rivers are a function of complexity and arrangement of the different forms of organic and inorganic material.

c. Movement of Coarse Woody Debris in Channels. The movement of CWD in channels involves periodic input and redistribution. Stand structure and composition, physical and biological processes leading to the input of CWD, and the geometry of the site where CWD comes to rest all influence the size distribution and stability of CWD when it first enters a channel. Once in a channel or floodway, pieces of CWD may move downstream. The potential for downstream movement is controlled by a variety of factors determining the stability of CWD pieces: size of piece relative to size of stream (Bilby, 1984; Lienkaemper and Swanson, 1980; Likens and Bilby, 1982); degree of burial; degree of stabilization by rooting into substrate by the piece itself or by other trees extending roots through the CWD; stage of decay, which influences potential for fragmentation; and position and orientation, including considerations such as proportion of mass supported outside the flooded area and tendency of streamflow to force CWD pieces against stable features.

The movement of CWD in streams has been examined by dendrochronologic analysis of the residence time of CWD pieces (Swanson and Lienkaemper, 1978; Swanson *et al.*, 1976), repeated mapping of CWD (Bryant, 1980; Lienkaemper and Swanson, 1980; Megahan, 1982), and repeatedly relocating tagged pieces of CWD (Bilby, 1984). CWD pieces whose length substantially exceeded channel width remained in place for many decades and in some cases several centuries (Keller and Tally, 1979; Swanson *et al.*, 1976).

VI. CONCLUSIONS

A. Comparison of Coarse Woody Debris in Terrestrial and Aquatic Ecosystems

The behavior of coarse woody debris (CWD) in terrestrial and aquatic environments can be viewed along a continuum, starting with dry forests at one extreme, progressing to mesic forests, and finally ending with freshwater aquatic environments at the other end of the gradient (Table 14). This represents a progression from conditions in which CWD is rarely, if ever, completely saturated with water to those in which water dominates the external as well as internal CWD environment.

In extremely wet environments, low amounts of oxygen restrict biological processing to a thin outer shell. Few invertebrates and no social insects occupy submerged CWD. Most aquatic invertebrates do not penetrate deeply into sound wood, and this, in combination with low oxygen concentrations, restricts microbial colonization and decomposition rates.

The absence of gallery-forming invertebrates in freshwater environments is a biological anomaly (Cummins *et al.*, 1983). Marine environments have a variety of organisms that bore into wood, provided temperature conditions are favorable. These include the crustacean genera *Limnoria* and *Sphaeroma* and the molluscan genera *Bankia*, *Martesia*, *Teredo*, and *Xylophaga* (Hochman, 1973). Wood often disappears quickly in marine environments as a consequence of aggressive utilization by these organisms. We know of no obvious explanation why similar organisms have not evolved in freshwater habitats that are typically much richer in wood substrates.

Fungal and bacterial species carrying out decomposition also change along the moisture gradient. Basidiomycetes dominate at the dry end of the gradient, ascomycetes and fungi imperfecti midway along the gradient, and actinomycetes and other bacteria predominate in streams and other very wet environments. Rates of respiration and, apparently, the ability to

Table 14 Contrasts in Behavior of CWD in Terrestrial and Freshwater Ecosystems

Feature	Environment		
	← Terrestrial →		← Freshwater →
	Dry	Mesic	Wet
O_2 levels	High O_2 →		Low O_2
Invertebrates	Abundant deep gallery and social →		Some shallow, no social
Type of animal use	Internal and external →		External only
Decomposers	Basidiomycetes →	Ascomycetes →	Actinomycetes
Decomposition	Rapid rate, lignin breakdown →		Slow rate, little lignin breakdown
Fragmentation	Biotic processes dominate (animals) →		Physical processes dominate
Spatial pattern	Nonaggregated →		Aggregated
Plant-habitat function	Low, lichens →	High, lichens, mosses, vascular plants ←	Moderate, algae and mosses

decompose lignin decline as one progresses from terrestrial to aquatic environments.

Although biological processing is slower in aquatic than in terrestrial environments, the length of time a piece of CWD plays a structural role is probably similar for both environments. This is because physical fragmentation caused by the abrasion and battering associated with flowing water systems shortens the life of CWD in streams and rivers in spite of slow rates of biological processing.

CWD is more mobile in aquatic than in terrestrial environments. CWD transport in streams and rivers creates log jams, producing a more aggregated spatial pattern than found in forests. CWD has a much greater influence on geomorphology of stream channels than hillslopes, in part because of the larger flux of organic and inorganic matter through flowing water systems.

The function of CWD also changes along the dry–wet continuum (Table 14). Use as autotrophic habitat is minor in the dry forest, reaches a maximum in diversity and size of life forms in the moist forest, and returns to low amounts in aquatic ecosystems. Vertebrate use is greatest in terrestrial ecosystems and involves both the exterior and interior of CWD, whereas vertebrate use in aquatic environments is almost entirely external, primarily for cover.

In summary, behavior of CWD varies along the moisture gradient. The gradient can be viewed as one in which moisture becomes limiting for decomposition at the dry end and oxygen becomes limiting in the aquatic environment. Decomposition tends to be dominated by external processes at the wettest and driest extremes of the gradient. In the center of the gradient, there is a balance of internal and external processes.

B. Effects of Human Activities on Coarse Woody Debris

The intensity and ubiquity of human influence on the amounts, dynamics, and functional importance of CWD have been tremendous. CWD has been eliminated from landscapes by forest clearing and from streams to improve navigation. CWD has been drastically modified in forest lands by harvest and salvage of trees and alteration of natural disturbance regimes.

Human impacts on CWD are so pervasive and have occurred for so long that few appreciate their magnitude and implications. Modifications of CWD in streams and rivers exemplify the temporal and geographic extent of human influences. Today, most ecologists do not fully appreciate the importance of CWD because deliberate removal of CWD from river channels has occurred for centuries. Huge accumulations of CWD up to 8 km long were common and blocked navigation on most of the large rivers in the United States. The U.S. Congress in 1776 made appropriations to clean rivers and streams of driftwood to maintain navigation. These navigational improvements began on the Mississippi River and its major tributaries before 1830. Over 800,000 pieces of CWD averaging 1.6 m in diameter at the base, 0.6 m at the top, and 40 m long were removed between 1830 and 1880 along the lower 1650 km of this river. Between 1878 and 1910, the U.S. Army Corps of Engineers expanded river navigation improvements to removing CWD from rivers in all parts of the country (Sedell and Frogatt, 1984; Triska, 1984). A conservative estimate of the number of pieces pulled from several rivers of the United States is provided in Table 15. These data illustrate the former abundance of CWD in these rivers and the extent to which large-river habitat was altered from a structurally diverse system to an aquatic highway.

On smaller streams, a series of "splash" dams were constructed to provide holding ponds for logs, and when these dams were opened ("splashed"), the large volume of released water moved logs downstream. While erect, the dams barred fish migration and when broken, the resulting floods eroded streambanks, damaged riparian vegetation, and cleared channels of CWD. Splash dams were common on many small streams. By the late 1880s, there were ~70 sites that were repeatedly used for splash dams on the St. Croix River, 41 in the Menominee Valley, and ~25 on the 150-km-long Red Cedar River of the Upper Mississippi River (Rector, 1949). Over 150 major dams existed in coastal rivers of Washington State (Wendler and Deschamps, 1955), and more than 160 splash dams were used on coastal rivers and Oregon tributaries of the Columbia River (Sedell and Luchessa, 1982). In ancient times, many old-world rivers were used to transport timbers, including the Euphrates, Haliacman, Axius, and Tiber Rivers, which no doubt altered the amount of CWD present. Many other rivers in Europe and Asia were used in the same manner during the fourteenth and fifteenth centuries (Albion, 1926; Totman, 1983). This historical analysis indicates that much of the biological integrity of streams and rivers was lost hundreds of years ago in Europe and Asia and by 1910 in much of North America. Consequently, even the oldest citizens and scientists have generally seen nothing but highly altered river and stream ecosystems.

In the case of forests, management practices, especially intermediate and final harvests of timber, have resulted in drastic contrasts between managed and natural forests. As with streams and rivers, these practices have gone on for so long in some places that there is little or no awareness that CWD is missing. In other areas, constant removal of materials, as small as twigs in some Eurasian forests, may have altered productivity and biological diversity.

Table 15 Summary of Snags Pulled from Rivers in the United States for Navigation Improvement from 1867 to 1912[a,b]

Rivers by region	Period of snagging	Kilometers snagged	Snags removed	Streamside trees cut	Logs pulled	Drift piles removed
Southeast Region						
Pamunkey R., Virginia	1880–1912	50	3,677	369	67	—[c]
North Landing R., North Carolina and Virginia	1879–1897	28	9,012	9	1,685	—
Pamlico and Tar R., North Carolina	1879–1912	81	29,260	7,625	728	—
Contentnia Cr., North Carolina	1881–1912	116	10,372	5,223	1,320	2
Black R., North Carolina	1887–1912	116	11,685	785	6,789	30
Edisto R., South Carolina	1882–1906	124	26,512	8,447	1,896	164
Savannah R. to Augusta, Georgia	1881–1912	409	37,812	1,167	9,766	—
Oconee R., Georgia	1877–1912	163	44,840	16,480	1,742	—
Noxubee R., Alabama and Mississippi	1890–1901	114	143,700	—	—	13
Pearl R., Mississippi	1879–1912	744	294,300	—	—	39
Tombigbee R., Mississippi	1892–1912	794	286,220	243	—	1,076
Guyandot R., West Virginia	1890–1899	134	8,060	—	—	—
Cumberland R., above Nashville, Tennessee	1879–1908	591	38,828	38,273	—	—
Choctawhatchee R., Florida and Alabama	1874–1912	350	177,599	—	—	—
Oklawaha R., Florida	1891–1911	102	9,089	1,080	984	—
Caloosahatchee R., Florida	1886–1911	36	7,874	6,860	1,192	—
Central Region						
Grand R., Michigan	1905–1911	67	2,019	—	—	—
Minnesota R., Minnesota	1867–1912	396	13,740	13,613	—	—
Red R., North Dakota and Minnesota	1877–1912	528	3,600	4,160	335	—
Red Lake R., North Dakota and Minnesota	1877–1912	248	1,500	—	—	—
Wabash R., Illinois and Indiana	1872–1906	79	7,700	154	—	109

(*Continued*)

Table 15 Continued

Rivers by region	Period of snagging	Kilometers snagged	Snags removed	Streamside trees cut	Logs pulled	Drift piles removed
Missouri R.	1879–1901	2,888	25,030	330	—	82
Arkansas R.	1879–1912	1,980	139,214	53,246	—	130
White R., Arkansas	1880–1912	495	22,500	37,118	—	177
Cache R., Arkansas	1888–1912	162	26,030	7,918	—	319
St. Francis and L'Anguille R., Arkansas	1902–1912	363	6,700	21,800	—	115
Southwest Region						
Guadalupe R., Texas	1907–1912	86	70,583	—	—	—
West Coast Region						
Sacramento R., California	1886–1920	380	33,545	—	—	—
Chehalis R., Washington	1884–1910	25	4,838	—	—	2
Willamette R., above Albany, Oregon	1870–1880	91	5,362	—	—	10

[a] From Secretary of War (1915).
[b] Most rivers in the United States lost significant amounts of fish habitat by the year 1910.
[c] Not available.

Effects of forest management on CWD include modification of the input rate, direct effects on biomass, alterations in size and species of CWD, and effects on the rates and patterns of decomposition. Input is affected at both the landscape and stand level by removing potential sources of CWD or altering disturbance regimes. Harvesting procedures that selectively remove trees with a high potential for death, as has been done in many *Pinus* forests, reduce the rate CWD is added. Introduction of fire or a pathogen can drastically accelerate CWD formation, while control programs reduce input rates below natural conditions. Harvest of timber affects CWD biomass when existing dead material is removed. Management affects the path and rate of decomposition by changing size and species of CWD, by felling snags, and by modifying the physical environment.

The consequences of human removal of CWD may be great because of the numerous functions of CWD. CWD introduces complexity to ecosystems, and when it is eliminated, ecosystems are simplified—organisms, structures, pathways, and functions are reduced. CWD removal may also reduce long-term site productivity, although current model simulations (Section V.C) indicate nutrient losses caused by CWD removal will probably not reduce productivity greatly on sites of average or better quality. Another concern has been the effects of CWD removal on biological diversity. CWD provides shelter and food for a large number of organisms, and elimination of CWD may lead to decreases in the populations of many organisms in both terrestrial and aquatic environments. For example, cavity-nesting bird populations have been substantially reduced in European forests because of long-term intensive forestry (Haapanen, 1965).

Our expanding knowledge of CWD provides an increased scientific appreciation of CWD, and we hope it will also lead to more enlightened management of this important ecological resource. Understanding the behavior and functional importance of CWD is still rudimentary, however, and deserves greater scientific attention given the incredible diversity of processes and relationships associated with it.

VII. SUMMARY

1. Coarse woody debris (CWD) is an important component of temperate stream and forest ecosystems. We have reviewed the rates at which CWD is added and removed from ecosystems, the biomass found in streams and forests, and many functions that CWD serves.

2. CWD is added to ecosystems by numerous mechanisms, including wind, fire, insect attack, pathogens, competition, and geomorphic processes. Despite the many long-term studies on tree mortality, there are few published rates of CWD input on mass-area^{-1} time^{-1} basis. Most ecological studies have not measured CWD input over a long enough period or a large enough area to give accurate estimates. Input rates measured in temperate ecosystems range from 0.12 to 14.9 Mg ha^{-1} year^{-1} and vary greatly over time and space.

3. Once CWD enters the detrital food web, it is decomposed by a large array of organisms and physical processes. Although respiration-caused losses have been the focus of many studies, CWD is also significantly transformed physically and chemically. Movement of CWD, especially in streams, is also an important but poorly documented mechanism whereby CWD is lost from ecosystems. Many factors control the rate at which CWD decomposes, including temperature, moisture, the internal gas composition of CWD, substrate quality, the size of the CWD, and the types of organisms involved. However, the importance of many of these factors has yet to be established in field experiments.

4. The mass of CWD in an ecosystem ideally represents the balance between addition and loss. In reality, slow decomposition rates and erratic variations in input of CWD cause the CWD mass to deviate markedly from steady-state projections. The mass of CWD in stream and forest ecosystems varies widely, ranging between 1 and 269 Mg ha^{-1}. Many differences correspond to forest type, with deciduous-dominated systems having generally lower biomass than conifer-dominated systems. However, conifer-dominated systems with low productivity also have low CWD mass. Stream size also influences CWD mass in lotic ecosystems, while successional stage dramatically influences CWD mass in boat aquatic and terrestrial settings.

5. CWD performs many functions in ecosystems, serving as auto-trophic and heterotrophic habitat and strongly influencing geomorphic processes, especially in streams. It is also a major component of nutrient cycles in many ecosystems. We have reviewed these many functions and conclude that CWD is an important functional component of stream and forest ecosystems.

6. Humans have greatly affected the amount of CWD found in temperate ecosystems by removing CWD and by changing the rate of input and the rate and pattern of loss. In some cases, human influences have been so pervasive that natural conditions are difficult to define. Management practices concerning CWD often have not been based on the numerous beneficial roles this material plays in ecosystems. Better scientific understanding of these functions and the natural factors influencing CWD dynamics should lead to more enlightened management practices.

ACKNOWLEDGEMENTS

Many individuals and institutions contributed to the preparation of this article. This work is a legacy of the U.S. Biological Programme Coniferous Forest Biome Project. Current financial support has been provided by National Science Foundation Grants DEB 78-10594, BSR 80-22190, DEB 80-12162, DEB 80-04652, DEB-8112455, and BSR-8508356, and by the U.S. Forest Service, Pacific Northwest Forest and Range Experiment Station, and the U.S. Department of the Interior, Bureau of Land Management. J. Means, K. Luchessa, A. Lehre, T. Fahey, and H. McDade were generous is sharing their unpublished data with us. Numerous individuals contributed intellectually to development of this manuscript, including A. McKee, F. Triska, C. Maser, J. Trappe, T. Spies, R. Graham, R. Fogel, E. Keller, and R. Waring. We also thank D. Ford for his helpful review and editorial work on this article. We particularly thank T. Blinn for the outstanding job of editing and orchestrating the production of the manuscript. We also thank J. Brenneman for her help on the manuscript, including her excellent drafting of the figures. R. Anderson and C. Iwai assisted with the typing, and we are grateful for their efforts.

REFERENCES

Abbott, D.T. and Crossley, D.A. (1982) Woody litter decomposition following clear-cutting. *Ecology* **63**, 35–42.

Adams, C.C. (1915) An ecological study of prairie and forest invertebrates. *Bull. Ill. State Lab. Nat. Hist.* **11**, 30–280.

Adler, E. (1977) Lignin chemistry—past, present and future. *Wood Sci. Technol.* **11**, 169–218.

Ahlgren, C.E. (1966) Small mammals and reforestation following prescribed burning. *J. For.* **64**, 614–618.

Aho, P.E., Seidler, R.J., Evans, H.J. and Raju, P.N. (1974) Distribution, enumeration, and identification of nitrogen-fixing bacteria associated with decay in living white fir trees. *Phytopathology* **64**, 1413–1420.

Alban, D.H., Perala, D.A. and Schaegel, B.E. (1978) Biomass and nutrient distribution in aspen, pine and spruce stands on the same soil type in Minnesota. *Can. J. For. Res.* **8**, 290–299.

Albion, R.G. (1926) *Forests and Sea Power: The Timber Problem of the Royal Navy 1652–1862.* Harvard Economic Studies, Vol. 29. Harvard Univ. Press, Cambridge, Massachusetts.

Aldrich, J.W. (1943) Biological survey of the bogs and swamps in northeastern Ohio. *Am. Midl. Nat.* **30**, 346–402.

Alexander, C.P. (1931) Deutsche Lininologische Sunda-Expedition. The craneflies (Tipulidae, Diptera). *Arch. Hydrobiol.* **9**, 135–191.

Alexander, R.R. (1954) Mortality following partial cutting in virgin lodgepole pine. USDA For. Serv. Paper 16. Rocky Mt. For. and Range Exp. Stn., Fort Collins, Colorado.

Alexander, R.R. (1956) A comparison of growth and mortality following cutting in oldgrowth mountain spruce-fir stands. USDA For. Serv. Res. Note RM-20. Rocky Mt. For. and Range Exp. Stn., Fort Collins, Colorado.

Allsopp, A. and Misra, P. (1940) The constitution of the cambium, the new wood and the mature sapwood of the common ash, the common elm and the Scotch pine. *Biochem. J.* **34**, 1078–1084.

Anderson, N.H. (1982) A survey of aquatic insects associated with wood debris in New Zealand streams. *Mauri Ora* **10**, 21–33.

Anderson, N.H. and Sedell, J.R. (1979) Detritus processing by macroinvertebrates in stream ecosystems. *Annu. Rev. Entomol.* **24**, 351–377.

Anderson, N.H., Sedell, J.R., Roberts, L.M. and Triska, F.J. (1978) The role of aquatic invertebrates in processing of wood debris in coniferous forest streams. *Am. Midl. Nat.* **100**, 64–82.

Anderson, N.H., Steedman, R.J. and Dudley, T. (1984) Patterns of exploitation by stream invertebrates of wood debris (xylophagy). *Verh. Int. Ver. Limnol.* **22**, 1847–1852.

Anderson, S.H. (1972) Seasonal variation in forest birds of western Oregon. *Northwest Sci.* **46**, 194–206.

Armson, K.A. and Fessenden, R.J. (1973) Forest windthrow and their influence on soil morphology. *Soil Sci. Soc. Am. Proc.* **37**, 781–783.

Atkinson, S.W. (1971) BOD and toxicity of log leachates. M.S. thesis, Oregon State Univ., Corvallis.

Attiwill, P.M. (1980) Nutrient cycling in a *Eucalyptus obliqua* (L'Hérit.) forest. IV. Nutrient uptake and return. *Aust. J. Bot.* **28**, 199–222.

Aumen, N.G. (1985) Characterization of lignocellulose decomposition in stream wood samples using ^{14}C and ^{15}N techniques. Ph.D. dissertation, Oregon State Univ., Corvallis.

Aumen, N.G., Bottomley, P.J., Ward, G.M. and Gregory, S.V. (1983) Microbial decomposition of wood in streams: Distribution of microflora and factors affecting [^{14}C]lignocellulose mineralization. *Appl. Environ. Microbial.* **46**, 1409–1416.

Ausmus, B.S. (1977) Regulation of wood decomposition rates by arthropod and annelid populations. *Ecol. Bull.* **25**, 180–192.

Avery, C.C., Larson, F.R. and Schubert, G.H. (1976) Fifty-year records of virgin stand development in southwestern ponderosa pine. USDA For. Serv. Gen. Tech. Rep. RM-22. Rocky Mt. For. and Range Exp. Stn., Fort Collins, Colorado.

Ayre, G.L. (1963) Laboratory studies on the feeding habits of seven species of ants (Hymenoptera:Formicidae) in Ontario. *Can. Entamol.* **95**, 712–715.

Baker, C.O. (1979) The impacts of log jam removal on fish populations and stream habitat in western Oregon. M.S. thesis, Oregon State Univ., Corvallis.

Baker, J.H., Morita, R.Y. and Anderson, N.H. (1983) Bacterial activity of woody substrates in a stream sediment. *Appl. Environ. Microbiol.* **45**, 516–521.

Balch, R.E., Clark, J. and Bonga, J.M. (1964) Hormonal action in production of tumors and compression wood by an aphid. *Nature (London)* **202**, 721–722.

Baldwin, H.I. (1927a) The spruce reproduction problem in Norway. *Ecology* **8**, 139–141.

Baldwin, H.I. (1927b) A humus study in Norway. *Ecology* **8**, 380–383.

Banerjee, B. (1967) Seasonal changes in the distribution of the millipede *Cylindroiulus punctatus* (Leach) in decaying logs and soil. *J. Anim. Ecol.* **36**, 171–177.

Barnes, D.P. and Sinclair, S.A. (1983) Time-related changes in specific gravity and moisture content of spruce budworm-killed balsam fir. *Can. J. For. Res.* **13**, 257–263.

Basham, J.T. (1951) The pathological deterioration of balsam fir killed by the spruce budworm. *Pulp Pap. Mag. Can.* **52**, 120–134.

Beattie, R.K. and Diller, J.P. (1954) Fifty years of chestnut blight in America. *J. For.* **52**, 323–329.

Benke, A.C., Van Arsdall, T.C., Gillespie, D.M. and Parrish, F.K. (1984) Invertebrate productivity in a subtropical blackwater river: The importance of habitat and life history. *Ecol. Monogr.* **54**, 25–63.

Benner, R., Maccubbin, A.E. and Hodson, R.E. (1984a) Preparation, characterization and microbial degradation of specifically radiolabelled [^{14}C]lignocelluloses from marine and freshwater macrophytes. *Appl. Environ. Microbiol.* **47**, 381–389.

Benner, R., Maccubbin, A.E. and Hodson, R.E. (1984b) Anaerobic biodegradation of the lignin and polysaccharide components of lignocellulose and synthetic lignin by sediment microfora. *Appl. Environ. Microbiol.* **47**, 998–1004.

Berryman, A.A. (1982) Mountain pine beetle outbreaks in Rocky Mountain lodgepole pine forests. *J. For.* **80**, 410–413, 419.

Beschta, R.L. (1979) Debris removal and its effects on sedimentation in an Oregon Coast Range stream. *Northwest Sci.* **53**, 71–77.

Bidwell, A. (1979) Observations on the nymphs of *Povilla adusta* Navas (Ephemeroptera: Polymitarcyidae) in Lake Kainji, Nigeria. *Hydrobiologia* **67**, 161–172.

Bilby, R.E. (1981) Role of organic debris dams in regulating the export of dissolved and particulate matter from a forested watershed. *Ecology* **62**, 1234–1243.

Bilby, R.E. (1984) Post-logging removal of woody debris affects stream channel stability. *J. For.* **82**, 609–613.

Bilby, R.E. and Likens, G.E. (1980) Importance of organic debris dams in the structure and function of stream ecosystems. *Ecology* **61**, 1107–1113.

Bisson, P.A., Nielsen, J.L., Pahmason, R.A. and Grove, L.E. (1982) A system of naming habitat types in small streams, with examples of habitat utilization by salmonids during low streamflow. In: *Acquisition and Utilization of Aquatic Habitat Inventory Information* (Ed. by N.B. Armantrout), pp. 62–73. Western Div., Am. Fish. Soc., Portland, Oregon.

Blackman, M.W. and Stage, H.H. (1918) Notes on insects bred from bark and wood of the American larch. Tech. Publ. 10, pp. 11–115. N.Y. State Coll. For., Syracuse Univ., Syracuse.

Blackman, M.W. and Stage, H.H. (1924) On the succession of insects living in the bark and wood of dying, dead and decaying hickory. Tech. Publ. 17, pp. 1–269. N.Y. State Coll. For., Syracuse Univ., Syracuse.

Blanchett, R.A. and Shaw, C.G. (1978) Associations among bacteria, yeasts and basidiomycetes during wood decay. *Phytopathology* **68**, 631–637.

Bock, C.E. and Lynch, J.F. (1970) Breeding bird populations of burned and unburned conifer forest in the Sierra Nevada. *Condor* **72**, 182–189.

Boddy, L. (1983) Microclimate and moisture dynamics of wood decomposing in terrestrial ecosystems. *Soil Biol. Biochem.* **15**, 149–157.

Borkent, A. (1984) The systematics and phylogeny of the *Stenochironomus* complex (*Xestochironomus*, *Harrisius*, and *Stenochironomus*) (Diptera: Chironomidae). *Mem. Entomol. Soc. Can.* **128**, 269.

Boussu, M.F. (1954) Relationship between trout populations and cover on a small stream. *J. Wildl. Manage.* **18**, 229–239.

Bowen, G.D. (1973) Mineral nutrition of ectomycorrliizae. In: *Ectomycorrhizae* (Ed. by G.C. Marks and T.T. Kozlowski), pp. 151–205. Academic Press, New York.

Boyce, J.S. (1923) The deterioration of felled western yellow pine on insect control projects. USDA Agric. Bull. 1140. U.S. Gov. Print. Off., Washington, D.C.

Boyce, J.S. (1929) Deterioration of windthrown timber on the Olympic Peninsula, Washington. USDA Tech. Bull. 104. U.S. Gov. Print. Off., Washington, D.C.

Brackebusch, A.P. (1975) Gain and loss of moisture in large forest fuels. USDA For. Serv. Res. Pap. INT-173. Intermt. For. and Range Exp. Stn., Ogden, Utah.

Breznak, J.A. (1982) Intestinal microbiota of termites and other xylophagous insects. *Annu. Rev. Microbiol.* **36**, 323–343.

Brokaw, N.V. (1982) Treefalls: Frequency, timing and consequences. In: *The Ecology of a Tropical Forest: Seasonal Rhythms and Long-Term Changes* (Ed. by E.G. Leigh, A.S. Rand and D.M. Windsor), pp. 101–108. Smithsonian Institution Press, Washington, D.C.

Brown, J.K. (1974) Handbook for inventorying downed woody material. USDA For. Serv. Gen. Tech. Rep. INT-16. Intermt. For. and Range Exp. Stn., Ogden, Utah.

Brown, J.K. and See, T.E. (1981) Downed dead woody fuel and biomass in the northern Rocky Mountains. USDA For. Serv. Gen. Tech. Rep. INT-117. Intermt. For. and Range Exp. Stn., Ogden, Utah.

Brush, T. (1981) Response of secondary cavity-nesting birds to manipulation of nest-site availability. M.S. thesis, Arizona State Univ., Tempe.

Bryant, M. (1980) Evolution of large, organic debris after timber harvest: Maybeso Creek, 1949 to 1978. USDA For. Serv. Gen. Tech. Rep. PNW-101. Pacific Northwest For. and Range Exp. Stn., Portland, Oregon.

Bryant, M.D. (1982) Organic debris in salmonid habitat in southeast Alaska: Measurement and effects. In: *Proceedings of the Symposium on Acquisition and Utilization of Aquatic Habitat Inventory Information, October 28–30, 1981* (Ed. by N.B. Armantrout), pp. 259–265. Western Div., Am. Fish. Soc., Portland, Oregon.

Buchanan, D.V., Tate, P.S. and Moring, J.R. (1976) Acute toxicities of spruce and hemlock bark extracts to some estuarine organisms in southeastern Alaska. *J. Fish. Res. Board Can.* **33**, 1188–1192.

Buchanan, T.S. and Englerth, G.H. (1940) Decay and other volume losses in windthrown timber on the Olympic Peninsula, Washington. USDA Tech. Bull. 733. U.S. Gov. Print. Off., Washington, D.C.

Buckley, B.M. and Triska, F.J. (1978) Presence and ecological role of nitrogen-fixing bacteria associated with wood decay in streams. *Verh. Int. Ver. Limnol.* **20**, 1333–1339.

Bull, E.L. (1975) Habitat utilization of the pileated woodpecker, Blue Mountains, Oregon. M.S. thesis, Oregon State Univ., Corvallis.

Bull, E.L. (1983) Longevity of snags and their use by woodpeckers. In: *Snag Habitat Management* (Ed. by J.W. Davis, G.A. Goodwin and R.A. Ockenfels), pp. 64–67. USDA For. Serv. Gen. Tech. Rep. RM-99. Rocky Mt. For. and Range Exp. Stn., Fort Collins, Colorado.

Bustard, D.R. and Narver, D.W. (1975a) Preferences of juvenile coho salmon (*Oncorhynchus kisutch*) and cutthroat trout (*Salmo clarki*) relative to simulated alteration of winter habitat. *J. Fish. Res. Board Can.* **32**, 681–687.

Bustard, D.R. and Narvar, D.W. (1975b) Aspects of the winter ecology of juvenile coho salmon (*Oncorhynchus kisutch*) and steelhead trout (*Salmo gairdnen*). *J. Fish. Res. Board Can.* **32**, 667–680.

Campbell, R.N. and Clark, J.W. (1960) Decay resistance of baldcypress heartwood. *For. Prod. J.* **10**, 250–253.

Carey, A.B. (1983) Cavities in trees in hardwood forests. In: *Snag Habitat Management; Symposium Proceedings* (Ed. by J.W. Davis, G.A. Goodwin and R.A. Ockenfels), pp. 167–184. USDA For. Serv. Gen. Tech. Rep. RM-99. Rocky Mt. For. and Range Exp. Stn., Fort Collins, Colorado.

Carpenter, S.R. (1981) Decay of heterogenous detritus: A general model. *J. Theor. Biol.* **89**, 539–547.

Christensen, O. (1977) Estimation of standing crop and turnover of dead wood in a Danish oak forest. *Oikos* **28**, 177–186.

Christensen, O. (1984) The states of decay of woody litter determined by relative density. *Oikos* **42**, 211–219.

Christy, E.J. and Mack, R.N. (1984) Variation in demography of juvenile *Tsuga heterophylla* across the substratum mosaic. *J. Ecol.* **72**, 75–91.

Christy, E.J., Sollins, P. and Trappe, J.M. (1982) First-year survival of *Tsuga heterophylla* without mycorrhizae and subsequent ectomycorrhizal development of decaying logs and mineral soil. *Can. J. Bot.* **60**, 1601–1605.

Churchill, G.B., John, H.H., Duncan, D.P. and Hodson, A.C. (1964) Long-term effects of defoliation of aspen by the forest tent caterpillar. *Ecology* **45**, 630–633.

Clark, J.W. (1957) Comparative decay resistance of some common pines, hemlock, spruce and true fir. *For. Sci.* **3**, 314–320.

Clermont, L.P. and Schwartz, H. (1951) The chemical composition of Canadian woods. *Pulp Pap. Mag. Can.* **52**, 103–105.

Cline, S.P., Berg, A.B. and Wight, H.M. (1980) Snag characteristics-and dynamics in Douglas-fir forests, western Oregon. *J. Wlidl. Manage.* **44**, 773–786.

Collins, B.D., Dunne, T. and Lehre, A.K. (1983) Erosion of tephra-covered hillslopes north of Mount St. Helens, Washington: May 1980–May 1981. *Z. Geomorphol.* **46**, 103–121.

Conner, R.N., Miller, O.K. Jr. and Adkisson, C.S. (1976) Woodpecker dependence on trees infected by fungal heart rots. *Wilson Bull.* **88**, 575–581.

Corbet, G.B. and Southern, H.N. (Eds.) (1977) *The Handbook of British Mammals*. Blackwell, Oxford.

Cornaby, B.W. and Waide, J.B. (1973) Nitrogen fixation in decaying chestnut logs. *Plant Soil* **39**, 445–448.

Côté, W.A. (1977) Wood ultrastructure in relation to chemical composition. *Recent Adv. Phytochem.* **11**, 1–14.

Coulson, R.N. and Witter, J.A. (1984) *Forest Entomology, Ecology and Management*. Wiley, New York.

Cowling, E.B. and Merrill, W. (1966) Nitrogen in wood and its role in wood deterioration. *Can. J. Bot.* **44**, 1539–1554.

Cranston, P.S. (1982) The metamorphosis of *Symposiocladius lignicola* (Kieffer) n. gen., n. comb., a wood-mining Chironomidae (Diptera). *Entomol. Scand.* **13**, 419–429.

Crawford, D.L. and Crawford, R.L. (1976) Microbial degradation of lignocellulose: The lignin component. *Appl. Environ. Microbiol.* **31**, 714–717.

Crawford, D.L. and Sutherland, J.B. (1979) The role of actinomycetes in the decomposition of lignocellulose. *Dev. Ind. Microbiol.* **20**, 143–151.

Crawford, D.L., Crawford, R.L. and Pometto, A.L.III (1977a) Preparation of specifically labeled ^{14}C [Lignin] and [Glucan]-lignocelluloses and their decomposition by the microflora of soil. *Appl. Environ. Microbiol.* **33**, 1247–1251.

Crawford, D.L., Floyd, S., Pometto, A.L. III and Crawford, R.L. (1977b) Degradation of natural and Kraft lignins by the microflora of soil and water. *Can. J. Microbiol.* **23**, 434–440.

Crawford, R.L. (1981) *Lignin Biodegradation and Transformation*. Wiley, New York.

Crawford, R.L. and Crawford, D.L. (1978) Radioactive methods for the study of lignin biodegradation. *Dev. Ind. Microbiol.* **19**, 35–49.

Crawford, R.L., Robinson, L.E. and Cheh, A.M. (1980) ^{14}C-labeled lignins as substrates for the study of lignin biodegradation and transformation. In: *Lignin Biodegradation: Microbiology, Chemistry, and Potential Applications* (Ed. by T.K. Kirk, T. Higuchi and H.M. Chang), Vol. 1, pp. 61–76. CRC Press, Boca Raton, Florida.

Creighton, W.S. (1950) The ants of North America. *Bull. Mus. Comp. Zool.* **104**, 1–585.

Cromack, K. Jr. and Monk, C.D. (1975) Litter production, decomposition, and nutrient cycling in a mixed hardwood watershed and a white pine watershed. In: *Mineral Cycling in Southeastern Ecosystems* (Ed. by F.G. Howell, J.B. Gentry and M.H. Smith), pp. 609–624. National Technical Information Service, U.S. Dept. of Commerce, Springfield, Virginia.

Cudney, M.D. and Wallace, J.B. (1980) Life cycles, microdistribution and production dynamics of six species of net-spinning caddisflies in a large Southeastern (U.S.A.) river. *Holarctic Ecol.* **3**, 169–182.

Cummins, K.W. (1979) The multiple linkages of forests to streams. In: *Forests: Fresh Perspectives from Ecosystem Analysis. Proceedings 40th Biology Colloquium (1979)* (Ed. by R.H. Waring), pp. 191–198. Oregon State Univ. Press, Corvallis.

Cummins, K.W. and Klug, M.J. (1979) Feeding ecology of stream invertebrates. *Annu. Rev. Ecol. Syst.* **10**, 147–172.

Cummins, K.W., Sedell, J.R., Swanson, F.J., Minshall, G.W., Fisher, S.G., Cushing, C.E., Petersen, R.C. and Vannote, R.L. (1983) Organic matter budgets for stream ecosystems: Problems in their evaluation. In: *Stream Ecology, Application and Testing of General Ecological Theory* (Ed. by J.R. Barnes and G.W. Minshall), pp. 299–353. Plenum, New York.

Curtis, J.D. (1943) Some observations on wind damage. *J. For.* **41**, 877–882.

Dahms, W.G. (1949) How long do ponderosa pine snags stand? USDA For. Serv. Res. Note PNW-57. Pacific Northwest For. And Range Exp. Stn., Portland, Oregon.

David, K.A.V. and Fay, P. (1977) Effects of long-term treatment with acetylene on nitrogen fixing microorganisms. *Appl. Environ. Microbiol.* **34**, 640–646.

Davis, J.W., Goodwin, G.A. and Ockenfels, R.A. (1983) Snag habitat management: Proceedings of the symposium. USDA For. Serv. Gen. Tech. Rep. RM-99. Rocky Mt. For. And Range Exp. Stn., Fort Collins, Colorado.

Dennis, W.M. and Batson, W.T. (1974) The floating log and stump communities in the Santee Swamp of South Carolina. *Castanea* **39**, 166–170.

Denny, C.S. and Goodlett, J.C. (1956) Microrelief resulting from fallen trees. In: *Surficial Geology and Geomorphology of Potter County, Pennsylvania* (Ed. by C.S. Denny), pp. 59–66. USGS Prof. Pap. 288, Washington Off., Washington, D.C.

Deverall, B.J. (1965) The physical environment for fungal growth. 1. Temperature. In: *The Fungi. I. The Fungi Cell* (Ed. by G.C. Ainsworth and A.S. Sussman), pp. 543–550. Academic Press, New York.

Deyrup, M.A. (1975) The insect community of dead and dying Douglas-fir. 1. The Hymenoptera. Coniferous For. Biome, Ecosyst. Anal. Stud. Bull. 6. Univ. of Washington, Seattle.

Deyrup, M.A. (1976) The insect community of dead and dying Douglass-fir: Diptera, Coleoptera, and Neuroptera. Ph.D. thesis, Univ. of Washington, Seattle.

Deyrup, M.A. (1981) Deadwood decomposers. *Nat. Hist.* **90**, 84–91.

Dietrich, W.E., Dunne, T., Hymphrey, N.F. and Reid, L.M. (1982) Construction of sediment budgets for drainage basins. In: *Sediment Budgets and Routing in Forested Drainage Basins* (Ed. by F.J. Swanson, R.J. Dunne and D.N. Swanston), pp. 5–23. USDA For. Serv. Res. Pap. PNW-141. Pacific Northwest For. and Range Exp. Stn., Portland, Oregon.

Dimbley, G.W. (1953) Natural regeneration of pine and birch on the heather moors of northeast Yorkshire. *Forestry* **26**, 41–52.

Dudley, T.L. (1982) Population and production ecology of Lipsothrix spp. (Diptera: Tipulidae) M.S. thesis, Oregon State Univ., Corvallis.

Dudley, T. and Anderson, N.H. (1982) A survey of invertebrates associated with wood debris in aquatic habitats. *Melanderia* **39**, 1–21.

Duncan, C.G. (1961) Relative aeration requirements by soft rot and basidiomycete wood-destroying fungi. USDA For. Prod. Lab. Rep. 2218, Madison, Wisconsin.

Duvigneand, P. and Denaeyer-DeSmet, S. (1970) Biological cycling of minerals in temperate deciduous forests. In: *Analysis of Temperate Forest Ecosystems* (Ed. by D.E. Reichle), pp. 199–225. Springer-Verlag, Berlin and New York.

Edwards, N.T., coord. (1972) Terrestrial decomposition. *Ecol. Sci. Div. Annu. Prog. Rep. (1971)*, USAEC Rep. No. ORNL-4759, pp. 74–85. Oak Ridge Natl. Lab., Oak Ridge, Tennessee.

Edwards, P.J. and Grubb, P.J. (1977) Studies of mineral cycling in a montane rain forest in New Guinea. *J. Ecol.* **65**, 943–969.

Eidmann, H. (1929) Zur Kenntnis der Biologie der Rossameise (*Camponotus herculeanus* (L.)). *Z. Angew. Entomol.* **14**, 229–253.

Elliot, S. (1978) Ecology of rearing fish. Annu. Rep., Proj. F-9-10, 19(D-I), 39–52. Alaska Dep. Fish and Game, Fed. Aid in Fish Restoration, Juneau, Alaska.

Ellis, E.L. (1965) Inorganic elements in wood. In: *Cellular Ultrastructure of Woody Plants* (Ed. by W.A. Côté), pp. 181–189. Syracuse Univ. Press, Syracuse, New York.

Ellwood, E.L. and Ecklund, B.A. (1959) Bacterial attack of pine logs in pond storage. *For. Prod. J.* **9**, 283–292.

Elton, C.S. (1966) Dying and dead wood. In: *The Pattern of Animal Communities*, pp. 279–305. Wiley, New York.

Eslyn, W.E. and Highley, T.L. (1976) Decay resistance and susceptibility of sapwood of fifteen tree species. *Phytopathology* **66**, 1010–1017.

Essig, E.O. (1942) *College Entomology*. Macmillan, New York.

Everest, F.H. and Meehan, W.R. (1981) Forest management and anadromous fish habitat productivity. *Trans. North Am. Wildl. Nat. Resour. Conf.* **46**, 521–530.

Everitt, B.L. (1968) Use of the cottonwood in an investigation of the recent history of a floodplain. *Am. J. Sci.* **266**, 417–439.

Eyre, F.H. and Longwood, F.R. (1951) Reducing mortality in old-growth northern hardwoods through partial cutting. USDA For. Serv. Stn. Pap. 25. Lake States For. Exp. Stn., St. Paul, Minnesota.

Fager, E.W. (1968) The community of invertebrates in decaying oak wood. *J. Anim. Ecol.* **37**, 121–142.

Fahey, T.J. (1983) Nutrient dynamics of aboveground detritus in lodgepole pine (Pinus contorta sp. latifolia) ecosystems, southeastern Wyoming. *Ecol. Monogr.* **53**, 51–72.

Falinski, J.B. (1978) Uprooted trees, their distribution and influences in the primeval forest biotope. *Vegetatio* **38**, 175–183.

Federle, T.W. and Vestal, J.R. (1980) Lignocellulose mineralization by Arctic lake sediments in response to nutrient manipulation. *Appl. Environ. Microbiol.* **40**, 32–39.

Felt, E.P. (1906) Insects affecting park and woodland trees. N.Y. State Mus. Memoir No. 8, Vol. 2, New York.

Fischer, W.C. and McClelland, B.R. (1983) A cavity-nesting bird bibliography—including related titles on forest snags, fire, insects, disease, and decay. USDA For. Serv. Gen. Tech. Rep. INT-144. Intermt. For. and Range Exp. Stn., Ogden, Utah.

Fogel, R. and Cromack, K. Jr. (1977) Effect of habitat and substrate quality on Douglas-fir litter decomposition in western Oregon. *Can. J. Bot.* **55**, 1632–1640.

Fogel, R., Ogawa, M. and Trappe, J.M. (1973) Terrestrial decomposition: A synopsis. Coniferous For. Biome Rep. 135. Univ. of Washington, Seattle.

Forward, C.D. (1984) Organic debris complexity and its effect on small scale distribution and abundance of coho (*Oncorhynchus kisutch*) fry populations in

Carnation Creek, British Columbia. B.S.F. thesis, Univ. of British Columbia, Vancouver.

Foster, F.R. and Lang, G.E. (1982) Decomposition of red spruce and balsam fir boles in the White Mountains of New Hampshire. *Can. J. For. Res.* **12**, 617–626.

Fowler, H.G. and Roberts, R.B. (1980) Foraging behavior of the carpenter ant, *Camponotus pennsylvanicus* (Hymenoptera: Formicidae), in New Jersey. *J. N.Y. Entomol. Soc.* **53**, 295–304.

Frank, R.M. and Blum, B.M. (1978) The selection system of silviculture in spruce-fir stands-procedures, early results, and comparisons with unmanaged stands. USDA For. Serv. Res. Pap. NE-425. Northeast. For. Exp. Stn., Broomall, Pennsylvania.

Frankland, J.C., Hedger, J.N. and Swift, M.J. (Eds.) (1982) *Decomposer Basidiomycetes, Their Biology and Ecology*. Cambridge Univ. Press, London and New York.

Franklin, J.F. and Dyrness, C.T. (1973) Natural vegetation of Oregon and Washington. USDA For. Serv. Gen. Tech. Rep. PNW-8. Pacific Northwest For. and Range Exp. Stn., Portland, Oregon.

Franklin, J.F. and Waring, R.H. (1980) Distinctive features of the Northwestern coniferous forest: Development, structure and function. In: *Forests: Fresh Perspectives from Ecosystem Analysis. Proceedings 40th Biology Colloquium (1979)* (Ed. by R.H. Waring), pp. 59–85. Oregon State Univ. Press, Corvallis.

Franklin, J.F., Cromack, K. Jr., Denison, W., McKee, A., Maser, C., Sedell, J., Swanson, F. and Juday, G. (1981) Ecological characteristics of old-growth Douglas-fir forests. USDA For. Serv. Gen. Tech. Rep. PNW-118. Pacific Northwest For. and Range Exp. Stn., Portland, Oregon.

Franklin, J.F., Swanson, F.J. and Sedell, J.R. (1982) Relationships within the valley floor ecosystems in western Olympic National Park: A summary. In: *Ecological Research in National Parks of the Pacific Northwest. Proceedings of the Second Conference on Scientific Research in the National Parks, San Francisco, California*, pp. 43–45. Oregon State Univ., For. Res. Lab., Corvallis.

Franklin, J.F., Klopsch, M. and Luchessa, K.J. (1984) Timing and causes of mortality in natural coniferous forests of Oregon and Washington. *Bull. Ecol. Soc. Am.* **65**, 206–207.

Froehlich, H.A. (1973) Natural and man-caused slash in headwater streams. Loggers Handb. XXXIII. Pacific Logging Congr., Portland, Oregon.

Furniss, R.L. (1936) Bark beetles active following Tillamook Fire. *Timberman* **37**, 21–22.

Furniss, R.L. and Carolin, V.M. (1977) Western forest insects. USDA For. Serv. Misc. Publ. 1339. U.S. Gov. Print. Off., Washington, D.C.

Gammon, J.R. (1970) The effect of inorganic sediment on stream biota. Water Pollut. Control Res. Ser. 18050DWC12/70. Environ. Prot. Agency, Water Qual. Off., Washington, D.C.

Gardner, J.A.F. and Barton, G.M. (1960) The distribution of dihydroquercetin in Douglas-fir and western larch. *For. Prod. J.* **10**, 171–173.

Gashwiler, J.R. (1970) Plant and mammal changes on a clearcut in west-central Oregon. *Ecology* **51**, 1018–1026.

Gauch, H. (1982) *Multivariate Analysis in Community Ecology*. Cambridge Univ. Press, London and New York.

Gentry, J.B. and Whitford, W.G. (1982) The relationship between wood litter infall and relative abundance and feeding activity of subterranean termites *Riticulitermes* spp. in three southeastern coastal plain habitats. *Oecologia* **54**, 63–67.

Gessel, S.P. and Turner, J. (1976) Litter production in western Washington Douglas-fir stands. *Forestry* **49**, 63–72.

Golley, F.B., Petrusewicz, K. and Ryszkowski, L. (Eds.) (1975) *Small Mammals: Their Productivity and Population Dynamics*. Cambridge Univ. Press, London and New York.

Gosz, J.R. (1980) Biomass distribution and production budget for a nonaggrading forest ecosystem. *Ecology* **61**, 507–514.

Gosz, J.R., Likens, G.E. and Bormann, F.H. (1972) Nutrient content of litter fall on the Hubbard Brook Experimental Forest. *Ecology* **53**, 769–784.

Gosz, J.R., Likens, G.E. and Bormann, F.H. (1973) Nutrient release from decomposing leaf and branch litter in the Hubbard Brook Forest, New Hampshire. *Ecol. Monogr.* **43**, 173–191.

Gottfried, G.L. (1978) Five-year growth and development in a virgin Arizona mixed conifer stand. USDA For. Serv. Res. Pap. RM-203. Rocky Mt. For. and Range Exp. Stn., Fort Collins, Colorado.

Gotwald, W.H. Jr. (1968) Food gathering behavior of the ant *Camponotus noveboracensis* (Fitch) (Hymenoptera: Formicidae). *J. N.Y. Entomol. Soc.* **76**, 278–296.

Graham, R.L. (1982) Biomass dynamics of dead Douglas-fir and western hemlock boles in mid-elevation forests of the Cascade Range. Ph.D. thesis, Oregon State Univ., Corvallis.

Graham, R.L. and Cromack, K. Jr. (1982) Mass, nutrient content and decay rate of dead boles in rain forests of Olympic National Park. *Can. J. For. Res.* **12**, 511–521.

Graham, S.A. (1925) The felled tree trunk as an ecological unit. *Ecology* **6**, 397–416.

Gratkowski, H.J. (1956) Windthrow around staggered settings in old-growth Douglas-fir. *For. Sci.* **2**, 60–74.

Green, G.W. and Sullivan, C.R. (1950) Ants attacking larvae of the forest tent caterpillar, *Malacosoma disstria* Hbn. (Lepidoptera: Lasiocampidae). *Can. Entomol.* **82**, 194–195.

Greene, S.E. (1984) Forest structure and dynamics in an Oregon coast *Tsuga heterophylia–Picea sitchensis* forest. *Bull. Ecol. Soc. Am.* **65**, 207.

Greenland, D.J. and Kowal, J.M.L. (1960) Nutrient content of the moist tropical forest of Ghana. *Plant Soil* **12**, 154–174.

Grier, C.C. (1976) Biomass, productivity and nitrogen–phosphorus cycles in hemlock–spruce stands of the central Oregon coast. In: *Proceedings: Western Hemlock Management Conference* (Ed. by W.A. Atkinson and R.J. Zasoski), pp. 71–81. Coll. For. Resour., Univ. of Washington, Seattle.

Grier, C.C. (1978) A *Tsuga heterophylla–Picea sitchensis* ecosystem of coastal Oregon: Decomposition and nutrient balances of fallen logs. *Can. J. For. Res.* **8**, 198–206.

Grier, C.C. and Logan, R.S. (1977) Old-growth *Pseudotsuga menziesii* communities of a western Oregon watershed: Biomass distribution and production budgets. *Ecol. Monogr.* **47**, 373–400.

Grier, C.C., Vogt, K.A., Keyes, M.R. and Edmonds, R.L. (1981) Biomass distribution and above- and below-ground production in young and mature *Abies amabilis* zone ecosystems of the Washington Cascades. *Can. J. For. Res.* **11**, 155–167.

Griffin, D.M. (1977) Water potential and wood-decay fungi. *Annu. Rev. Phytopathol.* **15**, 319–329.

Griffith, B.G. (1931) The natural regeneration of spruce in central B.C. *For. Chron.* **7**, 204–219.

Grodzinski, W. (1971) Energy flow through populations of small mammals in the Alaskan Taiga Forest. *Acia Theriot.* **16**, 231–275.

Grodzinski, W., Bobek, B., Drozdz, A. and Gorecki, A. (1970) Energy flow through small rodent populations in a beech forest. In: *Energy Flow through Small Mammal Populations* (Ed. by K. Petrusewicz and L. Ryszkowski), pp. 291–298. PWN-Polish Scientific Publ., Warsaw.

Gross, C.G. (1980) The biochemistry of lignifcation. *Adv. Bar. Res.* **8**, 26–63.

Gunther, P.M., Horn, B.S. and Babb, G.D. (1983) Small mammal populations and food selection in relation to timber harvest practices in the western Cascade Mountains. *Northwest Sci.* **57**, 32–44.

Haapanen, A. (1965) Bird fauna of the Finnish forests in relation to forest succession. 1. *Ann. Zool. Fenn.* **2**, 153–196.

Hack, J.T. and Goodlett, J.C. (1960) Geomorphology and forest ecology of a mountain region in the Central Appalachians. USGS Prof. Pap. 347. Washington Off., Washington, D.C.

Hager, D.C. (1960) The interrelationships of logging, birds, and timber regeneration in the Douglas-fir region of northwestern California. *Ecology* **41**, 116–125.

Hall, J.D. and Baker, C.O. (1982) Rehabilitating and enhancing stream habitat: 1. Review and evaluation. In: *Influence of Forest and Rangeland Management on Anadromous Fish Habitat in Western North America* (Ed. by W.R. Meehan). USDA For. Serv. Gen. Tech. Rep. PNW-138. Pacific Northwest For. and Range Exp. Stn., Portland, Oregon.

Hall, J.D. and Lantz, R.L. (1969) Effects of logging on the habitat of coho salmon and cutthroat trout in coastal streams. In: *Symposium on Salmon and Trout in Streams* (Ed. by T.G. Northcote), pp. 355–375. Inst. Fish., Univ. of British Columbia, Vancouver.

Hall, T.F. and Penfound, W.T. (1943) Cypress-gum communities in the Blue Girth Swamp near Selma, Alabama. *Ecology* **24**, 208–217.

Hancock, W.V. (1957) The distribution of dihydroquercetin and leucoanthocyanidin in a Douglas-fir tree. *For. Prod. J.* **7**, 335–338.

Hansson, L. (1971) Estimates of the productivity of small mammals in a south Swedish spruce plantation. *Ann. Zool. Fenn.* **8**, 118–126.

Harcombe, P.A. (1984) Stand development in a 130-yr-old spruce–hemlock forest in the Oregon coast. *Bull. Ecol. Soc. Am.* **65**, 150.

Harcombe, P.A. and Marks, P.L. (1983) Five years of tree death in a *Fagus–Magnolia* forest, southeast Texas, USA. *Oecologia* **57**, 49–64.

Harmon, M.E. (1980) The distribution and dynamics of forest fuels in the low elevation forests of Great Smoky Mountains National Park. Res./Resour. Manage. Rep. 32. Uplands Field Res. Lab., Gatlinburg, Tennessee.

Harmon, M.E. (1982) Decomposition of standing dead trees in the southern Appalachians. *Oecologia* **52**, 214–215.

Harmon, M.E. (1985) Logs as sites of tree regeneration in *Picea sitchensis–Tsuga heterophylla forests* of Washington and Oregon. Ph.D. thesis, Oregon State Univ., Corvallis.

Harper, J.L. (1977) *Population Biology of Plants*. Academic Press, New York.

Harris, R.D. (1983) Decay characteristics of pileated woodpecker nest trees. In: *Snag Habitat Management* (Ed. by J.W. Davis, G.A. Goodwin and R.A. Ockenfels), pp. 125–129. USDA For. Serv. Gen. Tech. Rep. RM-99. Rocky Mt. For. and Range Exp. Stn., Fort Collins, Colorado.

Harris, W.F. (1976) Nutrient release from decaying wood In: *Environmental Sciences Division Annual Progress Report* (Ed. by S.I. Auerbach, D.E. Reichle and E.G. Struxness), pp. 169–170. Environ. Sci. Div. Publ. 1145. Oak Ridge Natl. Lab., Oakridge, Tennessee.

Harris, W.F., coord. (1978) Terrestrial ecology section. In: *Environmental Sciences Division Annual Progress Report* (Ed. by S.I. Auerbach, D.E. Reichle and E.G. Struxness), pp. 156–175. ORNL-5365, Oak Ridge Nail. Lab., Oakridge, Tennessee.

Harris, W.F., Henderson, G.S. and Todd, D.E. (1972) Measurement of turnover of biomass and nutrient elements from the woody component of forest fitter on Walker Branch Watershed. East. Deciduous For. Biome Memo. Rep. 72–146. Oak Ridge Natl. Lab., Oak Ridge, Tennessee.

Harris, W.F., Goldstein, R.A. and Henderson, G.S. (1973) Analysis of forest biomass pools, annual primary production and turnover of biomass for a mixed deciduous forest watershed. In: *IUFRO Biomass Studies* (Ed. by H.E. Young), pp. 43–64. Univ. of Maine Press, Orono.

Hart, J.H. (1968) Morphological and chemical differences between sapwood; discolored sapwood and heartwood in black locust and Osage orange. *For. Sci.* **14**, 334–338.

Hartley, C. (1958) Evaluation of wood decay in experimental work. USDA For. Prod. Lab. Rep. 2119, Madison, Wisconsin.

Hartman, G.F. (1965) The role of behavior in the ecology and interaction of underyearliag coho salmon (*Oncorhynchus kisutch*) and steelhead trout (*Salmo gairdneri*). *J. Fish. Res. Board Can.* **22**, 1035–1081.

Harvey, A.E., Jurgensen, M.F. and Lanen, M.J. (1976) Seasonal distribution of ectomycorrhizae in a mature Douglas-fir/larch forest soil in western Montana. *For. Sci.* **24**, 203–208.

Harvey, A.E., Larsen, M.J. and Jurgensen, M.F. (1979) Comparative distribution of ectomyconrhizae in soils of three western Montana forest habitat types. *For. Sci.* **25**, 350–358.

Hawley, L.F. and Wise, L.E. (1929) *The Chemistry of Wood*. Chemical Catalog Co, New York.

Hawley, L.F., Fleck, L.C. and Richards, C.A. (1924) The relation between durability and chemical composition of wood. *Ind. Eng. Chem.* **16**, 699–706.

Hayes, G.L. (1940) The moisture content of large sized fuels as an index of intraseasonal and seasonal fire danger severity. M. S. thesis, Yale Univ., New Haven, Connecticut.

Heede, B.H. (1972a) Influences of a forest on the hydraulic geometry of two mountain streams. *Water Resour. Bull.* **8**, 523–530.

Heede, B.H. (1972b) Flow and channel characteristics of two high mountain streams. USDA For. Serv. Gen. Tech. Rep. RM-96. Rocky Mt. For. and Range Exp. Stn., Fort Collins, Colorado.

Heede, B.H. (1977) Influence of forest density on bedload movement in a small mountain stream. In: *Hydrology and Water Resources in Arizona and the Southwest. Vol. 7. Proceedings of the 1977 Meetings of the Arizona Section of the American Water Resources Association and the Hydrology Section of the Arizona Academy of Science*, pp. 103–107. Tucson, Arizona.

Heger, L. (1974) Longitudinal variation of specific gravity in stems of black spruce, balsam fir and lodgepole pine. *Can. J. For. Res.* **4**, 321–326.

Henderson, G.S. and Harris, W.F. (1975) An ecosystem approach to characterization of the nitrogen cycle in a deciduous forest watershed. In: *Proceedings of the Fourth North American Forest Sods Conference* (Ed. by B. Bernier and C.H. Winget), pp. 179–193. Les Presses de l'Université Laval, Quebec, Canada.

Henderson, G.S., Swank, W.T., Waide, J.B. and Grier, C.C. (1978) Nutrient budgets of Appalachian and Cascade region watersheds: A comparison. *For. Sci.* **24**, 385–397.

Henry, J.D. and Swan, J.M.A. (1974) Reconstructing forest history from live and dead plant material—an approach to the study of forest succession in southwest New Hampshire. *Ecology* **55**, 772–783.

Hepting, G.H. (1971) Diseases of forest and shade trees of the United States. USDA For. Serv. Agric. Handb. 386. U.S. Gov. Print. Off., Washington, D.C.

Hickin, N.E. (1963) *The Insect Factor in Wood Decay*. Hutchinson, London.

Hillis, W.E. (1962) *Wood Extractives and Their Significance to the Pulp and Paper Industry*. Academic Press, New York.

Hillis, W.E. (1977) Secondary changes in wood. *Recent Ado. Phytochem.* **11**, 247–309.

Hinds, T.E., Hawksworth, F.G. and Davidson, R.W. (1965) Beetle-killed Engelmann spruce: Its deterioration in Colorado. *J. For.* **63**, 536–542.

Hines, W.W. (1971) Plant communities in the old-growth forests of north coastal Oregon. M.S. thesis, Oregon State Univ., Corvallis.

Hintikka, V. (1973) Passive entry of spores into wood. *Karstenia* **13**, 5–8.

Hirth, H.F. (1959) Small mammals in old field succession. *Ecology* **40**, 417–425.

Hochman, H. (1973) Degradation and protection of wood from marine organisms. In: *Wood Deterioration and Its Prevention by Preservative Treatments. Vol. 1: Degradation and Protection of Wood* (Ed. by D.D. Nicholas), pp. 247–275. Syracuse Univ. Press, Syracuse, New York.

Hodkinson, I.D. (1975) Dry weight loss and chemical changes in vascular plant litter of terrestrial origin, occurring in a beaver pond ecosystem. *J. Ecol.* **63**, 131–142.

Hogan, D. (1985) The influence of large organic debris on channel morphology in Queen Charlotte Island streams. In: *Proceedings of the Western Division of the American Fisheries Society, Victoria, British Columbia, July 17–20, 1984*. In press.

Hölldobler, K. (1944) Ueber die forstlich wichtigen Ameisen des nordostkarelischen Urwaldes. Teil I. *Z. Angew. Entomol.* **30**, 606–622.

Holmes, R.T. and Sturges, F.W. (1975) Bird community dynamics and energetics in a northern hardwoods ecosystem. *J. Anim. Ecol.* **44**, 175–200.

Holt, J.P. (1974) Bird populations in the hemlock sere on the Highlands Plateau, North Carolina, 1946 to 1972. *Wilson Bull.* **86**, 397–406.

Holtam, B.W. (Ed.) (1971) *Windblow of Scottish forests in January 1968*. [British] For. Comm. Bull. 45. Her Majesty's Stationery Off., Edinburgh.

Hooven, E.F. and Black, H.C. (1976) Effects of some clearcutting practices on small-mammal populations in western Oregon. *Northwest Sci.* **50**, 189–208.

Howden, H.F. and Vogt, G.B. (1951) Insect communities of standing dead pine (*Pinus virginiana* Mill.). *Ann. Entomol. Sec. Am.* **44**, 581–595.

Huff, M.H. (1984) Post-fire succession in the Olympic Mountains, Washington: Forest vegetation, fuels, and avifauna. Ph.D. dissertation, Univ. of Washington, Seattle.

Hulme, M.A. and Shields, J.K. (1970) Biological control of decay fungi in wood by competition for non-structural carbohydrates. *Nature (London)* **227**, 300–301.

Humphrey, C.J. (1916) Laboratory tests on durability of American woods. I. Flask test on conifers. *Mycologia* **8**, 80–92.

Hunt, R.L. (1969) Effects of habitat alteration on production, standing crops and yield of brook trout in Lawrence Creek, Wisconsin. In: *Proceedings of the Symposium on Salmon and Trout in Streams, February 22–24* (Ed. by T.G. Northcote), pp. 281–312. Univ. of British Columbia, Vancouver.

Imeson, A.C. and Von Zon, H. (1979) Erosion processes in small forested catchments in Luxemborg. In: *Geographical Approaches to Fluvial Processes* (Ed. by A.F. Pitty), pp. 93–107. Geo Abstracts, Norwich.

International Pacific Salmon Fisheries Commission (1966) Effects of log driving on the salmon and trout populations in the Stellako River, Vancouver, British Columbia. Prog. Rep. 14, Vancouver, British Columbia.

Jackman, S. (1974) Some characteristics of cavity nesters: Can we ever leave enough snags? Oregon Coop. Wildl. Res. Unit, Oregon State Univ., Corvallis.

[Japanese] Forestry Agency. (1955) A memoir of the scientific investigations of the primeval forests in the headwaters of the River Ishikari, Hokkaido, Japan, 1952–1954. For. Agency, Tokyo, Japan.

Jenny, H., Gessel, S.P. and Bingham, T. (1949) Comparative study on decomposition rates of organic matter in temperate and tropical regions. *Soil Sci.* **68**, 419–432.

Jensen, K.F. (1967) Oxygen and carbon dioxide affect the growth of wood-decaying fungi. *For. Sci.* **13**, 384–389.

Jensen, R.A. and Zasada, Z.A. (1977) Growth and mortality in an old-age jack pine stand. Minn. For. Res. Note 261, Univ. of Minnesota, St. Paul.

Johnston, D.W. and Odum, E.P. (1956) Breeding bird populations in relation to plant succession on the Piedmont of Georgia. *Ecology* **37**, 50–62.

Jones, E.W. (1945) The structure and reproduction of the virgin forest of the North Temperate Zone. *New Phytol.* **44**, 130–148.

June, J. (1981) Life history and habitat utilization of cutthroat trout (*Salmo clarki*) in a headwater stream on the Olympic Peninsula, Washington. M.S. thesis, Univ. of Washington, Seattle.

Käärik, A.A. (1974) Decomposition of wood. In: *Biology of Plant Litter Decomposition* (Ed. by C.H. Dickson and G.J.E. Pugh), Vol. 1, pp. 129–174. Academic Press, London.

Kaufman, M.G. (1983) Life history and feeding ecology of *Xylotopus par* (Coquillett) (Diptera: Chironomidae) M.S. thesis, Central Michigan Univ., Mount Pleasant.

Keen, F.P. (1929) How soon do yellow pine snags fall? *J. Ecol.* **27**, 735–737.

Keen, F.P. (1955) The rate of natural falling of beetle-killed ponderosa pine snags. *J. For.* **53**, 720–723.

Keller, E.A. and Swanson, F.J. (1979) Effects of large organic material on channel form and fluvial processes. *Earth Surf. Process.* **4**, 361–380.

Keller, E.A. and Tally, T. (1979) Effects of large organic debris on channel form and fluvial processes in the coastal redwood environment. In: *Adjustments of the Fluvial System. 1979 Proceedings of the Tenth Annual Geomorphology Symposium* (Ed. by D.D. Rhodes and G.P. Williams), pp. 169–197. State Univ. of New York, Binghamton.

Keller, E.A., MacDonald, A., Tally, T. and Merritt, N.J. (1985) Effects of large organic debris on channel morphology and sediment storage in selected tributaries of Redwood Creek. In: *Geomorphic Processes and Aquatic Habitat in the Redwood Creek Drainage Basin*. USGS Prof. Pap. Washington Off, Washington, D.C. In press.

Kendeigh, S.C. (1948) Bird populations and biotic communities in northern lower Michigan. *Ecology* **29**, 101–114.

Ker, M.F. (1980) Tree biomass equations for tea major tree species in Cumberland County, Nova Scotia. Can. For. Serv. Inf. Rep. M-X-108. Maritimes For. Res. Cent., Fredericton, New Brunswick.

Kimmey, J.W. (1955) Rate of deterioration of fire-killed timber in California. USDA Circ. 962, Washington Off., Washington, D.C.

Kimmey, J.W. and Furniss, R.L. (1943) Deterioration of fire-killed Douglas-fir. USDA Tech. Bull. 851. Washington Off., Washington, D.C.

Kimmins, J. and Scoullar, K. (1979) FORCYTE: A computer simulation approach to evaluating the effect of whole-tree harvesting on the nutrient budget in Northwest forests. In: *Forest Fertilization Conference* (Ed. by S.P. Gessel, R.M. Kenady and W.A. Atkinson), pp. 266–273. Inst. For. Resour. Contrib. No. 40. Univ. of Washington, Seattle.

Kimmins, J. and Scoullar, K. (1981) FORCYTE-10. In: *Proceedings of the 3rd Bioenergy R&D Seminar*, pp. 55–59. Nat. Res. Counc. of Canada, Ottawa.

Kirk, R. (1966) *The Olympic Rain Forest*. Univ. of Washington Press, Seattle.

Kirk, T.K. (1973) The chemistry and biochemistry of decay. In: *Wood Deterioration and Its Prevention by Preservative Treatments* (Ed. by D.A. Nicholas), pp. 149–181. Syracuse Univ. Press, Syracuse, New York.

Kirk, T.K., Schultz, E., Conners, W.J., Lorenz, L.F. and Zeikus, J.G. (1978) Influence of culture parameters on lignin metabolism by *Phanerochaete chrysosporium*. *Arch. Microbiol.* **117**, 277–285.

Kirkland, G.L. Jr. (1977) Responses of small mammals to the clearcutting of northeastern Appalachian forests. *J. Mamm.* **58**, 600–609.

Knechtel, A. (1903) Natural reproduction in the Adirondack forests. *For. Q.* **1**, 50–55.

Knuth, D.T. (1964) Bacteria associated with wood products and their effect on certain chemical and physical properties of wood. *Diss. Abstr.* **25**, 2175.

Knuth, D.T. and McCoy, E. (1962) Bacterial deterioration of pine logs in pond storage. *For. Prod. J.* **12**, 437–442.

Korstian, C.F. (1937) Perpetuation of spruce on cut-over and burned lands in the higher Southern Appalachian Mountains. *Ecol Monogr.* **7**, 125–167.

Kotarba, A. (1970) The morphogenic role of foehn wind in the mountains. *Stud. Geomorphol. Carpatho-Balancia* **4**, 171–186.

Kozak, A. and Yang, R.C. (1981) Equations for estimating bark volume and thickness of commercial trees in British Columbia. *For. Chron.* **57**, 112–115.

Kozlowski, T.T. and Cooley, J.H. (1961) Natural root grafting in northern Wisconsin. *J. For.* **59**, 105–107.

Kropp, B.R. (1982) Fungi from decayed wood as ectomycorrhizal symbiont of western hemlock. *Can. J. For. Res.* **12**, 36–39.

Lachaussée, E. (1947) La régéneration de l'épicea en haute Montague. Rev. Eaux For. 85, 281–302. (English summary in *For. Abstr.* **9,** 168.)

Lambert, R.C., Lang, G.E. and Reiners, W.A. (1980) Loss of mass and chemical change in decaying boles of a subalpine balsam fir forest. *Ecology* **61**, 1460–1473.

Lang, G.E. and Forman, R.T.T. (1978) Detritus dynamics in a mature oak forest: Hutchenson Memorial Forest, New Jersey. *Ecology* **59**, 580–595.

Larsen, M.J., Jurgensen, M.F. and Harvey, A.F. (1978) N_2 fixation associated with wood decayed by some common fungi in western Montana. *Can. J. For. Res.* **8**, 341–345.

Larsen, M.J., Jurgensen, M.F. and Harvey, A.F. (1982) N_2 fixation in brown-rotted soil wood in an intermountain cedar–hemlock ecosystem. *For. Sci.* **28**, 292–296.

Larsson, S., Oren, R., Waring, R.H. and Barrett, J.W. (1983) Attacks of mountain pine beetle as related to tree vigor of ponderosa pine. *For. Sci.* **29**, 395–402.

Leach, J.G., Orr, L.W. and Christensen, C. (1934) The interrelationships of bark beetles and blue-staining fungi in felled Norway pine timber. *J. Agric. Res.* (*Washington, D.C.*) **49**, 315–341.

Leach, J.G., Orr, L.W. and Christensen, C. (1937) Further studies on the interrelationship of insects and fungi in the deterioration of felled Norway pine logs. *J. Agric. Res.* (*Washington, D.C.*) **55**, 129–140.

LeBarron, R.K. (1950) Silvicultural management of black spruce in Minnesota. USDA Circ. 791. U.S. Gov. Print. Off., Washington, D.C.

Lemon, P.C. (1945) Wood as a substratum for perennial plants in the Southeast. *Am. Midl. Nat.* **34**, 744–749.

Leopold, L.B., Wolman, M.G. and Miller, J.P. (1964) *Fluvial Processes in Geomorphology*. Freeman, San Francisco.

LeSage, L. and Harper, P.P. (1976a) Notes on the life history of the toed-winged beetle *Anchytarsus bicolor* (Melsheimer) (Coleoptem: Ptilodactilidae). *Coleopt. Bull.* **30**, 233–238.

LeSage, L. and Harper, P.P. (1976b) Cycles biologiques d'Elmidae (Coleopteres) de ruisseaux des Laurentides, Quebec. *Ann. Limnol.* **12**, 139–174.

Lestelle, L.C. (1978) The effects of forest debris removal on a population of resident cutthroat trout in a small headwater stream. M.S. thesis, Univ. of Washington, Seattle.

Levi, J. (1965) The soft rot fungi: Their mode of action and significance in the degradation of wood. *Adv. Bot. Res.* **2**, 323–357.

Levy, J.F. (1982) The place of basidiomycetes in the decay of wood in contact with the ground. In: *Decomposer Basidiomycetes: Their Biology and Ecology* (Ed. by J.C. Frankland, J.N. Hedger and M.J. Swift), pp. 161–178. Cambridge Univ. Press, London and New York.

Lienkaemper, G.W. and Swanson, F. (1980) Changes in large organic debris in forested streams, western Oregon. *Abstr. Annu. Meet. Geol. Soc. Am. 76th* **12**, 116.

Likens, G.E. and Bilby, R.E. (1982) Development, maintenance and role of organic-debris dams in New England streams. In: *Sediment Budgets and Routing in Forested Drainage Basins* (Ed. by F.J. Swanson, R.J. Janda, T. Dunne and D.N. Swanston), pp. 122–128. USDA For. Serv. Res. Pap. PNW-141. Pacific Northwest For. and Range Exp. Stn., Portland, Oregon.

Likens, G.E. and Bormann, F.H. (1970) Chemical analysis of plant tissues from the Hubbard Brook Ecosystem in New Hampshire. Bull. 79, Yale Sch. For., New Haven, Connecticut.

Likens, G.E., Bormann, F.H., Pierce, R.S., Eaton, J.S. and Johnson, N.M. (1977) *Biogeochemistry of a Forested Ecosystem.* Springer-Verlag, Berlin and New York.

Lilly, V.G. (1965) Chemical constituents of the fungal cell. In: *The Fungi. I. The Fungal Cell* (Ed. by G.C. Ainsworth and A.S. Sussman), pp. 163–177. Academic Press, New York.

Lisle, T.E. and Kelsey, H.M. (1982) Effects of large roughness elements on the thalweg course and pool spacing. In: *American Geomorphological Field Group Field Trip Guidebook, 1982 Conference, Pinedale, Wyoming* (Ed. by L.B. Leopold), pp. 134–135. Leopold, Berkeley, California.

Lister, D.B. and Genoe, H.S. (1979) Stream habitat utilization by cohabiting underyearlings of chinook (*Oncorhynchus tshawyacha*) and coho salmon (*O. kisutch*) in the Big Qualicum River, British Columbia. *J. Fish. Res. Board Can.* **27**, 1215–1224.

Lloyd, M. (1963) Numerical observations on movements of animals between beech litter and fallen branches. *J. Anim. Ecol.* **32**, 157–163.

Lobeck, A.K. (1939) *Geomorphology*. McGraw-Hill, New York.

Lowdermilk, W.C. (1925) Factors affecting the reproduction of Engelmann spruce. *J. Agric. Res.* **30**, 995–1009.

Lyell, Sir Charles. (1969) *Principles of Geology*, Vol. 2. The Sources of Science No. 84. Johnson Reprint Corp., New York and London.

Lyford, W.H. and MacLean, D.W. (1966) Mound and pit microrelief in relation to soil disturbance and tree distribution in New Brunswick, Canada. Harvard For. Pap. 15. Harvard Univ., Cambridge, Massachusetts.

Lyon, L.J. (1977) Attrition of lodgepole pine snags on the Sleeping Child Burn, Montana. USDA For. Serv. Res. Note INT-219. Intermt. For. and Range Exp. Stn., Ogden, Utah.

McArdle, R.W. (1931) Overtopping of Douglas-fir snags by reproduction. USDA For. Serv. Res. Note PNW-8. Pacific Northwest For. and Range Exp. Stn., Portland, Oregon.

McCauley, K.J. and Cook, S.A. (1980) *Phellinus wierii* infestations of two mountain hemlock forests in the Oregon Cascades. *For. Sci.* **26**, 23–29.

McClelland, B.R. (1977) Relationships between hole-nesting birds, forest snags, and decay in western larch–Douglas-fir forests of the northern Rocky Mountains. Ph.D. dissertation, Univ. of Montana, Missoula.

Maccubbin, A.E. and Hodson, R.E. (1980) Mineralization of detrital lignocelluloses by salt marsh sediment microflora. *Appl. Environ. Microbiol.* **40**, 735–740.

McCullough, H.A. (1948) Plant succession on decaying lags in a virgin spruce-fir forest. *Ecology* **29**, 508–513.

MacDonald, A. and Keller, E. (1983) Large organic debris and anadromous fish habitat in the coastal redwood environment: The hydrologic system. Tech. Completion Rep., OWRT Proj. B-213-CAL, Water Resour. Cent., Univ. of California, Davis.

McFee, W.W. and Stone, E.L. (1966) The persistence of decaying wood in humus layers of northern forests. *Soil Sci. Soc. Am. Proc.* **30**, 513–516.

McKee, A., LeRoi, G. and Franklin, J.F. (1982) Structure, composition and reproductive behavior of terrace forests, South Fork Hoh River, Olympic National Park. In: *Ecological Research in National Parks of the Pacific Northwest* (Ed. by E.E. Starkey, J.F. Franklin and J.W. Matthews), pp. 22–29. Natl. Park Serv. Coop. Stud. Unit, Corvallis, Oregon.

McLachlan, A.J. (1970) Submerged trees as a substrate for benthic fauna in the recently created Lake Kariba (Central Africa). *J. Appl. Ecol.* **7**, 253–266.

MacLean, H. and Gardner, J.A.F. (1956) Distribution of fungicidal extractives (thujaplicin and water-soluble phenols) in western red cedar heartwood. *For. Prod. J.* **6**, 510–516.

MacLean, J.S. (1941) Thermal conductivity of wood. *Heating, Piping Air Condition* **13**, 380–391.

MacMillan, P.C. (1981) Log decomposition in Donaldson's Woods, Spring Mill State Park, Indiana. *Am. Midl. Nat.* **106**, 335–344.

McMullen, L.H., Fiddick, R.L. and Wood, R.O. (1981) Bark beetles, *Pseudohylesinus* spp. (Coleoptera:Scolytidae), associated with *Amabilis* fir defoliated by *Neodiprion* sp. (Hymenoptera: Diprionidae). *J. Entomol. Soc. B.C.* **78**, 43–45.

Mannan, R.W. (1977) Use of snags by birds, Douglas-fir region, western Oregon. M.S. thesis, Oregon State Univ., Corvallis.

Mannan, R.W., Meslow, E.C. and Wight, H.M. (1980) Use of snags by birds in Douglas-fir forests, western Oregon. *J. Wildl. Manage.* **44**, 787–797.
Manuwal, D.A. and Zarnowitz, J. (1981) *Cavity Nesting Birds of the Olympic National Forest*. Coll. For. Resour., Univ. of Washington, Seattle.
Manville, R.H. (1949) A study of small mammal populations in northern Michigan. Misc. Publ. 73, pp. 1–83. Mus. Zoos., Univ. of Michigan, Ann Arbor.
Marikovskii, P.I. (1956) Observations on the biology of the destructive carpenter ant and wood ants inhabiting the montane forests of Kirghizia. Trudy Inst. Zool. i Parasitologii A N Kergizskoi SSR, No. 5, Frunze. (In Russian, as cited in Rozhkov, 1966, English translation, 1970.)
Marston, RA. (1982) The geomorphic significance of log steps in forest streams. *Ann. Assoc. Am. Geogr.* **72**, 99–108.
Martell, A.M. and Radvanyi, A. (1977) Changes in small mammal populations after clearcutting of northern Ontario black spruce forest. *Can. Field Nat.* **91**, 41–46.
Martin, N.D. (1960) An analysis of bird populations in relation to forest succession in Algonquin Provincial Park, Ontario. *Ecology* **41**, 126–140.
Maser, C. and Trappe, J.M., tech. eds. (1984) The seen and unseen world of the fallen tree. USDA For. Serv. Gen. Tech. Rep. PNW-164. Pacific Northwest For. and Range Exp. Stn., Portland, Oregon.
Maser, C., Trappe, J.M. and Nussbaum, R.A. (1978) Fungal–small mammal interrelationships with emphasis on Oregon coniferous forests. *Ecology* **59**, 799–809.
Maser, C., Anderson, R., Cromack, K. Jr., Williams, J.T. and Martin, R.E. (1979) Dead and down woody material. In: *Wildlife Habitats in Managed Forests, the Blue Mountains of Oregon and Washington* (Ed. by J. Thomas), Ch. 6. USDA For. Serv. Agric. Handb. 553, Washington Off., Washington, D.C.
Maser, C., Mate, B.R., Franklin, J.F. and Dyrness, C.T. (1981) Natural history of Oregon coast mammals. USDA For. Serv. Gen. Tech. Rep. PNW-133. Pacific Northwest For. and Range Exp. Stn., Portland, Oregon, in cooperation with USDI Bur. Land Manage.
Matson, K.G. and Swank, W.T. (1984) Carbon dioxide fluxes from conventional and whole tree harvested watersheds—effects of woody residue. *Bull. Ecol. Soc. Am.* **65**, 123.
Meehan, W.R., Farr, W.A., Bishop, D.M. and Patric, J.H. (1969) Some effects of clearcutting on salmon habitat on two southeastern Alaska streams. USDA For. Serv. Gen. Tech. Rep. PNW-82. Pacific Northwest For. and Range Exp. Stn., Portland, Oregon.
Meehan, W.R., Swanson, F.J. and Sedell, J.R. (1977) Influences of riparian vegetation on aquatic ecosystems with particular references to salmonid fishes and their food supply. In: *Importance, Preservation and Management of Riparian Habitat, a Symposium* (R.R. Johnson and D.A. Jones, tech. coords.). USDA For. Serv. Gen. Tech. Rep. RM-43. Rocky Mt. For. and Range Exp. Stn., Fort Collins, Colorado.
Megahan, W.F. (1982) Channel sediment storage behind obstructions in forested drainage basins draining the granitic bedrock of the Idaho Batholith. In: *Sediment Budgets and Routing in Forested Drainage Basins* (Ed. by F.J. Swanson, R.J. Janda,

T. Dunne and D.N. Swanston), pp. 114–121. USDA For. Serv. Res. Pap. PNW-141. Pacific Northwest For. and Range Exp. Stn., Portland, Oregon.

Megahan, W.F. and Nowlin, R.A. (1976) Sediment storage in channels draining small forested watersheds in the mountains of central Idaho. In: *Proceedings of the small the 3rd Federal Interagency Sedimentation Conference*, Denver, Colorado, pp. 4–115/4–126. Sediment. Comm. of Water Resour. Counc., Washington, D.C.

Melillo, J.M., Aber, J.D. and Muratore, J.F. (1982) Nitrogen and lignin control of hardwood leaf litter decomposition dynamics. *Ecology* **63**, 621–626.

Melillo, J.M., Naiman, R.J., Aber, J.D. and Eshleman, K.N. (1983) The influence of substrate quality and stream size on wood decomposition dynamics. *Oecologia* **38**, 281–285.

Melillo, J.M., Naiman, R.J., Aber, J.D. and Linkins, A.E. (1985) Factors controlling mass loss and nitrogen dynamics of plant litter decaying in northern streams. *Bull. Mar. Sci. Spec. Detritus Symp.*, 1984.

Merrill, W. and Cowling, E.B. (1966) Role of nitrogen in wood deterioration: Amounts and distribution in tree stems. *Can. J. Bot.* **44**, 1555–1580.

Michigan Historical Society, Manistee County. (1883) *History of Manistee County, Michigan.* Page, Chicago.

Mielke, J.L. (1950) Rate of deterioration of beetle-killed Engelmann spruce. *J. For.* **48**, 882–888.

Miller, E. and Miller, D.R. (1980) Snag use by birds. In: *Proceedings of the Workshop on Management of Western Forests and Grasslands for Nongame Birds, 1980 February 11–14; Salt Lake City* (R.M. DeGraaf, tech. coord.), pp. 337–356. USDA For. Serv. Gen. Tech. Rep. INT 86. Intermt. For. and Range Exp. Stn., Ogden, Utah.

Miller, J.M. and Keen, K.P. (1960) Biology and control of the western pine beetle. USDA For. Serv. Misc. Publ. 800, Washington Off, Washington, D.C.

Miller, W.E. (1983) Decomposition rates of aspen bole and branch litter. *For. Sci.* **29**, 351–356.

Minderman, G. (1968) Addition, decomposition and accumulation of organic matter in forests. *J. Ecol.* **56**, 355–362.

Minore, D. (1972) Germination and early survival of coastal tree species on organic seed beds. USDA For. Serv. Res. Pap. PNW-135. Pacific Northwest For. and Range Exp. Stn., Portland, Oregon.

Mintzer, A. (1979) Colony foundation and pleometrosis in *Camponotus* (Hymenoptera:Formicidae). *Pan-Pac. Entomol.* **55**, 81–89.

Mitchell, R.G., Waring, R.H. and Pitman, G.B. (1983) Thinning lodgepole pine increases tree vigor and resistance to mountain pine beetle. *For. Sci.* **29**, 204–211.

Monserud, R.A. (1976) Simulation of forest tree mortality. *For. Sci.* **22**, 438–444.

Montgomery, A.P. (1982) The role of polysaccharidase enzymes in the decay of wood by basidiomycetes. In: *Decomposer Basidiomycetes: Their Biology and Ecology* (Ed. by J.C. Frankland, J.N. Hedger and M.J. Swift), pp. 51–65. Cambridge Univ. Press, London and New York.

Mooney, H.A., Bonnicksen, T.M., Christenson, N.L., Lotan, J.E. and Reiners, W.A. (1981) Fire regimes and ecosystem properties. USDA Gen. Tech. Rep. WO-26, Washington Off., Washington, D.C.

Morgan, F.D. (1968) Bionomics of Siricidae. *Annu. Rev. Entomol.* **13**, 239–256.

Morris, R.F. (1955) Population studies on some small forest mammals in eastern Canada. *J. Mamm.* **36**, 21–35.

Mortensen, E. (1977) Density-dependent mortality of trout fry (*Salmo tritta* L.) and its relationship to the management of small streams. *J. Fish. Biol.* **11**, 613–617.

Mosley, M.P. (1981) The influence of organic debris on channel morphology and bedload transport in a New Zealand forest stream. *Earth Surf. Process. Landforms* **6**, 572–579.

Murgatroyd, A.L. and Ternan, J.L. (1983) The impact of afforestation on stream bank erosion and channel form. *Earth Surf. Process. Landforms* **8**, 357–369.

Murphy, M.L., Koski, K.V., Heifetz, J., Johnson, S.W., Kirchhofer, D. and Thedinga, J.F. (1985) Role of large organic debris as winter habitat for juvenile salmonids in Alaska streams. Proc. Western Assoc. Fish and Wildl. Agencies, Vol. 64. Victoria, B.C. In press.

Myers, J.H. and Campbell, B.J. (1976) Predation by carpenter ants: A deterrent to the spread of cinnabar moth. *J. Entomol. Soc. B.C.* **73**, 7–9.

Naiman, R.J. and Sedell, J.R. (1980) Relationships between metabolic parameters and stream order in Oregon. *Can. J. Fish. Aquat. Sci.* **37**, 834–847.

Narver, D.W. (1971) Effects of logging debris on fish production. In: *Proceedings of the Symposium on Forest Land Uses and Stream Environment, Ootober 19–21, 1970* (Ed. by J.T. Krygier and J.D. Hall), pp. 100–111. Oregon State Univ., Corvallis.

Nelson, N.D. (1975) Extractives produced during heartwood formation in relation to mounts of parenchyma in *Juglans nigra* and *Quercus rubra*. *Can. J. For. Res.* **5**, 291–301.

Nilsen, H.C. and Larimore, R.W. (1973) Establishment of invertebrate communities on log substrates in the Kaskaskia River, Illinois. *Ecology* **54**, 366–374.

Odum, E.P. (1949) Small mammals of the Highlands (North Carolina) Plateau. *J. Mamm.* **30**, 179–192.

Odum, E.P. (1950) Bird populations of the Highlands (North Carolina) Plateau in relation to plant succession and avian invasion. *Ecology* **31**, 587–605.

Ogden, P.S. (1961) *Peter Skene Ogden's Snake Country Journal 1826–27* (Ed. by K.G. Davies), Vol. 23, pp. lvii–lxi, 143–163. Hudson's Bay Re card Soc., London.

Okkonen, E.A., Wahlgren, H.E. and Maeglin, R.R. (1972) Relationship of specific gravity to tree height in commercially important species. *For. Prod. J.* **22**, 37–42.

Oliver, C.D. and Stephens, E.P. (1977) Reconstruction of a mixed forest in central New England. *Ecology* **58**, 562–572.

Olson, J.S. (1963) Energy storage and the balance of producers and decomposers in ecological systems. *Ecology* **44**, 322–331.

Orr, P.W. (1963) Windthrown timber survey in the Pacific Northwest 1962. USDA For. Serv., Pacific Northwest Reg., Portland, Oregon.

Osborn, J.G. (1981) The effects of logging on cutthroat trout (*Salmo clarki*) in small headwater streams. Fish. Res. last. FRI-VW-8113, Univ. Washington, Seattle.

Packard, A.S. Jr. (1890) Insects injurious to forest and shade trees. USDA, U.S. Entomol. Comm. Fifth Rep., Washington Off., Washington, D.C.

Paim, U. and Becker, W.E. (1963) Seasonal oxygen and carbon dioxide content of decaying wood as a component of *Orthosoma brunneum* (Forster) (Coleoptera: Cerambycidae). *Can. J. For. Res.* **41**, 1133–1147.

Panshin, A.J. and deZeeuw, C. (1980) *Textbook of Wood Technology*. McGraw-Hill, New York.

Parkin, E.A. (1940) The digestive enzymes of some wood-boring beetle larvae. *J. Exp. Biol.* **17**, 364–377.

Pastor, J. and Bockheim, J.G. (1984) Distribution and cycling of nutrients in an aspen-mixed-hardwood-spodosol ecosystem in Wisconsin. *Ecology* **65**, 339–353.

Pearce, A.J. and Watson, A. (1983) Medium term effects of two landsliding episodes on channel storage of sediment. Earth Surf. Process. *Landforms* **8**, 29–39.

Pearson, P.G. (1959) Small mammals and old field succession on the piedmont of New Jersey. *Ecology* **40**, 249–255.

Peck, E.C. (1953) The sap or moisture in wood. For. Prod. Lab. Rep. D768. USDA For. Serv., Madison, Wisconsin.

Peet, R.K. (1984) Twenty-six years of change in a *Pinus strobus–Acer saccharum* forest, Lake Itasca, Minnesota. *Bull. Torrey Bot. Club* **111**, 61–68.

Penney, M.M. (1967) Studies on the ecology of *Feronia oblongopunctata* (F.) (Coleoptera: Carabidae). *Trans. Soc. Br. Entomol.* **17**, 129–139.

Pereira, C.R.D. (1980) Life history studies of *Cinygma integrum* Eaton (Ephemeroptera: Heptageniidae) and other mayflies associated with wood substrates in Oregon streams. M.S. thesis, Oregon State Univ., Corvallis.

Pereira, C.R.D. and Anderson, N.H. (1982) Observations on the life histories and feeding of *Cinygma integrum* Eaton and *Ironodes nitidus* (Eaton) (Ephemeroptera: Heptageniidae). *Melanderia* **39**, 35–45.

Pereira, C.R.D., Anderson, N.H. and Dudley, T. (1982) Gut content analysis of aquatic insects from wood substrates. *Melanderia* **39**, 23–33.

Peters, G.B., Dawson, H.J., Hrutfiord, B.F. and Whitney, R.R. (1976) Aqueous leachate from western red cedar: Effects on some aquatic organisms. *J. Fish Res. Board Can.* **33**, 2703–2709.

Petersen, R.C. and Cummins, K.W. (1974) Leaf processing in a woodland stream. *Freshwater Biol.* **4**, 343–368.

Peterson, N.P. (1980) The role of spring ponds in the winter ecology and natural production of coho salmon (*Oncorhynchus kisutch*) on the Olympic Peninsula, Washington. M.S. thesis, Univ. of Washington, Seattle.

Peterson, R.T. (1961) *A Field Guide to Western Birds*. Houghton-Mifflin, Boston.

Petr, T. (1970) Macroinvertebrates of flooded trees in the man-made Volta Lake (Ghana) with special reference to the burrowing mayfly, *Pouilla adusta* Navas. *Hydrobiologia* **36**, 353–360.

Pfankuch, D.J. (1978) Stream reach inventory and channel stability evaluation. USDA For. Serv., Missoula, Montana.

Pickford, S.G. and Hazard, J.W. (1978) Simulation studies of line intersect sampling of forest residue. *For. Sci.* **24**, 469–483.

Pielou, E.C. (1977) *Mathematical Ecology*. Wiley (Interscience), New York.

Place, I.C.M. (1955) The influence of seedbed conditions on the regeneration of spruce and balsam fir. For. Branch Bull. 117, Dep. North. Aff. and Nad. Resour., Ottawa, Ontario.

Pneumaticos, S.M., Jacger, T.A. and Perem, E. (1972) Factors influencing the weight of black spruce and balsam fir stems. *Can. J. For. Res.* 2, 427–433.

Ponce, S.L. (1974) The biochemical oxygen demand of finely divided logging debris in stream water. *Water Resour. Res.* **10**, 983–988.

Pricer, J.L. (1908) The life history of the carpenter ant. *Biol. Bull.* **14**, 177–218.

Putz, F.E., Coley, P.D., Lu, K., Montalvo, A. and Aiello, A. (1983) Uprooting and snapping of trees: Structural determinants and ecological consequences. *Can. J. For. Res.* **13**, 1011–1020.

Raphael, M.G. (1980) Utilization of standing dead trees by breeding birds at Sagehen Creek, California. Ph.D. dissertation, Univ. of California, Berkeley.

Raphael, M.G. (1983) Analysis of habitat requirements of amphibians, reptiles, and mammals in three early-successional stages of Douglas-fir forest. USDA For. Serv. Final Rep., Agreement No. PSW-82–0068. Pacific Southwest Reg., San Francisco, California.

Raphael, M.G. and White, M. (1984) Use of snags by cavity-nesting buds in the Sierra Nevada. *Wildl. Monogr.* **86**, 66.

Raphael, M.G., Rosenberg, K.V. and Taylor, C.A. (1982) Administrative study of relationships between wildlife and old-growth forest stands, phase III. USDA For. Serv. Interim Rep. 3, Suppi. RO-44, Master Agreement No. 21–395. Pacific Southwest Reg., San Francisco, California.

Rayner, A.D.M. and Todd, N.K. (1979) Population and community structure and dynamics of fungi in decaying wood. *Adv. Bot. Res.* **7**, 333–420.

Rayner, A.D.M. and Todd, N.K. (1982) Population structure in wood-decomposing basidiomycetes. In: *Decomposer Basidiomycetes: Their Biology and Ecology* (Ed. by J.C. Frankland, J.N. Hedger and M.J. Swift), pp. 109–129. Cambridge Univ. Press, London and New York.

Rector, W.G. (1949) From woods to sawmill: Transportation problems in logging. *Agric. Hist.* **23**, 239–244.

Reid, L.R. (1981) Sediment production from gravel-surfaced roads, Clearwater basin. Final Rep. FR1-UW-8108. Wash. Fish. Res. Inst., Univ. of Washington, Seattle.

Reiners, N.M. and Reiners, W.A. (1965) Natural harvesting of trees. *William L. Hutcheson Mem For. Bull.* **2**, 9–17.

Reis, M.S. (1972) Decay resistance of six wood species from the Amazon Basin of Brazil. *Holzforschung* **27**, 103–111.

Riley, C.G. and Skolko, A.J. (1942) Rate of deterioration in spruce killed by the European spruce sawfly. *Pulp Pap. Mag. Can.* **43**, 521–524.

Roskoski, J.P. (1980) Nitrogen fixation in hardwood forests of the northeastern United States. *Plant Soil* **54**, 33–44.

Roskoski, J.P. (1981) Comparative C_2H_2 reduction and $^{15}N_2$ fixation in deciduous wood litter. *Soil Biol. Biochem.* **13**, 83–85.

Rust, H.J. (1933) Many bark beetles destroyed by predaceous mites. *J. Econ. Entomol.* **26**, 733–734.

Ruth, R.H. and Yoder, R.A. (1953) Reducing wind damage in the forests of the Oregon Coast Range. USDA For. Serv. Stn. Pap. 7. Pacific Northwest For. and Range Exp. Stn., Portland, Oregon.

Rydholm, S.A. (1965) *Pulping Processes*. Wiley (Interscience), New York.

Ryszkowski, L. (1969) Estimates of consumption of rodent populations in different pine forest ecosystems. In: *Energy Flow Through Small Mammal Populations* (Ed. by K. Petrusewcz and L. Ryszkowski), pp. 281–289. PWN-Polish Scientific Publ, Warsaw.

Sacket, S.S. (1979) Natural fuel loadings in a ponderosa pine and mixed conifer forests of the Southwest. USDA For. Serv. Res. Pap. RM-213. Rocky Mt. For. and Range Exp. Stn., Fort Collins, Colorado.

Salt, G.W. (1957) An analysis of avifaunas in the Teton Mountains and Jackson Hole, Wyoming. *Condor* **59**, 373–393.

Sanders, C.J. (1964) The biology of carpenter ants in New Brunswick. *Can. Entomol.* **96**, 894–909.

Sanders, C.J. (1970) The distribution of carpenter ant colonies in the spruce-fir forests of northwestern Ontario. *Ecology* **51**, 865–873.

Sanders, C.J. (1972) Seasonal and daily activity patterns of carpenter ants (*Camponotus* spp.) in northwestern Ontario (Hymenoptera:Formicidae). *Can. Entomol.* **104**, 1681–1687.

Savely, H.E. Jr. (1939) Ecological relations of certain animals in dead pine and oak logs. *Ecol. Monogr.* **9**, 322–385.

Savory, J.G. (1954a) Breakdown of timber by ascomycetes and fungi imperfecti. *Ann. Appl. Biol.* **41**, 336–347.

Savory, J.G. (1954b) Damage to wood caused by microorganisms. *J. Appl. Bacterial.* **17**, 213–218.

Scheffer, T.C. and Cowling, E.B. (1966) Natural resistance of wood to microbial deterioration. *Annu. Rev. Phytopathol* **4**, 147–170.

Scheffer, T.C. and Duncan, C.G. (1947) The decay resistance of certain Central American and Ecuadorian woods. *Trop. Woods* **92**, 1–24.

Scheffer, T.C. and Englerth, G.H. (1952) Decay resistance of second-growth Douglas-fir. *J. For.* **50**, 439–442.

Scheffer, T.C. and Hopp, H. (1949) Decay resistance of black locust heartwood. USDA Tech. Bull. 984, Washington Off., Washington, D.C.

Scheffer, T.C., Englerth, G.H. and Duncan, C.G. (1949) Decay resistance of seven native oaks. *J. Agric. Res.* **78**, 129–152.

Scott, J.T., Siccama, T.G., Johnson, A.H. and Breisch, A.R. (1984) Decline of red spruce in the Adirondacks, New York. *Bull. Torrey Bot. Club* **111**, 438–444.

Scott, V.E., Crouch, G.L. and Whelan, J.A. (1982) Responses of birds and small mammals to clearcutting in a subalpine forest in central Colorado. USDA For. Serv. Res. Note RM-422. Rocky Mr. For. and Range Exp. Stn., Fort Collins, Colorado.

Secretary of War (1875) Report of the Chief of Engineers. In: *House Executive Documents: 1st Session 44th Congress, 1875–1876*, Vol. 2, Pt. 2. U.S. Gov. Print. Off., Washington, D.C.

Secretary of War (1881) Letter from the Secretary of War. Letter from Chief of Engineers. In: *Senate Executive Documents: 3rd Session 46th Congress*, Vol. 3, Doc. No. 39, Ser. No. 1943. U.S. Gov. Print. Off., Washington, D.C.

Secretary of War (1915) Index to the reports of the Chief of Engineers, U.S. Army, 1866–1912. In: *House Documents: 2nd Session 63rd Congress, 1913–1914*, Vol. 20, Pt. 2. U.S. Gov. Print. Off., Washington, D.C.

Sedell, J.R., Everest, F.H. and Swanson, F.J. (1982) Fish habitat and streamside management: Past and present. In: *Proceedings of the Society of American Foresters Annual Meeting, September 27–30, 1981*, pp. 244–255. Soc. Am. For., Bethesda, Maryland.

Sedell, J.R. and Frogatt, J.L. (1984) Importance of streamside forests to large rivers: The isolation of the Willamette River, Oregon, U.S.A., from its floodplain by snagging and streamside forest removal. *Verh. Int. Ver. Limnol.* **22**, 1828–1834.

Sedell, J.R. and Luchessa, K.J. (1982) Using the historical record as an aid to salmonid habitat enhancement. In: *Proceedings of a Symposium on Acquisition and Utilization of Aquatic Habitat Inventory Information, October 28–30, 1981, Portland, Oregon* (Ed. by N.B. Armantrout), pp. 210–223. West. Div. Am. Fish. Soc., Bethesda, Maryland.

Sedell, J.R. and Swanson, F.J. (1982) Fish habitat and streamside management: Past and present. In: *Proceedings of the Society of American Foresters Annual Meeting*, pp. 244–255. Soc. Am. For., Bethesda, Maryland.

Sedell, J.R., Yuska, J.E. and Speaker, R.W. (1984) Habitats and salmonid distribution in pristine sediment-rich river valley systems: S. Fork Hoh and Queets River, Olympic National Park. In: *Proceedings of a Symposium on Fish and Wildlife Relationships in Old-Growth Forests, April 12–15, 1982, Juneau, Alaska* (Ed. by W.R. Meeham, T.R. Merrell Jr. and T. Hanley), pp. 33–46. Am. Inst. Fish. Res. Biol., Juneau, Alaska.

Servizi, J.A., Martens, D.W. and Gordon, R.W. (1970) Effects of decaying bark on incubating salmon eggs. Int. Pacific Salmon Fish. Comm. Prog. Rep. 24, New Westminster, British Columbia.

Sharp, R.F. and Milbank, J.W. (1973) Nitrogen fixation in deteriorating wood. *Experimentia* **29**, 895–896.

Sharpe, G.W. (1956) A taxonomical–ecological study of vegetation by habitats in eight forest types of the Olympic Rain Forest, Olympic National Park, Washington. Ph.D. thesis, Univ. of Washington, Seattle.

Shugart, H.H. and James, D. (1973) Ecological succession of breeding bird populations in northwestern Arkansas. *Auk* **90**, 62–77.

Shugart, H.H., Smith, T.M., Kitchings, J.T. and Kroodsma, R.L. (1978) The relationship of non-game birds to southern forest types and successional stages. In: *Management of Southern Forests for Nongame Birds*, Workshop Proc., Atlanta, Georgia, Jan. 24–26, 1978. USDA For. Serv. Gen. Tech. Rep. SE-14. Southeast. For. Exp. Stn., Asheville, North Carolina.

Siccam, T.G., Bliss, M. and Vogelmann, H.W. (1982) Decline of red spruce in the Green Mountains of Vermont. *Bull. Torrey Bot. Club* **109**, 163–168.

Silsbee, D.G. and Larson, G.L. (1983) A comparison of streams in logged and unlogged areas of Great Smoky Mountains National Park. *Hydrobiologia* **102**, 99–111.

Silvester, W.B., Sollins, P., Verhoeven, T. and Cline, S.P. (1982) Nitrogen fixation and acetylene reduction in decaying conifer boles: Effects of incubation time, aeration, and moisture content. *Can. J. For. Res.* **12**, 646–652.

Singer, S. and Swanson, M.L. (1983) Soquel Creek storm damage recovery plan. USDA Soil Conserv. Serv., Aptos, California.

Singh, J.S. and Gutpka, S.R. (1977) Plant decomposition and soil respiration in terrestrial ecosystems. *Bot. Rev.* **43**, 449–528.

Skaar, C. (1972) *Water in Wood.* Syracuse Univ. Press, Syracuse, New York.

Smith, J.H.G. (1955) Some factors affecting reproduction of Engelmann spruce and alpine fir. Tech. Rep. T.43, Dep. Lands and For., British Columbia For. Serv., Victoria.

Smith, J.H.G. and Clark, M.B. (1960) Growth and survival of Engelmann spruce and alpine fir on seeds spots at Bolean Lake, British Columbia 1954–58. *For. Chronol.* **36**, 46–49.

Smith, L.V. and Zavarin, E. (1960) Free mono- and oligosaccharides of some California conifers. *Tappi* **43**, 218–221.

Smith, R.D. (1985) Sediment routing in a small watershed in the blast zone at Mount St. Helens, Washington. M.S. thesis, Oregon State Univ., Corvallis.

Smythe, R.V. and Carter, F.L. (1969) Feeding responses to sound wood by the eastern subterranean termite, *Reticulitermes flavipes. Ann. Entomol. Soc. Am.* **62**, 335–337.

Snyder, D.P. (1950) Bird communitiep in the coniferous forest biome. *Condor* **52**, 17–27.

Sollins, P. (1982) Input and decay of coarse woody debris in coniferous stands in western Oregon and Washington. *Can. J. For. Res.* **12**, 18–28.

Sollins, P., Reichle, D.E. and Olson, J.S. (1973) Organic matter budget and model for a southern Appalachian *Liriodendron* forest. USAEC Rep. EDFB-1BP-73-2. Oak Ridge Natl. Lab., Oak Ridge, Tennessee.

Sollins, P., Brown, A.T. and Swartzman, G. (1979) CONIFER: A model of carbon and water flow through a coniferous forest (revised documentation). *Conif. For. Biome Bull.* **15**, Univ. of Washington, Seattle.

Sollins, P., Grier, C.C., McCorison, F.M., Cromack, K. Jr., Fogel, R. and Fredriksen, R.L. (1980) The internal element cycles of an old-growth Douglas-fir ecosystem in western Oregon. *Ecol. Monogr.* **50**, 261–285.

Spano, S.D., Jurgensen, M.F., Larsen, M.J. and Harvey, A.E. (1982) Nitrogen-fixing bacteria in Douglas-fir residue decayed by *Fomitopsis pinicola. Plant Soil* **68**, 117–123.

Speaker, R.W., Moore, K.M. and Gregory, S.V. (1984) Analysis of the process of retention of organic matter in stream ecosystems. *Verh. Int. Ver. Limnol.* **22**, 1835–1841.

Sprugel, D.G. (1976) Dynamic structure of wave-generated *Abies balsamea* forests in the Northeast United States. *J. Ecol.* **64**, 889–911.

Spurr, S.H. and Hsuing, W. (1954) Growth rate and specific gravity in conifers. *J. For.* **52**, 191–200.

Stark, R.W. (1982) Generalized ecology and life cycle of bark beetles. In: *Bark Beetles in North American Conifers* (Ed. by J.B. Milton and K.B. Sturgeon), pp. 21–45. Univ. Texas Press, Austin.

Starker, T.J. (1934) Fire resistance in the forest. *J. For.* **33**, 595–598.

Steedman, R.J. (1983) Life history and feeding role of the xylophagous aquatic beetle, *Lara avara* LeConte (Dryopoidea: Elmidae) M.S. thesis, Oregon State Univ., Corvallis.

Stephens, E.P. (1955) The historical–developmental method of determining forest trends. Ph.D. thesis, Harvard Univ., Cambridge, Massachusetts.

Stephens, E.P. (1956) The uprooting of trees: A forest process. *Soil Sci. Soc. Am. Proc.* **20**, 113–116.

Stewart, R.E. and Aldrich, J.W. (1949) Breeding bird populations in the spruce region of the central Appalachians. *Ecology* **30**, 75–82.

Stiles, E.W. (1980) Bird community structure in alder forests in Washington. *Condor* **82**, 20–30.

Stone, E.L. (1975) Windthrow influences on spatial heterogeneity in a forest soil. *Bibliogr. Agric. Mitt. Eidg. Anst. Forstl. Versuchswes.* **51**, 77–87 (Int. Assoc. of Sci. Hydrol., Gentbrugge, Belgium).

Storer, T.I., Evans, F.C. and Palmer, F.G. (1944) Some rodent populations in the Sierra Nevada of California. *Ecol. Monogr.* **14**, 165–192.

Stupka, A. (1964) *Trees, Shrubs and Woody Vines of Great Smoky Mountains National Paris.* Univ. of Tennessee Press, Knoxville.

Sutherland, J.B., Blanchette, R.A., Crawford, D.L. and Pometto, A.L. III (1979) Breakdown of Douglas-fir phloem by a lignocellulose-degrading *Streptomyces. Curr. Microbial.* **2**, 123–126.

Swanson, F.J. and Lienkaemper, G.W. (1978) Physical consequences of large organic debris in Pacific Northwest streams. USDA For. Serv. Gen. Tech. Rep. PNW-69. Pacific Northwest For. and Range Exp. Stn., Portland, Oregon.

Swanson, F.J. and Lienkaemper, G.W. (1982) Interactions among fluvial processes, forest vegetation, and aquatic ecosystems, South Fork Hoh River, Olympic National Park. In: *Ecological Research in National Parks of the Pacific Northwest* (Ed. by J.F. Franklin, E.E. Starkey and J.W. Matthews), pp. 30–34. Oregon State Univ., For. Res. Lab., Corvallis.

Swanson, F.J., Lienkaemper, G.W. and Sedell, J.R. (1976) History, physical effects, and management implications of large organic debris in western Oregon streams. USDA For. Serv. Gen. Tech. Rep. PNW-56. Pacific Northwest For. and Range Exp. Stn., Portland, Oregon.

Swanson, F.J., Fredricksen, R.L. and McCorison, F.M. (1982a) Material transfer in a western Oregon forested watershed. In: *Analysis of Coniferous Forest Ecosystems in the Western United States* (Ed. by R.L. Edmonds), pp. 233–266. US/IBP Synth. Ser. 14, Hutchinson Ross, Stroudsburg, Pennsylvania.

Swanson, F.J., Gregory, S.V., Sedell, J.R. and Campbell, A.G. (1982b) Land–water interactions: The riparian zone. In: *Analysis of Coniferous Forest Ecosystems in the Western United States* (Ed. by R.L. Edmonds), pp. 267–291. US/IBP Synth. Ser. 14, Hutchinson Ross, Stroudsburg, Pennsylvania.

Swanson, F.J., Janda, R.J., Dunne, T. and Swanston, D.N. (1982c) Sediment budgets and routing in forested drainage basins. USDA For. Serv. Res. Pap. PNW-141. Pacific Northwest For. and Range Exp. Stn., Portland, Oregon.

Swanson, F.J., Collins, B., Dunne, T. and Wicherski, B.P. (1983) Erosion of tephra from hillslopes near Mount St. Helens and other volcanoes. In: *Proceedings of the Symposium on Erosion Control in Volcanic Areas, July 6–9, 1982, Seattle and Vancouver, Washington*, pp. 361–371. Sabo (Erosion Control) Div., Erosion Control Dep., Tsukuba Science City, Japan.

Swanson, F.J., Bryant, M.S., Lienkaemper, G.W. and Sedell, J.R. (1984) Organic debris in small streams, Prince of Wales Island, Southeast Alaska. USDA For. Serv. Gen. Tech. Rep. PNW-166. Pacific Northwest For. and Range Exp. Stn., Portland, Oregon.

Swift, M.J. (1973) The estimation of mycelial biomass by determination of the hexoamine content of wood tissue decayed by fungi. *Soil Biol. Biochem.* **5**, 321–332.

Swift, M.J. (1977a) The ecology of wood decomposition. *Sci. Prog.* **64**, 175–199.

Swift, M.J. (1977b) The roles of fungi and animals in the immobilization and release of nutrient elements from decomposing branch-wood. In: *Soil Organisms as Components of Ecosystems* (Ed. by V. Lohm and T. Persson), *Ecol. Bull.* (*Stockholm*) 25, pp. 193–202.

Swift, M.J., Healey, I.N., Hibberd, J.K., Sykes, J.M., Bampoe, V. and Nesbitt, M.E. (1976) The decomposition of branch-wood in the canopy and floor of a mixed deciduous woodland. *Oecologta* **26**, 139–149.

Swift, M.J., Heal, O.W. and Anderson, J.M. (1979) *Decomposition in Terrestrial Ecosystems*. Univ. of California Press, Berkeley.

Szaro, R.C. and Balda, R.P. (1979) Effects of harvesting ponderosa pine on nongame bird populations. USDA For. Serv. Res. Pap. RM-212. Rocky Mt. For. and Range Exp. Stn., Fort Collins, Colorado.

Tabak, H.H. and Cooke, W.B. (1968) The effects of gaseous environments on the growth and metabolism of fungi. *Bot. Rev.* **34**, 126–252.

Tarkow, H. and Stamm, A.J. (1960) Diffusion through air-filled capillaries of softwoods. Part I. Carbon dioxide. *For. Prod. J.* **10**, 247–254.

Tarzwell, C.M. (1936) Experimental evidence on the value of trout stream improvement in Michigan. *Trans. Am. Fish. Soc.* **66**, 177–187.

Taylor, R.F. (1935) Available nitrogen as a factor influencing the occurrences of Sitka spruce and western hemlock seedlings in the forests of southeastern Alaska. *Ecology* **16**, 580–642.

Teskey, H.J. (1976) Diptera larvae associated with trees in North America. *Mem. Entomol. Soc. Can.* **100**, 53.

Thomas, J.W., tech. ed. (1979) Wildlife habitats in managed forests, the Blue Mountains of Oregon and Washington. USDA For. Serv. Agric. Handb. 553. Washington Off., Washington, D.C.

Thompson, J.N. (1980) Treefalls and colonization patterns of temperate forest herbs. *Am. Midl. Nat.* **104**, 176–184.

Thornburgh, D.A. (1969) Dynamics of the true fir–hemlock forests of the west slope of the Washington Cascade Range. Ph.D. thesis, Univ. of Washington, Seattle.

Tilles, D.A. and Wood, D.L. (1982) The influence of carpenter ant (*Camponotus modoc*) (Hymenoptera:Formicidae) attendance on the development and survival of aphids (*Cinara* spp.) (Homoptera:Aphididae) in a giant sequoia forest. *Can. Entomol.* **114**, 1113–1142.

Timell, T.E. (1957) Carbohydrate composition of ten Northern American species of wood. *Tappi* **40**, 568–572.

Timell, T. (1967) Recent progress in the chemistry of wood hemicelluloses. *Wood Sci. Technol.* **1**, 45–70.

Toews, D.A.A. and Moore, M.K. (1982) The effects of streamside logging on large organic debris in Carnation Creek. Land Manage. Rep. 11, Prov. British Columbia, Minist. For., Vancouver.

Tomlinson, G.H. (1973) Air pollutants and forest decline. *Environ. Sci. Technol.* **17**, 246A–256A.

Toole, E.R. (1969) Effect of decay on crushing strength. *For. Prod. J.* **19**, 36–37.

Toole, E.R. (1971) Interaction of mold and decay fungi on wood in laboratory tests. *Phytopathology* **61**, 124–125.

Totman, C.D. (1983) Logging the unloggable: Timber transport in early modern Japan. *J. For. Hist.* **27**, 180–191.

Townsend, C.H.T. (1886) Coleoptera found in dead trunks of *Tilia americana* L. in October. *Can. Entomol.* **18**, 65–68.

Triska, F.J. (1984) Role of wood debris in modifying channel geomorphology and riparian areas of a large lowland river under pristine conditions: A historical case study. *Verh. Int. Ver. Limnol.* **22**, 1876–1892.

Triska, F.J. and Cromack, K. (1980) The role of wood debris in forests and streams. In: *Forests: Fresh Perspectives from Ecosystem Analysis. Proceedings 40th Biology Colloquium (1979)* (Ed. by R.H. Waring), pp. 171–190. Oregon State Univ. Press, Corvallis.

Triska, F.J., Sedell, J.R. and Gregory, S.V. (1982) Coniferous forest streams. In: *Analysis of Coniferous Forest Ecosystems in the Western United States* (Ed. by R.L. Edmonds), pp. 292–332. US/IBP Synth. Ser. 14. Hutchinson Ross, Stroudsburg, Pennsylvania.

Triska, F.J., Sedell, J.R., Cromack, K., Gregory, S.V. and McCorison, F.M. (1984) Nitrogen budget for a small coniferous forest stream. *Ecol. Monogr.* **54**, 119–140.

Tritton, L.M. (1980) Dead wood in the northern hardwood forest ecosystem. Ph.D. dissertation, Yale Univ., New Haven, Connecticut.

Tritton, L.M. and Hornbeck, J.W. (1982) Biomass equations for major tree species of the Northeast. USDA For. Serv. Gen. Tech. Rep. NE-69. Northeast. For. Exp. Stn., Broomall, Pennsylvania.

Truszkowski, J. (1974) Utilization of nest boxes by rodents. *Acta Theriol.* **19**, 441–452.

Tschaplinski, P.J. and Hartman, G.F. (1983) Winter distribution of juvenile coho salmon (*Oncorhynchus kisutch*) before and after logging in Carnation Creek, British Columbia, and some implications for overwinter survival. *Can. J. Fish. Aquat. Sci.* **40**, 452–461.

Tukey, H.B. Jr. (1970) The leaching of substances from plants. *Annu. Rev. Ptant Physiol.* **21**, 305–324.

Ure, D.C. and Maser, C. (1982) Mycophagy of red-backed voles in Oregon and Washington. *Can. J. Zool.* **60**, 3307–3315.

U.S. Forest Service. (1965a) 1965 status report: Southern wood density survey. USDA For. Serv. Res. Pap. FPL-26. For. Prod. Lab., Madison, Wisconsin.

U.S. Forest Service. (1965b) Western wood density survey: Report number 1. USDA For. Serv. Res. Pap. FPL-27. For. Prod. Lab., Madison, Wisconsin.
U.S. Forest Products Laboratory. (1967) Comparative decay resistance of heartwood of native species. USDA For. Prod. Lab. Res. Note FPL-0153, Madison, Wisconsin.
U.S. Forest Products Laboratory. (1976) Wood handbook: Wood as an engineering material. USDA Agric. Handb. 72, Madison, Wisconsin.
Van Wagner, C.E. (1968) The line intersect method is forest fuel sampling. *For. Sci* **14**, 20–26.
Van Wagner, C.E. (1973) Height of crown scorch in forest fires. *Can. J. For. Res.* **3**, 373–378.
Von Haartman, L. (1957) Adaptation in hole-nesting birds. *Evolution* **11**, 339–347.
Wadsworth, F.H. and Englerth, G.H. (1959) Effects of the 1956 hurricane on forests in Puerto Rico. *Carrib. For.* **20**, 38–51.
Wahlgren, H.E. and Fassnacht, D.L. (1959) Estimating tree specific gravity from a single increment core. USDA For. Serv. Rep. 2146. For. Prod. Lab., Madison, Wisconsin.
Wahlgren, H.E., Bakar, G., Maeglin, R.R. and Hart, A.C. (1968) Survey of specific gravity of eight Maine conifers. USDA For. Serv. Res. Pap. FPL-95. For. Prod. Lab., Madison, Wisconsin.
Wallace, J.B. and Benke, A.C. (1984) Quantification of wood habitat in subtropical coastal plain streams. *Can. J. Fish. Aquat. Sci.* **41**, 1643–1652.
Wallwork, J.A. (1976) *The Distribution and Diversity of Soil Fauna.* Academic Press, London.
Ward, G.M., Cummins, K.W., Speaker, R.W., Ward, A.K., Gregory, S.V. and Dudley, T.L. (1982) Habitat and food resources for invertebrate communities in South Fork Hoh River, Olympic National Park. In: *Ecological Research in National Parks of the Pacific Northwest* (Ed. by E.E. Starkey, J.F. Franklin and J.W. Matthews), pp. 9–14. Oregon State Univ., For. Res. Lab., Corvallis.
Warren, W.G. and Olsen, P.F. (1964) A line intersect technique for measuring logging waste. *For. Sci.* **10**, 267–276.
Watling, R. (1982) Taxonomic status and ecological identity in the basidiomycetes. In: *Decomposer Basidiomycetes: Their Biology and Ecology* (Ed. by J.C. Frankland, J.N. Hedger and M.J. Swift), pp. 1–32. Cambridge Univ. Press, London and New York.
Weesner, F.M. (1960) Evolution and biology of the termites. *Annu. Rev. Entomol.* **5**, 153–170.
Weesner, F.M. (1970) Termites of the nearctic region. In: *Biology of Termites* (Ed. by K. Krishna and F.M. Weesner), Vol. 1, pp. 477–525. Academic Press, New York.
Wendler, H.O. and Deschamps, G. (1955) Logging dams on coastal Washington streams. Fish. Res. Pap. 1, pp. 27–38. Washington Dep. Fish., Olympia, Washington.
Wenzl, F.J. (1970) *The Chemical Technology of Wood.* Academic Press, New York.
Westveld, M. (1931) Reproduction on pulpwood lands in the Northeast. USDA Agric. Tech. Bull. 223, Washington Off., Washington, D.C.

Wetzel, R.M. (1958) Mammalian succession on midwestern floodplains. *Ecology* **39**, 262–271.

White, D.S. (1982) Elmidae. In: *Aquatic Insects and Oligochaetes of North and South Carolina* (Ed. by A.R. Brigham, W.U. Brigham and A. Gnilka), pp. 10.99–10.110. Midwest Aquat. Enterprises, Mahomel, Illinois.

White, P.S. (1979) Pattern, process and natural disturbance in vegetation. *Bot. Rev.* **45**, 229–299.

Whittaker, R.H., Likens, G.E., Bormann, F.H., Eaton, J.S. and Siccama, T.G. (1979) The Hubbard Brook ecosystem study: Forest nutrient cycling and element behavior. *Ecology* **60**, 203–220.

Wickman, B.E. (1978) Tree mortality and top-kill related to defoliation by the Douglas-fir tussock moth in the Blue Mountains outbreak. USDA For. Serv. Res. Pap. PNW-233. Pacific Northwest For. and Range Exp. Stn., Portland, Oregon.

Wieder, R.K. and Lang, G.E. (1982) A critique of the analytical methods used in examining decomposition data. *Ecology* **63**, 1636–1642.

Wiens, J.A. and Nussbaum, R.A. (1975) Model estimation of energy flow in northwestern coniferous forest bird communities. *Ecology* **56**, 547–561.

Wiggins, G.B. (1977) *Larvae of North American Caddisfly Genera (Trichoptera)*. Univ. of Toronto Press, Toronto, Ontario.

Wiggins, G.B. and Mackay, R.J. (1978) Some relationships between systematics and trophic ecology in nearctic aquatic insects, with special reference to Trichoptera. *Ecology* **59**, 1211–1220.

Wilcox, W.W. (1973) Degradation in relation to wood structure. In: *Wood Deterioration and Its Prevention by Preservative Treatments* (Ed. by D.D. Nicholas), pp. 107–147. Syracuse Univ. Press, Syracuse, New York.

Williams, A.B. (1936) The composition and dynamics of a beech–maple climax community. *Ecol. Monogr.* **6**, 318–408.

Willoughby, L.G. and Archer, J.F. (1973) The fungal spora of a freshwater stream and its colonization pattern on wood. *Freshwater Biol.* **3**, 219–240.

Wise, L.E. and Jahn, E.C. (Eds.) (1952) *Wood Chemistry*, 2nd Ed., Vol. 1. Van Nostrand-Reinhold, Princeton, New Jersey.

Wisseman, R.W. and Anderson, N.H. (1984) Mortality factors affecting Trichoptera eggs and pupae in an Oregon Coast Range watershed. In: *Proceedings of the 4th International Symposium on Trichoptera* (Ed. by J.C. Morse), pp. 455–460. Dr. W. Junk, The Hague.

Witkamp, M. (1966) Rates of carbon dioxide evolution from the forest floor. *Ecology* **47**, 492–494.

Witkamp, M. (1969) Cycles of temperature and carbon dioxide evolution from litter and soil. *Ecology* **50**, 922–924.

Wood, T.G. (1976) The role of termites (Isoptera) in decomposition processes. In: *The Role of Terrestrial and Aquatic Organisms in Decomposition Processes* (Ed. by J.M. Anderson and A. MacFadyen), pp. 245–168. Blackwell, Oxford.

Wood, T.G. (1978) Food and feeding habits of termites. In: *Production Ecology of Ants and Termites* (Ed. by M.V. Brian), pp. 55–80. International Biological Programme 13. Cambridge Univ. Press, London and New York.

Wood, T.G. and Sands, W.A. (1978) The role of termites in ecosystems. In: *Production Ecology of Ants and Termites* (Ed. by M.V. Brian), pp. 245–292. International Biological Programme 13. Cambridge Univ. Press, London and New York.

Woodwell, G.M., Whittaker, R.H. and Houghton, R.A. (1975) Nutrient concentrations in plants in the Brookhaven oak–pine forest. *Ecology* **56**, 318–332.

Wright, H.A. and Bailey, A.W. (1982) *Fire Ecology*. Wiiley (Interscience), New York.

Wright, K.H. and Harvey, G.H. (1967) The deterioration of beetle-killed Douglas-fir in western Oregon and Washington. USDA For. Serv. Res. Pap. PNW-50. Pacific Northwest For. and Range Exp. Stn., Portland, Oregon.

Wright, K.H. and Lauterbach, P.G. (1958) A 10-year study of mortality in a Douglas-fir sawtimber stand in Coos and Douglas Counties, Oregon. USDA For. Serv. Res. Pap. PNW-27. Pacific Northwest For. and Range Exp. Stn., Portland, Oregon.

Yavitt, J.B. and Fahey, T.J. (1982) Loss of mass and nutrient changes of decaying woody roots in lodgepole pine forests, southeastern Wyoming. *Can. J. For. Res.* **12**, 745–752.

Yoda, K., Kira, T. and Hozumi, K. (1963) Self thinning in overcrowded pure stands under cultivated and natural conditions. *J. Biol. Osaka Cy Univ.* **14**, 107–129.

Yoneda, T. (1975) Studies on the rate of decay of wood litter on the forest floor. Part 2: Dry weight loss and carbon dioxide evolution of decaying wood. *Jpn. J. Ecol.* **25**, 132–140.

Yoneda, T., Yoda, K. and Kira, T. (1977) Accumulation and, decomposition of big wood litter in Pasoh Forest. *West Malaysia. Jpn. J. Ecol.* **27**, 53–60.

Young, H.E. and Guinn, V.P. (1966) Chemical elements in complete mature trees of seven species in Maine. *Tappi* **49**, 190–197.

Zeikus, J.G. (1980) Fate of lignin and related aromatic substrates in anaerobic environments. In: *Lignin Biodegradation: Microbiology, Chemistry and Potential Applications* (Ed. by T.K. Kirk, T. Higuchi and H.M. Chang), Vol. 1, pp. 101–109. CRC Press, Boca Raton, Florida.

Zimmerman, R.C., Goodlett, J.C. and Comer, G.H. (1967) The influence of vegetation on channel form of small streams. In: *Symposium on River Morphology*. Int. Assoc. Sci. Hydrol. Publ. 75, pp. 255–275. Bern, West Germany.

Originally Published in Volume 18 (this series), pp 271–317, 1988

A Theory of Gradient Analysis

CAJO J.F. TER BRAAK AND I. COLIN PRENTICE

ADVANCES IN ECOLOGICAL RESEARCH VOL. 34
0065-2504/04 $35.00 DOI 10.1016/S0065-2504(03)34003-6

I. INTRODUCTION

All species occur in a characteristic, limited range of habitats; and within their range, they tend to be most abundant around their particular environmental optimum. The composition of biotic communities thus changes along environmental gradients. Successive species replacements occur as a function of variation in the environment, or (analogously) with successional time (Pickett, 1980; Peet and Loucks, 1977). The concept of niche space partitioning also implies the separation of species along "resource gradients" (Tilman, 1982). Gradients do not necessarily have physical reality as continua in either space or time, but are a useful abstraction for explaining the distributions of organisms in space and time (Austin, 1985). Austin's review explores the interrelationships between niche theory and the concepts of ecological continua and gradients.

Our review concerns data analysis techniques that assist the interpretation of community composition in terms of species' responses to environmental gradients in the broadest sense. Gradient analysis *sensu lato* includes direct gradient analysis, in which each species' abundance (or probability of occurrence) is described as a function of measured environmental variables; the converse of direct gradient analysis, whereby environmental values are inferred from the species composition of the community; and indirect gradient analysis, *sensu* Whittaker (1967), in which community samples are displayed along axes of variation in composition that can subsequently be interpreted in terms of environmental gradients. There are close relationships among these three types of analysis. Direct gradient analysis is a *regression* problem—fitting curves or surfaces to the relation between each species' abundance or probability of occurrence (the response variable) and one or more environmental variables (the predictor variable(s)) (Austin, 1971). Inferring environmental values from species composition when these relationships are known is a *calibration* problem. Indirect gradient analysis is an *ordination* problem, in which axes of variation are derived from the total community data. Ordination axes can be considered as latent variables, or hypothetical environmental variables, constructed in such a way as to optimize the fit of the species data to a particular (linear or unimodal) statistical model of how species abundance varies along gradients (Ter Braak, 1985, 1987a). These latent variables are constructed without reference to environmental measurements, but they can subsequently be compared with actual environmental data if available. To these three well-known types of gradient analysis we add a fourth, *constrained ordination*, which has its roots in the psychometric literature on multidimensional scaling (Bloxom, 1978; De Leeuw and Heiser, 1980; Heiser, 1981). Constrained ordination also constructs axes of variation in overall community composition, but does so in such a way as to explicitly optimize the fit to supplied environmental data (Ter Braak, 1986; Jongman *et al.*, 1987). Constrained ordination is thus a multivariate generalization of direct gradient analysis, combining aspects of regression, calibration and ordination. Table 1 gives an arbitrary selection of literature references, chosen simply to illustrate the wide range of ecological problems to which each of the four types of gradient analysis has been applied; the reader is also referred to Gauch (1982), who includes an extensive bibliography, and to Gittins (1985).

Standard statistical methods that assume linear relationships among variables exist for all four types of problems (regression, calibration, ordination and constrained ordination), but have found only limited application in ecology because of the generally non-linear, non-monotone response of species to environmental variables. Ecologists have independently developed a variety of alternative techniques. Many of these techniques are essentially heuristic, and have a less secure theoretical basis. These heuristic techniques can nevertheless give useful results, and can be understood as approximate solutions to statistical problems similar to those solved by standard methods, but formulated in terms of a *unimodal* (Gaussian or similar) response model instead of a linear one. We present here a theory of gradient analysis, in which the heuristic techniques are integrated with regression, calibration, ordination and constrained ordination as distinct, well-defined statistical problems.

The various techniques used for each type of problem are classified into families according to their implicit response model and the method used to estimate parameters of the model. We consider three such families (Table 2). First we treat the family of standard statistical techniques based on the linear response model, because these are conceptually the simplest and provide a basis for what follows, even though their ecological application is restricted. Second, we outline a family of somewhat more complex statistical techniques which are formal extensions of the standard linear techniques and incorporate unimodal (Gaussian-like) response models explicitly. Finally, we consider the family of heuristic techniques based on weighted averaging. These are not more complex than the standard linear techniques, but implicitly fit a simple unimodal response model rather than a linear one. Our treatment thus unites such apparently disparate data analysis techniques as linear regression, principal components analysis, redundancy analysis, Gaussian ordination, weighted averaging, reciprocal averaging, detrended correspondence analysis and canonical correspondence analysis in a single theoretical framework.

Table 1 Selected applications of gradient analysis

Type of problem	Taxa	Environmental variables	Purpose of study
Regression			
Alderdice (1972)	Marine fish	Salinity, temperature	Defining ranges
Peet (1978)	Trees	Elevation, moisture, latitude	Biogeography
Wiens and Rotenberry (1981)	Birds	Vegetation structure	Niche characterization
Austin *et al.* (1984)	*Eucalyptus* spp.	Climatic indices	Habitat characterization
Bartlein *et al.* (1986)	Plant pollen types	Temperature, precipitation	Quaternary palaeoecology
Calibration			
Chandler (1970)	Benthic macro-invertebrates	Water pollution	Water quality management
Imbrie and Kipp (1971)	Foraminifera	Sea surface temperature	Palaeoclimatic reconstruction
Sládecek (1973)	Freshwater algae	Organic pollution	Ecological monitoring
Balloch *et al.* (1976)	Benthic macro-invertebrates	Water pollution	Ecological monitoring
Ellenberg (1979)	Terrestrial plants	Soil moisture, N, pH	Bioassay from vegetation
van Dam *et al.* (1981)	Diatoms	pH	Acid rain effects
Böcker *et al.* (1983)	Terrestrial plants	Soil moisture, N, pH	Bioassay from vegetation
Bartlein *et al.* (1984)	Plant pollen types	Temperature, precipitation	Palaeoclimatic reconstruction
Battarbee (1984)	Diatoms	pH	Acid rain effects
Charles (1985)	Diatoms	pH	Acid rain effects
Atkinson *et al.* (1986)	Beetles	Summer temperature, annual range	Palaeoclimatic reconstruction

Ordination[a]			
van der Aart and Smeenk-Enserink (1975)	Spiders	Microenvironmental features	Habitat characterization
Kooijman and Hengeveld (1979)	Beetles	Lutum content, elevation	Habitat characterization
Wiens and Rotenberry (1981)	Birds	Vegetation structure	Niche characterization
Prodon and Lebreton (1981)	Birds	Vegetation structure	Niche characterization
Kalkhoven and Opdam (1984)	Birds	Habitat and ladscape features	Habitat characterization
Macdonald and Ritchie (1986)	Plant pollen types	Vegetation regions	Quaternary palaeoecology
Constrained ordination			
Webb and Bryson (1972)	Plant pollen types	Climate variables, airmass frequencies	Palaeoclimatic reconstruction
Gasse and Tekaia (1983)	Diatoms	pH classes	Palaeolimnology
Ås (1985)	Beetles	Vegetation types	Niche theory
Cramer and Hytteborn (1987)	Terrestrial plants	Time, elevation	Land uplift effects
Purata (1986)	Tropical trees	Successional boundary conditions	Study of secondary succession
Fängström and Willén (1987)	Phytoplankton	Physical/chemical variables	Environmental monitoring

[a] Excluding vegetation studies, where ordination is used routinely: see Gauch (1982) for a review.

Table 2 Classification of gradient analysis techniques by type of problem, response model and method of estimation

Type of problem	Linear response model	Unimodal response model	
	Least-squares, estimation	Maximum likelihood estimation	Weighted averaging estimation
Regression	Multiple regression	Gaussian regression	Weighted averaging of site scores (WA)
Calibration	Linear calibration; "inverse regression"	Gaussian calibration	Weighted averaging of species' scores (WA)
Ordination	Principal components analysis (PCA)	Gaussian ordination	Correspondence analysis (CA); detrended correspondence analysis (DCA)
Constrained ordination[a]	Redundancy analysis (RDA)[d]	Gaussian canonical ordination	Canonical correspondence analysis (CCA); detrended CCA
Partial ordination[b]	Partial components analysis	Partial Gaussian ordination	Partial correspondence analysis; partial DCA
Partial constrained ordination[c]	Partial redundancy analysis	Partial Gaussian canonical ordination	Partial canonical correspondence analysis; partial detrended CCA

[a] Constrained multivariate regression.
[b] Ordination after regression on covariables.
[c] Constrained ordination after regression on covariables = constrained partial multivariate regression.
[d] "Reduced-rank regression" = "PCA of y with respect to x".

II. LINEAR METHODS

Species abundances may seem to change linearly through *short* sections of environmental gradients, so a linear response model may be a reasonable basis for analysing quantitative abundance data spanning a narrow range of environmental variation.

A. Regression

If a plot of the abundance (y) of a species against an environmental variable (x) looks linear, or can easily be transformed to linearity, then it is appropriate to fit a straight line by linear regression. The formula $y = a + bx$ describes the linear relation, with a the intercept of the line on the y-axis and b the slope of the line, or regression coefficient. Separate regressions can be, carried out for each of m species.

We are usually most interested in how the abundance of each species changes with a change in the environmental variable, i.e. in the slopes b_k (the index k refers to species k). If we first centre the data—by subtracting the mean of each species' abundances from the species data and the mean of the environmental values from the environmental data—the intercept disappears. Then if y_{ki} denotes the centred abundance of species k in the ith out of n sites, and x_i, the centred environmental value for that site, the response model for fitting the straight lines becomes

$$y_{ki} = b_k x_i + e_{ki} \tag{1}$$

where e_{ki} is an error component with zero mean and variance v_{ki}. The standard estimator for the slope in Eq. (1) is

$$\tilde{b}_k = \sum_{i=1}^{n} y_{ki} x_i / s_x^2 \tag{2}$$

where $s_x^2 = \Sigma_{i=1}^{n} x_i^2$. This is the least-squares estimator, which is the best linear unbiased estimator when errors are uncorrelated and homogeneous across sites ($v_{ki} = v_k$). It is also the maximum likelihood (ML) estimator when the errors are normally distributed. The fitted lines can be used to predict the abundances of species in a site with a known value of the environmental variable simply by reading off the graph.

Species experience the effect of more than one environmental variable simultaneously, so more than one variable may be required to account for variation in species abundances. The joint effect of two or more environmental variables on a species can be analysed by multiple regression (see e.g. Montgomery and Peck, 1982). Standard computer packages are available to obtain least-squares (ML) estimates for the regression coefficients. Only when the environmental variables are uncorrelated will the partial regression coefficients be identical to the coefficients estimated by separate regressions using Eq. (1).

B. Calibration

We now turn to the inverse problem, calibration. When the relationship between the abundances of species and the environmental variable we are interested in is known, we can infer values of that environmental variable for new sites from the observed species abundances. If we took into account the abundance of only a single species, we could simply read off the graph, starting from a value on the vertical axis. However, another species may well give a different estimate. We therefore need a good and unambiguous estimator that combines the information from all m species. In terms of Eq. (1), the b_k are now assumed to be known and x_i is unknown. The role of the b_k and x_i have been interchanged. By interchanging their roles in Eq. (2) as well, we obtain

$$\tilde{x}_i = \sum_{k=1}^{m} y_{ki} b_k / s_b^2 \tag{3}$$

where $s_b^2 = \Sigma_{k=1}^{m} b_k^2$. This is the least-squares estimator (and the ML-estimator) when the errors follow a normal distribution and are independent and homogeneous across species ($v_{ki} = v_i$).

A problem with Eq. (3) is that these conditions are likely to be unrealistic, because effects of other environmental variables can cause correlation between the abundances of different species even after the effects of the environmental variable of interest have been removed. Further, the residual variance v_{ki} may be different for different species. If this occurs, we also need to take the residual correlations and variances into account. In practice, the residual correlations and variances are estimated from the residuals of the regressions used for estimating the b_k's. Searching for the maximum of the likelihood with respect to x_i then leads to a general weighted least-squares problem (Brown, 1979; Brown, 1982) that can be solved by using standard algorithms.

Inferring values of more than one environmental variable simultaneously has been given surprisingly little attention in the literature. However, Williams (1959) and Brown (1982) derived the necessary formulae from the ML-principle (Cox and Hinkley, 1974).

C. Ordination

After having fitted a particular environmental variable to the species data by regression, we might ask whether another environmental variable would provide a better fit. For some species one variable may fit better, and for other species another variable. To get an overall impression we might judge the goodness-of-fit (explanatory power) of an environmental variable by the total regression sum of squares (Jongman *et al.*, 1987). The question then arises: what is the best possible fit that is theoretically obtainable with the straight line model of Eq. (1)?

This question defines an ordination problem, i.e. to construct the single "hypothetical environmental variable" that gives the best fit to the species data according to Eq. (1). This hypothetical environmental variable is termed the *latent variable*, or simply the (first) ordination axis. Principal components analysis (PCA) provides the solution to this ordination problem. In Eq. (1), x_i is then the score of site i on the latent variable, b_k is the slope for species k with respect to the latent variable (also called the species loading or species score) and the eigenvalue of the first PCA axis is equal to the goodness-of-fit, i.e. the total sum of squares of the regressions of the species abundances on the latent variable. PCA provides the least-squares estimates of the site and species scores: these estimates are also ML estimates if the errors are independently and normally distributed with constant variance ($v_{ki} = v$).

PCA is usually performed using a standard computer package, but several different algorithms can be used to do the same job. The following algorithm, known as the power method (Gourlay and Watson, 1973), makes the relationship between PCA and regression and calibration clear in a way that the usual textbook treatment, in terms of singular value decomposition of inner product matrices, does not; it also facilitates comparison with correspondence analysis, which we discuss later. The power method shows that PCA can be obtained by an alternating sequence of linear regressions and calibrations:

Step 1 Start with some (arbitrary) initial site scores $\{x_i\}$ with zero mean.
Step 2 Calculate new species scores $\{b_k\}$ by linear regression (Eq. (2)).
Step 3 Calculate new site scores $\{x_i\}$ by linear calibration (Eq. (3)).
Step 4 Remove the arbitrariness in scale by standardizing the site scores as follows: new x_i = old $x_i/n/s_x$, with s_x as defined beneath Eq. (2).
Step 5 Stop on convergence, i.e. when the newly obtained site scores are close to the site scores of the previous cycle of iteration, else go to Step 2.

The final scores do not depend on the initial scores.

The ordination problem for a two-dimensional linear model turns out to be relatively simple, compared with the regression and calibration problems. The solution does not need an alternating sequence of *multiple* regressions and calibrations, because the latent variables can always be chosen in such a way that they are uncorrelated; and if the latent variables are uncorrelated, then the multiple regressions and calibrations reduce to a series of separate linear regressions and calibrations. PCA provides the solution to the linear ordination problem in any number of dimensions; one latent variable is derived first, as in the one-dimensional case of Eq. (1), and the second latent variable can be obtained next by applying the same algorithm again but with one extra step—after Step 3, the trial scores are made uncorrelated with the first latent variable. On denoting the scores of the first axis by x_{i1}, this orthogonalization is computed by

Step 3b Calculate $f = \sum_i x_i x_{i1}/n$,
Calculate new $x_i = \text{old } x_i - f x_{i1}$.

Further latent variables (ordination axes) may be derived analogously. As in the one-dimensional case, PCA provides the ML-solution to the multi-dimensional linear ordination problem if the errors are independently and normally distributed with constant variance across species and sites. Jolliffe (1986) reviews the theory and applications of PCA.

D. The Environmental Interpretation of Ordination Axes (Indirect Gradient Analysis)

In indirect gradient analysis the species data are first subjected to ordination, e.g. using PCA, to find a few major axes of variation (latent variables) with a good fit to the species data. These axes are then interpreted in terms of known variation in the environment, often by using graphical methods (Gauch, 1982). A more formal method for the latter step would be to calculate correlation coefficients between environmental variables and each of the ordination axes. This analysis is similar to performing a multiple regression of each separate environmental variable on the axes (Dargie, 1984), because the axes are uncorrelated. A joint analysis of all environmental variables can be carried out by multiple regression of each ordination axis on the environmental variables:

$$x_i = c_0 + \sum_{j=1}^{q} c_j z_{ji} \tag{4}$$

in which x_i is the score of site i on that one ordination axis, z_{ij} denotes the value at site i of the jth out of q actual environmental variables, and c_j is the corresponding regression coefficient. For later reference, the error term in Eq. (4) is not shown. The multiple correlation coefficient R measures how well the environmental variables explain the ordination axis.

E. Constrained Ordination (Multivariate Direct Gradient Analysis)

Indirect gradient analysis, as outlined above, is a *two-step* approach to relate species data to environmental variables. A few ordination axes that summarize the overall community variation are extracted in the first step; then in the second step one may calculate weighted sums of the environmental variables that most closely fit each of these ordination axes. However, the environmental variables that have been studied may turn out to be poorly related to the first few ordination axes, yet may be strongly related to other, "residual" directions of variation in species composition. Unless the first few ordination axes explain a very high proportion of the variation, this residual variation can be substantial, and strong relationships between species and environment can potentially be missed.

In constrained ordination this approach is made more powerful by combining the two steps into one. The idea of constrained ordination is to search for a few weighted sums of environmental variables that fit the data of all species best, i.e. that give the maximum total regression sum of squares. The resulting technique, redundancy analysis (Rao, 1964; van den Wollenberg, 1977), is an ordination of the species data in which the axes are constrained to be linear combinations of the environmental variables. These axes can be found by extending the algorithm of PCA described above with one extra step, to be performed directly after Step 3 (Jongman *et al.*, 1987):

Step 3a Calculate a multiple regression of the site scores $\{x_i\}$ on the environmental variables (Eq. (4)), and take as new site scores the fitted values of this regression.

The regression is thus carried out within the iteration algorithm, instead of afterwards. On convergence, the coefficients $\{c_j\}$ are termed canonical coefficients and the multiple correlation coefficient in Step 3a can be called the species-environment correlation.

Redundancy analysis is also known as reduced-rank regression (Davies and Tso, 1982), PCA of y with respect to x (Robert and Escoufier, 1976) and two-block mode C partial least-squares (Wold, 1982). It is intermediate between PCA and separate multiple regressions for each of the species: it is a constrained ordination, but it is also a constrained form of (multivariate) multiple regression (Davies and Tso, 1982; Israëls, 1984). By inserting Eq. (4) into Eq. (1), it can be shown that the "regression" coefficient of species k with respect to environmental variable j takes the simple form $b_k c_j$. With two ordination axes this form would be, in obvious notation, $b_{k1}c_{j1} + b_{k2}c_{j2}$. With two ordination axes, redundancy analysis thus uses $2(q+m)+m$ parameters to describe the species data, whereas the multiple regressions use $m(q+1)$ parameters. One of the attractive features of redundancy analysis is that it leads to an ordination diagram that simultaneously displays (i) the main pattern of community variation as far as this variation can be explained by the environmental variables, and (ii) the main pattern in the correlation coefficients between the species and each of the environmental variables. We give an example of such a diagram later on.

Redundancy analysis is much less well known than canonical correlation analysis (Gittins, 1985; Tso, 1981), which is the standard linear multivariate technique for relating two sets of variables (in our case, the set of species and the set of environmental variables). Canonical correlation analysis is very similar to redundancy analysis, but differs from it in the assumptions about the error component: uncorrelated errors with equal variance in redundancy analysis and correlated normal errors in canonical correlation analysis (Tso, 1981; Jongman *et al.*, 1987). The most important practical difference is that redundancy analysis can analyse any number of species whereas in canonical correlation analysis the number of species (m) must be less than $n-q$ (Griffins, 1985: 24); this restriction is often a nuisance.

Canonical variates analysis, or multiple discriminant analysis, is simply the special case of canonical correlation analysis in which the "environmental" variables are a series of dummy variables reflecting a single-factor classification of the samples. A similar restriction on the number of species thus also applies to canonical variates analysis. Redundancy analysis with dummy variables provides an alternative to canonical variates analysis, evading this restriction.

III. NON-LINEAR (GAUSSIAN) METHODS

A. Unimodal Response Models

Linear methods are appropriate to community analysis only when the species data are quantitative abundances (with few zeroes) and the range of environmental variation in the sample set is narrow. Alternative analytical methods can be derived from unimodal models.

A unimodal response model for one environmental variable can be obtained by adding a quadratic term (x_i^2) to the linear model, changing the response curve from a straight line into a parabola. But this quadratic model can predict large negative values, whereas species abundances are always zero or positive. A simple remedy for the problem of negative values is provided by the Gaussian response curve (Gauch and Whittaker, 1972) in which the *logarithm* of species abundance is a quadratic in the environmental variable:

$$\log y = b_0 + b_1 x + b_2 x^2$$
$$= a - \frac{1}{2}(x - u)^2/t^2 \tag{5a}$$

where $b_2 < 0$ (otherwise the curve would have a minimum instead of a mode). The coefficients b_0, b_1, and b_2 are most easily interpreted by transformation to u, t and a (Figure 1), u being the species' optimum (the value of x at the peak), t being its tolerance (a measure of response breadth or ecological amplitude), and a being a coefficient related to the height of the peak (Ter Braak and Looman, 1986).

A closely related model can describe species data in presence–absence form. In analysing presence–absence data, we want to relate probability of *occurrence* (p) to environment. Probabilities are never greater than 1, so rather than using Eq. (5a) we use the Gaussian logit model,

$$\log\left(\frac{p}{1-p}\right) = b_0 + b_1 x + b_2 x^2 \tag{5b}$$

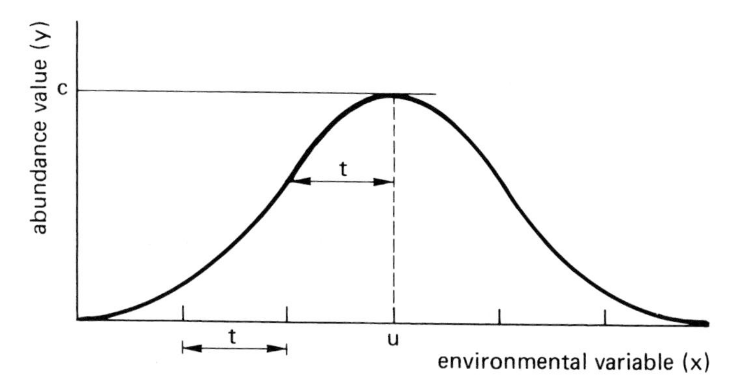

Figure 1 A Gaussian curve displays a unimodal relation between the abundance value (y) of a species and an environmental variable (x). (u = optimum or mode; t = tolerance; c = maximum = exp (a)).

which is very similar to the Gaussian model unless the peak probability is high (> 0.5); then Eq. (5b) gives a curve that is somewhat flatter on top. The coefficients b_0, b_1, and b_2 can be transformed as before into coefficients representing the species' optimum, tolerance and maximum probability value.

Although real ecological response curves are still more complex than implied by the Gaussian and Gaussian logit models, these models are nevertheless useful in developing statistical descriptive techniques for data showing mostly unimodal responses, just as linear models are useful in statistical analysis of data that are only approximately linear.

With two environmental variables, Eqs. (5a) and (5b) become full quadratic forms with both square and product terms (Alderdice, 1972). For example, the Gaussian model becomes

$$\log y = b_0 + b_1 x_1 + b_2 x_1^2 + b_3 x_2 + b_4 x_2^2 + b_5 x_1 x_2 \tag{6}$$

If $b_2 + b_4 < 0$, and $4b_2 b_4 - b_5^2 > 0$ then Eq. (6) describes a unimodal surface with ellipsoidal contours (Figure 2). If one of these conditions is not satisfied then Eq. (6) describes a surface with a minimum, or with a saddle point (e.g. Davison, 1983). Provided the surface is unimodal, its optimum (u_1, u_2) can be calculated from the coefficients in Eq. (6) by

$$\left.\begin{aligned} u_1 &= (b_5 b_3 - 2b_1 b_4)/d \\ u_2 &= (b_5 b_1 - 2b_3 b_2)/d \end{aligned}\right\} \tag{7}$$

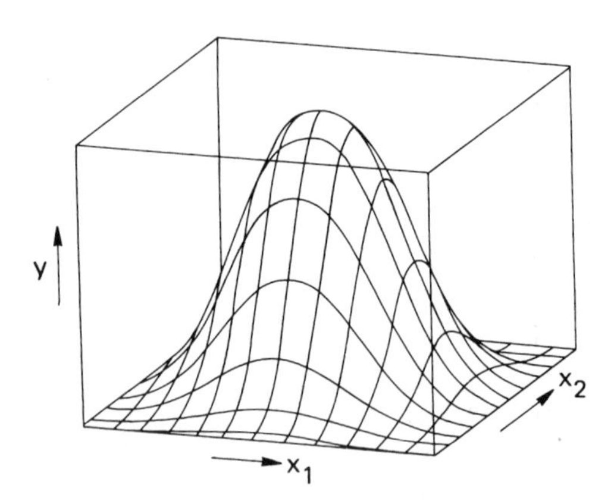

Figure 2 A Gaussian surface displays a unimodal relation between the abundance value (y) of a species and two environmental variables (x_1 and x_2).

where $d = 4b_2b_4 - b_5^2$. When $b_5 \neq 0$, the optimum with respect to x_1, depends on the value of x_2; the environmental variables are then said to show interaction in their effect on the species. In contrast, when $b_5 = 0$ the optimum with respect to x_1, does not depend on the value of x_2 (no interaction) and Eq. (7) simplifies considerably (Ter Braak and Looman, 1986).

The unknown parameters of non-linear response models in the context of regression, calibration or ordination can (at least in theory) be estimated by the maximum likelihood principle, however difficult this may be in a particular situation. Usually iterative methods are required, and initial parameter values must be specified. The likelihood function may have local maxima, so that different sets of initial parameter values may result in different final estimates. It cannot be guaranteed that the global maximum has been found. Furthermore, all kinds of numerical problems may occur. However, the special cases of Gaussian and Gaussian logit response models do allow reasonably practical solutions, which we consider now.

B. Regression

The regression problems of fitting Gaussian or Gaussian logit curves or surfaces are relatively straightforward, since these models can be fitted by Generalized Linear Modelling (GLM: McCullagh and Nelder, 1983; Dobson, 1983). An elementary introduction to GLM directed at ecologists is provided by Jongman *et al.* (1987). GLM is more flexible than ordinary multiple regression because one can specify "link functions" and error distributions other than the normal distribution. For example, the Gaussian models of Eqs. (5a) and (6) can be fitted with GLM to abundance data (which may include zeroes) by specifying the link function to be logarithmic and the error distribution to be Poissonian. The corresponding Gaussian logit models can be fitted with GLM to presence–absence data by specifying the link function to be logistic and the error distribution to be binomial-with-total-1. Alternatively, any statistical package that will do logit (= logistic) regression can be used to fit the Gaussian logic model. No initial estimates are needed and local maxima do not arise, so these techniques are quite practical for direct gradient analysis. For examples of the use of GLM in ecology see Austin and Cunningham (1981) and Austin *et al.* (1984).

The most common complications arise when the optimum for a species is estimated well outside the sampled range of environments, or if the fitted curve shows a minimum rather than a peak. These conditions suggest that the regression is ill-determined and that it might be better to fit a monotone curve by setting $b_2 = 0$ in Eq. (5); a statistical test can be used to determine whether this simplification is acceptable (Jongman *et al.*, 1987). Such cases are bound to arise in practice because any given set of samples will include some species that are near the edge of their range.

C. Calibration

The calibration problem of inferring environmental values at sites from species data and known Gaussian (logit) curves by ML is feasible by numerical optimization, but no computer programs are available at present that are easy to use (Jongman *et al.*, 1987). Local maxima may occur in the likelihood, when the tolerances of the species are unequal, and one needs to specify an initial estimate. The assumption of independence of species responses is required, but might not be tenable in practice; it remains to be studied how important this assumption is. Dependency among species could most obviously be caused by the effects of additional, unconsidered environmental variables, in which case the best remedy would be to identify these variables and include them in the analysis. Inferring the values of more than one environmental variable simultaneously on the basis of several Gaussian (logit) response surfaces is also possible in principle, but has not been done as far as we know.

D. Ordination

Ordination based on Gaussian (logit) curves aims to construct a latent variable such that these curves optimally fit the species data. This problem involves the ML estimation of site scores $\{x_k\}$ and the species' optima $\{u_k\}$, tolerances $\{t_k\}$ and maxima $\{a_k\}$, usually by an alternating sequence of Gaussian (logit) regressions and calibrations. This kind of ordination has been investigated by Gauch *et al.* (1974), Kooijman (1977), Kooijman and Hengeveld (1979), Goodall and Johnson (1982) and Ihm and Van Groenewoud (1975, 1984). The numerical methods required are computationally demanding, and in the general case, when the tolerances of the species are allowed to differ, the likelihood function typically contains many local maxima.

Kooijman (1977) and Goodall and Johnson (1982) reported numerical problems in their attempts to perform ML ordination using two-dimensional Gaussian-like models. A simple model with circular contours ($b_2 = b_4$ and $b_5 = 0$) may be amenable in practice, especially if b_2 is not allowed to vary among species (Kooijman, 1977). This model is equivalent to the "unfolding model" used by psychologists to analyse preference data (Coombs, 1964; Heiser, 1981; Davison, 1983; DeSarbo and Rao, 1984).

But with more than two latent variables the Gaussian (logit) model with a second-degree polynomial as linear predictor contains so many parameters that it is likely to be difficult to get reliable estimates of them, even if all the interaction terms are dropped.

E. Constrained Ordination

The constrained ordination problem for Gaussian-like response models is to construct ordination axes that are also linear combinations of the environmental variables, such that Gaussian (logit) surfaces with respect to these axes optimally fit the data. As in redundancy analysis (Section II.E), the joint effects of the environmental variables on the species are "channelled" through a few ordination axes which can be considered as composite environmental gradients influencing species composition. Ter Braak (1986) refers to this approach as Gaussian canonical ordination, the word canonical being chosen by analogy with canonical correlation analysis. The estimation problem is actually simpler than in unconstrained Gaussian ordination, and is more easily soluble in practice because the number of parameters to be estimated is smaller: instead of n site scores one has to estimate q canonical coefficients. Meulman and Heiser (1984) have applied similar ideas in the context of non-metric multidimensional scaling. Gaussian canonical ordination can also be viewed as multivariate Gaussian regression with constraints on the coefficients of the polynomial (Ter Braak, 1988). In multivariate Gaussian regression each species has its own optimum in the q-dimensional space formed by the environmental variables; the constraints imposed in Gaussian canonical ordination amount to a requirement that these optima lie in a low-dimensional subspace. If the optima lie close to a plane then the most important species–environment relationships can be depicted graphically in an ordination diagram.

IV. WEIGHTED AVERAGING METHODS

Ecologists have developed alternative, heuristic methods that are simpler but have essentially the same aims as the methods of the previous section based on Gaussian-type models. Each method in the Gaussian family has a counterpart in the family of heuristic methods based on weighted averaging (WA). These methods have been used extensively, and even re-invented in different branches of ecology.

A. Regression

WA can be used to estimate species' optima with respect to known environmental variables. When a species shows a unimodal relationship with environmental variables, the species' presences will be concentrated around the peak of this function. One intuitively reasonable estimate of the optimum is the average of the values of the environmental variable over those sites in which the species is present. With abundance data, WA applies weights proportional to species abundance; absences still carry zero weight. The estimate of the optimum for species k is thus

$$\tilde{u}_k = \sum_{i=1}^{n} y_{ki} x_i / y_{k+} \tag{8}$$

where y_{ki} is from now onwards the abundance (*not* centred) or presence/absence (1/0) of species k at site i, y_{k+} is the species total ($y_{k+} = \sum_i y_{ki}$) and x_i is the value of the environmental variable at site i. As a follow-up to an investigation of the theoretical properties of this estimator (Ter Braak and Barendregt, 1986), Ter Braak and Looman (1986) showed by simulation of presence–absence data that WA estimates the optimum of a Gaussian iogit curve as efficiently as the ML technique of Gaussian logic regression provided:

Condition *1a* The site scores $\{x_i\}$ are equally spaced over the whole range of occurrence of the species along the environmental variable.

WA also proved to be only a little less efficient whenever the distribution of the environmental variable among the sites was reasonably homogeneous (rather than strictly equally spaced) over the whole range of species occurrences, or more generally for species with narrow ecological amplitudes. But the estimate of the optimum of a rare species may be imprecise, because the standard error of the estimate is inversely proportional to the square root of the number of occurrences. So for efficiency, we also need

Condition 1b The site scores $\{x_i\}$ are closely spaced in comparison with the species' tolerance.

B. Calibration

WA is also used in calibration, to estimate environmental values at sites from species' optima—which in this context are often called indicator values

("Zeigerwerte", Ellenberg, 1979) or scores (Whittaker, 1956). When species replace one another along the environmental variable of interest, i.e. have unimodal response functions with optima spread out along that variable, then species with optima close to the environmental value of a site will naturally tend to be represented at that site. Intuitively, to estimate the environmental value at a site, one can average the optima of the species that are present. With abundance data, the corresponding intuitive estimate is the weighted average,

$$\tilde{x}_i = \sum_{k=1}^{m} y_{ki} u_k / y_{+i} \tag{9}$$

where y_{+i} is the site total ($y_{+i} = \sum_k y_{ki}$).

Ter Braak and Barendregt (1986) showed that WA estimates the value x_i of a site as well as the corresponding ML techniques if the species show Gaussian curves and Poisson-distributed abundance values (or, for presence–absence data, show Gaussian logit curves), and provided:

Condition 2a The species' optima are equally spaced along the environmental variable over an interval that extends for a sufficient distance in both directions from the true value x_i;
Condition 3 The species have equal tolerances;
Condition 4 The species have equal maximum values.

These conditions amount to a "species packing model" wherein the species have equal response breadth and equal spacing (Whittaker *et al.*, 1973). The conditions may be relaxed somewhat (Ter Braak and Barendregt, 1986) without seriously affecting the efficiency of the WA-estimate. When the optima are uniformly distributed instead of being equally spaced, the efficiency is still high if the maximum probabilities of occurrence are small (< 0.5). The species' maximum values may differ, but they must not show a trend along the environmental variable (for instance, leading to species-rich samples at one end of the gradient and species-poor samples at the other end). The efficiency of WA is less good if the tolerances substantially differ among species; a tolerance weighted version of WA, as suggested by Zelinka and Marvan (1961) and Goff and Cottam (1967), would be more efficient since it would give greater weight to species of narrower tolerance, which are more informative about the environment.

Under Conditions 2a–4 above, the standard error of the estimate of $\tilde{x}_i$ is approximately $t/\sqrt{y_{+i}}$, where t is the (common) species tolerance. For the weighted average to be practically useful, the number of species encountered in a site should therefore not be too small (not less than five). We therefore need the extra condition (cf. Section 5 in Ter Braak and Barendregt, 1986):

Condition 2b The species' optima must be closely spaced in comparison with their tolerances.

An alternative heuristic method of calibration is by "inverse regression". This is simply multiple linear regression of the environmental variable on the species abundances (Brown, 1982): the environmental variable is treated as if it were the response variable and the species abundances, possibly transformed, as predictor variables. The regression coefficients can be estimated from the training set of species abundances and environmental data, the resulting equations being applied directly to infer environmental values from further species abundance data. When applied to data on percentage composition, e.g. pollen spectra or diatom assemblages (Bartlein *et al.*, 1984; Charles, 1985), the method differs from WA calibration only in the way in which the species optima are estimated, since the linear combination of percentage values used to estimate the environmental value is by definition a weighted average of the regression coefficients.

C. Ordination

Hill (1973) turned weighted averaging into an ordination technique by applying alternating WA regressions and calibrations to a species-by-site data table. The algorithm of this technique of "reciprocal averaging" is similar to that given earlier for PCA:

Step 1 Start with arbitrary, but unequal, initial site scores $\{x_i\}$.
Step 2 Calculate new species scores $\{u_k\}$ by WA (Eq. (8)).
Step 3 Calculate new site scores $\{x_i\}$ by WA (Eq. (9)).
Step 4 Remove the arbitrariness in scale by standardizing the site scores by new $x_i = \{\text{old } x_i - z\}/s$ where $z = \sum_i y_{+i} x_i / \sum_i y_{+i}$ and

$$s^2 = \sum_i y_{+i}(x_i - z)^2 \Big/ \sum_i y_{+i} \tag{10}$$

Step 5 Stop on convergence, else go to Step 2.

As in PCA, the resulting site and species scores do not depend on the initial scores. The final scores produced by this reciprocal averaging algorithm form the first eigenvector or ordination axis of correspondence analysis (CA), an eigenvector technique that is widely used especially in the French-language literature (Laurec *et al.*, 1979; Hill, 1974). As with the power algorithm for PCA, the reciprocal averaging algorithm makes clear the relationship between CA and regression and calibration—this time, with WA regression and calibration. The method of standardization chosen in Step 4 is arbitrary, but chosen for later reference. On convergence, s in Step 4 is equal to the eigenvalue of the first axis, and lies between 0 and 1.

Correspondence analysis has many applications outside ecology. Nishisato (1980), Greenacre (1984) and Gifi (1981) provide a variety of different rationales for correspondence analysis, each adapted to a particular type of application. Heiser (1987) and Ter Braak (1985, 1987a) develop rationales for correspondence analysis that are particularly relevant to ecological applications.

Ter Braak (1985) showed that CA approximates ML Gaussian (logit) ordination under Conditions 1–4 listed above, i.e. under just these conditions for which WA is as good as ML-regression and ML-calibration. In practice CA can never be exactly equivalent to ML ordination, because Condition 1a implies that the range of site scores is broad enough to include the ranges of all of the species, whereas Condition 2a implies that there must be species with their optima situated beyond the edge of the range of site scores. These conditions cannot both be satisfied if the range of site scores is finite. As a result, CA shows an edge effect: the site scores near the ends of the axes become compressed relative to those in the middle (Gauch, 1982). This effect becomes less strong, however, as the range of site scores becomes wider and the spacing of the site scores and species scores becomes closer relative to the average species' tolerance.

Conditions 1–4 also disallow "deviant" sites and rare species. CA is sensitive to both (Hill, 1974; Feoli and Feoli Chiapella, 1980; Oksanen, 1983). This sensitivity may be useful in some applications, but is a nuisance if the aim is to detect major gradients. Deviant sites (and, possibly, the rarest species) should therefore ideally be removed from the data before analysis by CA.

As in PCA, further ordination axes can be extracted in CA by adding an extra step after Step 3, making the trial scores on the second axis uncorrelated with the (final) scores on the first axis. (In the calculation of f in Step 3b (see Section II.C) the sites are weighted proportional to the site total y_{+i}. This weighting is implicitly applied from now on.) However, there is a problem with the second and higher axes in CA. The problem is the well-known but hitherto not well-understood "arch effect" (Hill, 1974). If the species data come from an underlying one-dimensional Gaussian model the scores on the second ordination axis show a parabolic ("arch") relation with those of the first axis; if the species data come from a two-dimensional Gaussian model in which the true site and species scores are located homogeneously in a rectangular region in two-dimensional space (the extension to two dimensions of Conditions 1a and 2a), the scores of the second ordination axis lie not in a rectangle but in an arched band (Hill and

Gauch, 1980). The arch effect arises because the axes are extracted sequentially in order of decreasing "variance". Suppose CA has succeeded in constructing a first axis, such that species appear one after the other along that axis as in a species packing model. Then a possible second axis is obtained by folding the first axis in the middle and bringing the ends together. This axis is a superposition of two species packing models, each with half the gradient length of the first axis. It is a candidate for becoming the second axis, because it has *no linear correlation* with the first CA-axis yet has as much as half the gradient length of the first axis (Jongman *et al.*, 1987). The folded axis by itself thus "explains" a part of the variation in the species data, even though when taken jointly with the first axis it contributes nothing. Even if there is a strong second gradient, CA will not associate it with the second axis if it separates the species less than a folded first axis. As a result of the arch effect, the two-dimensional CA-solution is generally not a good approximation to the ML-solution (two-dimensional Gaussian ordination).

Hill and Gauch (1980) developed detrended correspondence analysis (DCA) as a heuristic modification of CA designed to remedy both the edge effect and the arch effect. The edge effect is removed in DCA by non-linear rescaling of the axis. Assuming a species packing model with randomly distributed species' optima, Hill and Gauch (1980) noted that the variance of the optima of the species present at a site (the "within-site variance") is an estimate of the average response curve breadth of those species (they used the standard deviation as a measure of breadth, which is about equal to tolerance as we define it). Because of the edge effect, the species' curves before rescaling are narrower near the ends of the axis than in the middle, and the within-site variance is correspondingly smaller in sites near the ends of the axis than in sites in the middle. The rescaling therefore attempts to equalize the within-site variance at all points along the ordination axis by dividing the axis into small segments, expanding the segments with sites with small within-site variance, and contracting the segments with sites with large within-site variance. The site scores are then calculated as weighted averages of the species scores and the scores are standardized such that the within-site variance is equal to 1.

Hill and Gauch (1980) defined the length of the ordination axis to be the range of the site scores. This length is expressed in "standard-deviation units" (SD). The tolerance of the species' curves along the rescaled axis are close to 1, and each curve therefore rises and falls over about 4 SD. Sites that differ by 4 SD can thus be expected to have no species in common. Even if non-linear resealing is not used, one can still set the average within-site variance of the species scores along a CA-axis equal to 1 by linear rescaling (Hill, 1979; Ter Braak, 1987b), so as to ensure that this useful interpretation of the length of the axis still approximately holds.

The arch effect, a more serious problem in CA, is removed in DCA by the heuristic method of "detrending-by-segments". This method ensures that at any point along the first ordination axis, the mean value of the site scores on subsequent axes is approximately zero. In order to achieve this, the first axis is divided into a number of segments and the trial site scores are adjusted within each segment by subtracting their mean after some smoothing across segments. Detrending-by-segments is built into the reciprocal averaging algorithm, and replaces Step 3b. Subsequent axes are derived similarly by detrending with respect to each of the existing axes.

DCA often works remarkably well in practice (Hill and Gauch, 1980; Gauch *et al.*, 1981). It has been critically evaluated in several recent simulation studies. Ter Braak (1985) showed that DCA gave a much closer approximation to ML Gaussian ordination than CA did, when applied to simulated data based on a two-dimensional species packing model in which species have identically shaped Gaussian surfaces and the optima and site scores are uniformly distributed in a rectangle. This improvement was shown to be mainly due to the detrending, not to the non-linear rescaling of axes. Kenkel and Orlóci (1986) found that DCA performed substantially better than CA when the two major gradients differed in length, but also noted that DCA sometimes "collapsed and distorted" CA results when there were: (a) few species per site, and (b) the gradients were long (we believe (a) to be the real cause of the collapse). Minchin (1987) further found that DCA can flatten out some of the variation associated with one of the underlying gradients. He ascribed this loss of information to an instability in the detrending-by-segments method. Pielou (1984, p. 197) warned that DCA is "overzealous" in correcting the "defects" in CA, and "may sometimes lead to the unwitting destruction of ecologically meaningful information". Minchin's (1987) results indicate some of the conditions under which such loss of information can occur.

DCA is popular among practical field ecologists, presumably because it provides an effective approximate solution to the ordination problem for a unimodal response model in two or more dimensions—given that the data are reasonably representative of sections of the major underlying environmental gradients. Two modifications might increase its robustness with respect to the problems identified by Minchin (1987). First, non-linear. rescaling aggravates these problems; since the edge effect is not too serious, we advise against the routine use of non-linear rescaling. Second, the arch effect needs to be removed (as Heiser, 1987, also noted), but this can be done by a more stable, less "zealous" method of detrending which was also briefly mentioned by Hill and Gauch (1980): namely detrending-by-polynomials. Under the one-dimensional Gaussian model, it can be shown that the second CA-axis is a quadratic function of the first axis, the third axis is a cubic function of the first axis, and so on (Hill, 1974; Iwatsubo, 1984).

Detrending-by-polynomials can be incorporated into the reciprocal averaging algorithm by extending Step 3b such that the trial scores are not only made uncorrelated with the previous axes, but are also made uncorrelated with polynomials of the previous axes. The limited experience so far suggests that detrending up to fourth-order polynomials should be adequate. In contrast with detrending-by-segments, the method of detrending-by-polynomials removes only specific defects of CA that are now theoretically understood.

D. Constrained Ordination

Just as CA/DCA is an approximation to ML Gaussian ordination, so is canonical correspondence analysis (CCA) an approximation to ML Gaussian canonical ordination (Ter Braak, 1986). CCA is a modification of CA in which the ordination axes are restricted to be weighted sums of the environmental variables, as in Eq. (4). CCA can be obtained from CA as redundancy analysis was obtained from PCA. An algorithm can be obtained by adding to the CA algorithm an extra multiple regression step. The only difference from Step 3a of redundancy analysis (see Section II.E) is that the sites must be weighted in the regression proportional to their site total y_{+i} (Ter Braak, 1986). CCA can also be obtained as the solution of an eigenvalue problem (Ter Braak, 1986). It is closely related to "redundancy analysis for qualitative variables" (Israëls, 1984) but has a different rationale and is applied to a different type of data.

In constrained ordination the constraints always become less strict as more environmental variables are included. If $q \geqslant n-1$, then there are no real constraints, and CA and CCA become equivalent. As in CA, the edge effect in CCA is a minor problem that is best left untreated. Detrending may sometimes be required to remove the arch effect, i.e. to prevent CCA from selecting weighted sums of environmental variables that are approximately polynomials of previous axes. Detrending-by-segments does not work very well here for technical reasons; detrending-by-polynomials is better founded and more appropriate (see Appendix and Ter Braak, 1987b). However, the arch effect in CCA can be eliminated much more elegantly, simply by dropping superfluous environmental variables (Ter Braak, 1987a). Variables that are highly correlated with the "arched" axis (often the second axis) are the most likely to be superfluous. If the number of environmental variables is small enough for the relationship of individual variables to the ordination axes to be significant, the arch effect is not likely to occur at all.

CCA can be sensitive to deviant sites, but only when they are outliers with regard to both species composition and environment. When realistically few environmental variables are included, CCA is thus more robust than CA in this respect too.

CCA leads to an ordination diagram that simultaneously displays (a) the main patterns of community variations, as far as these reflect environmental variation, and (b) the main pattern in the weighted averages (not correlations as in redundancy analysis) of each of the species with respect to the environmental variables (Ter Braak, 1986, 1987a). CCA is thus intermediate between CA and separate WA calculations for each species. Geometrically, the separate WA calculations give each species a point in the q-dimensional space of the environmental variables, which indicates the centre of the species' distribution. CCA attempts to provide a low-dimensional representation of these centres; CCA is thus also a constrained form of WA, in which the weighted averages are restricted to lie in a low-dimensional subspace.

Like redundancy analysis, CCA can be used with dummy "environmental" variables to provide an ordination constrained to show maximum separation among pre-defined groups of samples. This special case of CCA is described, for example, by Feoli and Orlóci (1979) under the name of "analysis of concentration", by Greenacre (1984, Section 7.1) and by Gasse and Tekaia (1983).

V. ORDINATION DIAGRAMS AND THEIR INTERPRETATION

The linear ordination techniques (PCA and redundancy analysis) and, the ordination techniques based on WA (CA/DCA and CCA) represent community data in substantially different ways. We focus on two-dimensional ordination diagrams, as these are the easiest to construct and to inspect, and illustrate the interpretation of each type of diagram with an example.

A. Principal Components: Biplots

PCA fits planes to each species' abundances in the space defined by the ordination axes. The species' point (b_{k1}, b_{k2}) may be connected with the origin (0,0) to give an arrow (Figure 3). Such a diagram, in which sites are marked by points and species by arrows is called a "biplot" (Gabriel, 1971). There is a useful symbolism in this use of arrows: the arrow points in the direction of maximum variation in the species' abundance, and its length is proportional to this maximum rate of change. Consequently, species on the edge of the diagram (far from the origin) are the most important for

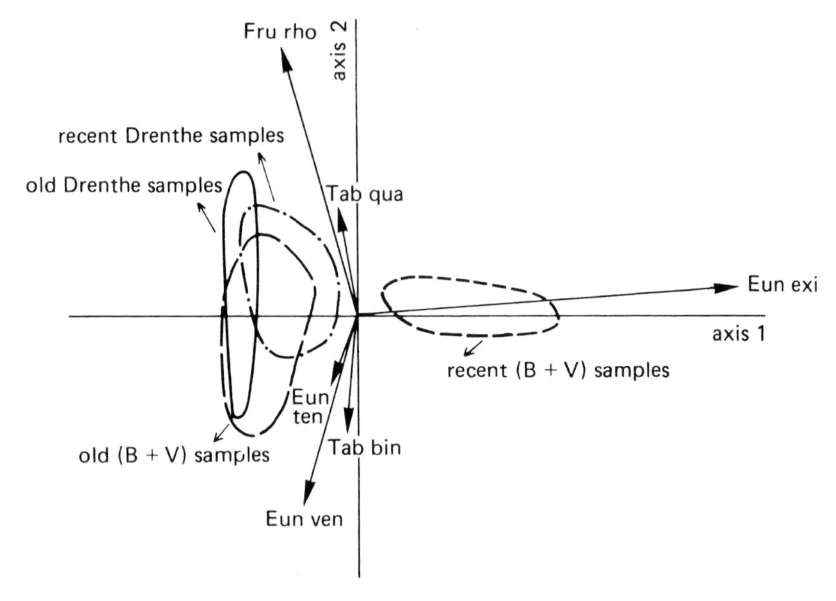

Figure 3 Biplot based on principal components analysis of diatom assemblages from Dutch moorland pools (schematic after van Dam *et al.*, 1981). The arrows for the six most frequent species and the regions where different categories of samples lie jointly display the approximate community composition in each of the regions (old, *c.* 1920; recent, 1978; B + V, from the province of Brabant and the Veluwe). Abbreviations: Eun exi, *Eunotia exigua*; Eun ten, *Eunotia tenella*; Eun ven, *Eunotia veneris*; Fru rho, *Frustulia rhomboides* var. *saxonica*; Tab bin, *Tabellaria binalis*; Tab qua, *Tabellaria quadriseptara*.

indicating site differences; species near the centre are of minor importance. Ter Braak (1983) provides more detailed, quantitative rules for interpreting PCA ordination diagrams.

van Dam *et al.* (1981) applied PCA to data consisting of diatom assemblages from 16 Dutch moorland pools, sampled in the 1920s and again in 1978, to investigate the impact of acidification on these shallow water bodies. Ten clearwater (non-humic) pools were situated in the province of Brabant and on the Veluwe and six brownwater (humic) pools in the province of Drenthe. Figure 3 displays the major variation in the data. The arrow of *Eunotia exigua* indicates that this species increases strongly along the first principal component: *E. exigua* is abundant in the recent Brabant and Veluwe samples, which lie on the right-hand side of the diagram, and rare in the remaining samples, which lie more to the left. The second axis accounts for some of the difference among the old and recent samples from Drenthe. These groups differ in the abundances of *Frustulia rhomboides* var. *saxonica*, *Tabellaria quadriseptata*, *Eunotia tenella*, *Tabellaria binalis*, and *Eunotia veneris*, as shown by the directions of the arrows for these species in Figure 3. As *E. exigua* is acidobiontic and the first principal component is strongly correlated with the sulphate concentration of the 1978 samples, this component clearly depicts the impact of acidification of the moorland pools in Brabant and the Veluwe (and to a smaller extent also in Drenthe). Thus van Dam *et al.* (1981) used PCA to summarize the changes in diatom composition between the 1920s and 1978. PCA helped them to detect that the nature of the change differed among provinces, hence stressing the importance for diatoms of the distinction between clearwater and brownwater pools.

B. Correspondence Analysis: Joint Plots

In CA and DCA both sites and species are represented by points, and each site is located at the centre of gravity of the species that occur there. One may therefore get an idea of the species composition at a particular site by looking at "nearby" species points. Also, in so far as DCA approximates the fitting of Gaussian (logit) surfaces (Figure 2), the species points are approximately the optima of these surfaces; hence the abundance or probability of occurrence of a species tends to decrease with distance from its location in the diagram.

Figure 4 illustrates this interpretation of the species' points as optima in ordination space. DCA was applied to presence–absence data on 51 bird species in 526 contiguous, 100 m × 100 m grid-cells in an area with pastures and scattered woodlots in the Rhine valley near Amerongen, the Netherlands (Opdam *et al.*, 1984). Figure 4 shows the DCA scores of the 20 most frequent species by small circles, and the outline (dashed) of the region in which the scores for the grid-cells fall (the individual grid-cells are not shown, to avoid crowding). Opdam *et al.* (1984) interpreted the first axis, of length 5.6 SD, as a gradient from open to closed landscape and the second axis, of length 5.3 SD, as a gradient from wet to drier habitats.

In order to test the interpretation of species' scores as optima, we fitted a response surface for each species by logit regression using Eq. (6) with the first and the second DCA-axes as the predictor variables x_1 and x_2. For 13 of the 20 bird species, the fitted surface had a maximum. The optimum was calculated for each of these species by Eq. (7) and plotted as a triangle in Figure 4. The fitted optima lie close to the DCA scores. The regression analysis also allowed us to estimate species' tolerances in ordination space: these are indicated in Figure 4 by ellipses representing the region within which each species occurs with at least half of its maximum probability, according to the fitted surface.

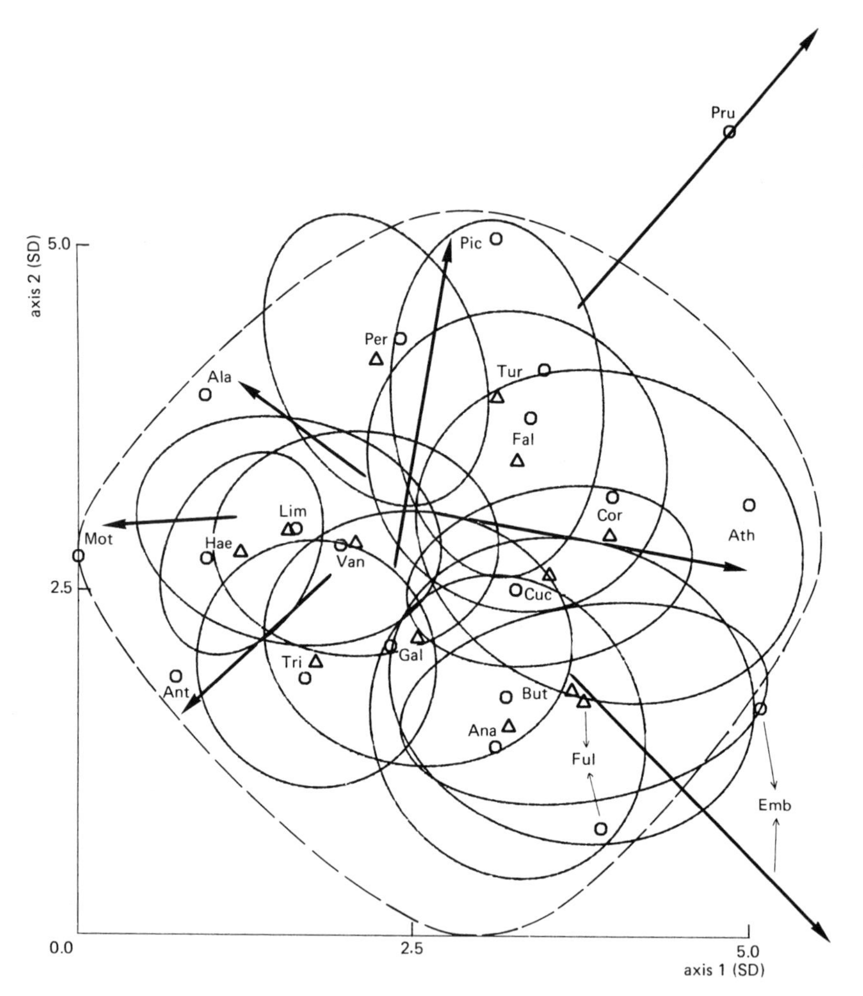

Figure 4 Joint plot based on detrended correspondence analysis (DCA) of bird species communities in the Rhine valley near Amerongen, the Netherlands (data from Opdam *et al.*, 1984), displaying the major variation in bird species composition across the landscape. This plot shows the DCA-scores (○) of the 20 most frequent species and the region in which the samples fall (— — —). Also shown are optima (Δ) and lines of equal probability for the 13 species whose probability surfaces had clear maxima (as fitted by Gaussian logit regression), and arrows representing directions of increase for the seven species whose probability surfaces were monotonic. Abbreviations: Ala, *Alauda arvensis*; Ana, *Anas platyrhynchos*; Ant, *Anthus pratensis*; Ath, *Athene noctua*; But, *Buteo buteo*; Cor, *Corvus corone*; Cuc, *Cuculus canorus*; Emb, *Emberiza schoeniclus*; Fal, *Fatco tinnunculus*; Ful, *Fulica atra*; Gal, *Gallinago gallinago*; Hae, *Haematopus ostralegus*; Lim, *Limosa limosa*; Mot, *Motacitla flava flava*; Per, *Perdix perdix*; Pic, *Pica pica*; Pru, *Prunella modularis*; Tri, *Tringa totanus*; Tur, *Turdus merula*; Van, *Vanellus vanellus*.

The fitted surfaces for the remaining seven species had a minimum or saddle point, suggesting that their optima are located well outside the sampled range. For these species we fitted a "linear" logit surface by setting b_2, b_4 and b_5 in Eq. (6) to zero. The direction of steepest increase of each of the fitted surfaces is indicated in Figure 4 by an arrow through the centroid of the site points; the beginning and end points of each arrow correspond to fitted probabilities of 0.1 and 0.9 respectively. As expected from our interpretation of DCA, these arrows point more or less in the same direction as the DCA scores of the corresponding species (Figure 4).

In contrast to the PCA-diagram, the species points on the edge of the CA- or DCA-diagram are often rare species, lying there either because they prefer extreme (environmental) conditions, or (very often) because their few occurrences by chance happen to fall in sites with extreme conditions; one cannot decide between these possibilities without additional data. Such peripheral species have little influence on the analysis and it is often convenient not to display them at all. Furthermore, species near the centre of the diagram may be ubiquitous, unrelated to the ordination axes, bimodal, or in some other way not fitting a unimodal response model—or they may be genuinely specific with a habitat-optimum near the centre of the sampled range of habitats. The correct interpretation may be found by the kind of secondary analysis shown in Figure 4, or more straightforwardly just by plotting the species' abundances in ordination space.

C. Redundancy Analysis

In redundancy analysis sites are indicated by points, and both species and environmental variables are indicated by arrows whose interpretation is similar to that of the arrows in the PCA biplot. The pattern of abundance of each species among the sites can be inferred in exactly the same way as in a PCA biplot, and so may the direction of variation of each environmental variable. One may also get an idea of the correlations between species' abundances and environmental variables. Arrows pointing in roughly the same direction indicate a high positive correlation, arrows crossing at right angles indicate near-zero correlation, and arrows pointing in opposite directions indicate high negative correlation. Species and environmental variables with long arrows are the most important in the analysis; the longer the arrows, the more confident one can be about the inferred correlation. (It is assumed here that for the purpose of the ordination diagram the environmental variables have been standardized to zero mean and unit variance, so as to make the lengths of arrows comparable.) Jongman *et al.* (1987) provide more quantitative rules for interpreting the ordination diagrams derived from redundancy analysis.

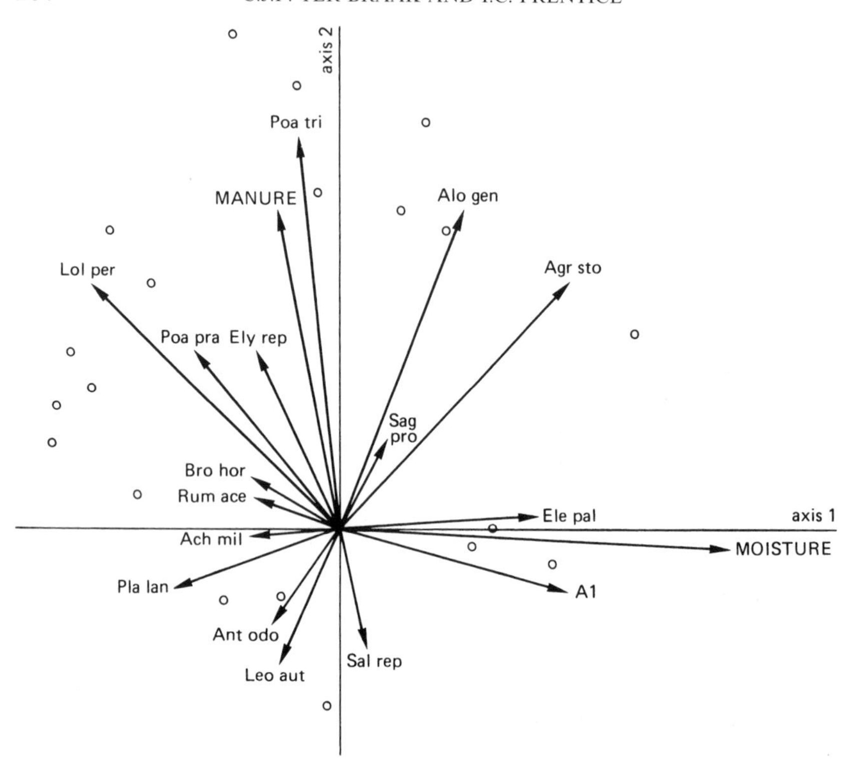

Figure 5 Biplot based on redundancy analysis of vegetation with respect to three environmental variables (quantity of manure, soil moisture and thickness of the A1 horizon) in dune meadows (○) on the island of Terschelling, The Netherlands. The arrows for plant species and environmental variables display the approximate (linear) correlation coefficients between plant species and the environmental variables. Abbreviations: Ach mil, *Achillea millefolium*; Agr sto, *Agrostis stolonifera*; Alo gen, *Alopecurus geniculatus*; Ant odo, *Anthoxanthum odoratum*; Bro hor, *Bromus hordaceus*; Ele pal, *Eleocharis palustris*; Ely rep, *Elymus repens*; Leo aut, *Leontodon autumnalis*; Loi per, *Lolium perenne*; Pla lan, *Plantago lanceolata*; Poa pra, *Poa pratensis*; Poa tri, *Poa trivialis*; Rum ace, *Rumex acetosa*; Sag pro, *Sagina procumbens*; Sal rep, *Salix repens*.

The data we use to illustrate redundancy analysis were collected to study the relation between the vegetation and management of dune meadows on the island of Terschelling, The Netherlands (M. Batterink and G. Wijffels, unpublished). Figure 5 displays the main variation in the vegetation in relation to three environmental variables (thickness of the A1 horizon, moisture content of the soil and quantity of manuring). The arrows for *Poa trivialis* and *Elymus repens* make small angles with the arrow for manuring;

these species are inferred to be positively correlated with manuring. *Salix repens* and *Leontodon autumnalis* have arrows pointing in directions roughly opposite to that of manuring, and are inferred to be negatively correlated with manuring. The former species are thus most abundant in the heavily manured meadows of standard farms (positioned at the top of the diagram), whereas the latter species are most abundant in the unmanured meadows (owned by the nature conservancy and positioned at the bottom of the diagram). The relationships of the species with moisture and thickness of the A1 horizon can be inferred in a similar way. The short arrows for *Bromus hordaceus* and *Sagina procumbens*, for example, indicate that their abundance is not so much affected by moisture, manure and thickness of the A1 horizon. Redundancy analysis can summarize the species–environment relationships in such an informative way, because the gradients are short ($\approx$2SD: Ter Braak, 1987b).

D. Canonical Correspondence Analysis

In CCA, since species are assumed to have unimodal response surfaces with respect to linear combinations of the environmental variables, the species are logically represented by points (corresponding to their approximate optima in the two-dimensional environmental subspace), and the environmental variables by arrows indicating their direction and rate of change through the subspace.

Purata (1986, and unpublished results) applied CCA to plant species abundance data from 40 abandoned cultivation sites within Mexican tropical rain forest. Data were available for 24 of these sites on the regrowth age (A), the length of the cropping period in the past (C), and the proportion of the perimeter that had remained forested (F). These three variables were used as environmental variables in CCA. The remaining 16 sites were entered as "passive" sites, to be positioned with respect to the CCA axes according to their floristic composition in relation to the "active" sites.

Figure 6 illustrates the results. The first axis, with length 4.7 SD, was interpreted as an indicator of the general trend of secondary succession. The direction of the arrow for regrowth age shows that this trend runs broadly from right to left. The species' locations are consistent with their life-history characteristics: the trend of succession runs from ruderals (to the right), through pioneer shrubs and trees, to late-secondary canopy dominants and shade-tolerant understorey species (to the left). The directions of the other two arrows in relation to axis 1 show that a long cropping period delays succession, while an extensive forested perimeter accelerates succession. Axis 2 (3.0 SD) may (more speculatively) differentiate species whose

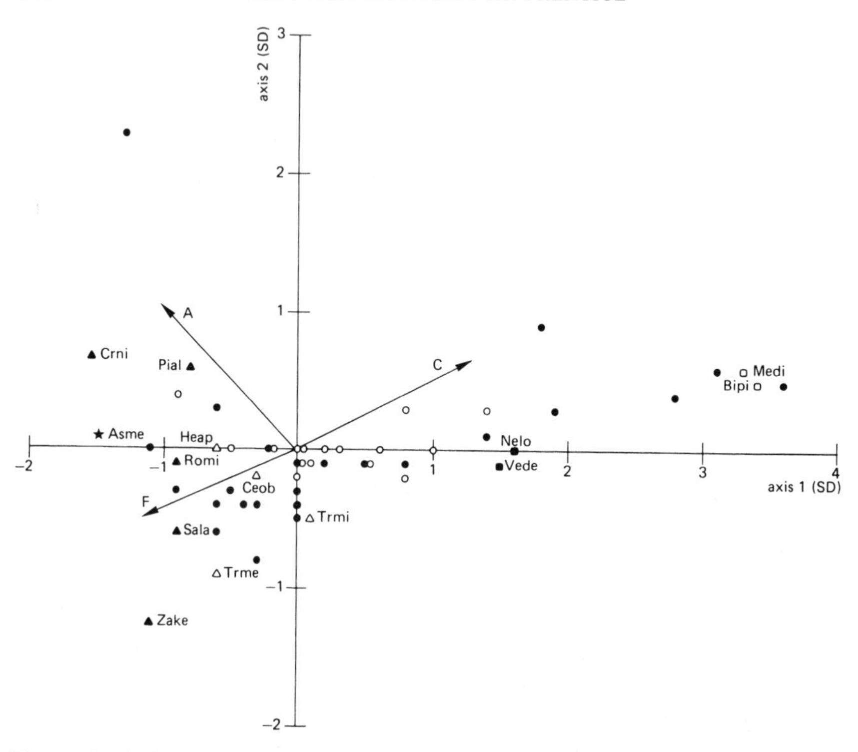

Figure 6 Ordination diagram based on canonical correspondence analysis of successional plant communities with respect to three environmental variables (regrowth age A, length of cropping period C, and extent of forested perimeter F) on abandoned cultivation sites within Mexican tropical rain forest (Purata, 1986 and unpublished). ●, sites with environmental data; ○, sites added "passively" on the basis of floristic composition. The species shown are a selection among the 285 included in the analysis. □, denotes ruderals; ■, pioneer shrubs; △, pioneer trees; ▲, late-secondary canopy trees; and ★, an understorey palm. Abbreviations: Asme, *Astrocaryum mexicanum*; Bipi, *Bidens pilosa*; Ceob, *Cecropia obtusifolia*; Crni, *Croton nitens*; Heap, *Heliocarpus appendicalatus*; Medi, *Melampodium divaricatum*; Nelo, *Neurolaena lobata*; Pial, *Piper amalago*; Romi, *Robinsonella mirandae*; Sala, *Sapium lateriflorum*; Trme, *Trichospermum mexicanum*; Trmi, *Trema micrantha*; Vede, *Vernonia deppeana*; Zake, *Zanthoxylum kellermanii*.

establishment is favoured by the presence of mature forest around the site from those that simply require a long time to grow.

CCA also allows the computation of unconstrained, "residual" axes summarizing floristic variation that remains after the effect of the environmental variables has been taken out. In Purata's study, the successive eigenvalues of the first three (constrained) CCA axes were 0.49, 0.34 and 0.18. (There cannot be more constrained axes than there are environmental variables). The first residual axis gave an eigenvalue of 0.74, showing that at least as much floristic variation was *not* explained by the environmental variables. In our experience, terrestrial community data commonly give a residual eigenvalue as large as the first constrained eigenvalue, however carefully the environmental variables are chosen. Thus DCA and CCA tend to give different ordinations, and CCA is more powerful in detecting relationships between species composition and environment.

VI. CHOOSING THE METHODS

A. Which Response Model?

Regression methods can fit response models with a wide variety of shapes. The linear and Gaussian-like models are convenient starting points; more complex shapes can be fitted by adding further parameters, if the data are sufficiently detailed to support it. Other species may be used as additional explanatory variables if the specific aim is to detect species interactions (Fresco, 1982). The shapes of the response functions may be made even more general by applying Box–Cox transformations to the explanatory variables (Bartlein *et al.*, 1986) or still more general by fitting splines (Smith, 1979). Even with all these modifications, regression can still be done with standard packages for Generalized Linear Modelling.

After species response curves or surfaces have been fitted by regression, calibration based on the maximum likelihood principle can be used to make inferences about the environment from community data. If the surfaces fitted by regression have complex shapes, then calibration by *numerical* maximization of the likelihood may be problematic. But even then, if there are only a few environmental variables involved, the "most likely" combination of environmental values can be searched for on a grid across the environmental space (Atkinson *et al.*, 1986; Bartlein *et al.*, 1986). So the type of response model used in both regression and calibration should generally be guided by the characteristics and resolution of the data, and inspection of the data and the residuals after regression should show whether the model being used is adequate for the purpose.

In contrast to regression and calibration, the ordination problem requires the simultaneous estimation of large numbers of parameters and cannot be solved practically without some constraints on the structure one wants to fit. That these constraints may seem unduly restrictive simply shows that there are limits to what ordination can achieve. The number of ordination axes to be extracted must be small, and the type of response model must be restricted, in order to permit a solution. For example, it seems necessary to disregard the possibility of bimodal species distributions (Hill, 1977). Certainly bimodal distributions sometimes occur, but ordination has to assume that species "on average" have simple distributions—otherwise, the problem would be insoluble; the utility of ordination techniques depends on them being robust with respect to departures from the simple models on which they are based. The Gaussian model seems to be of the right order of complexity for ordination of ecological data, but the full second-degree model of Eq. (6) is already difficult to fit (Kooijman, 1977; Goodall and Johnson, 1982). The Gaussian model with circular contour lines and equal species tolerances, i.e. the unfolding model, might provide a good compromise between practical solubility and realism in ordination. Promising algorithms for unfolding are developed by Heiser (1987) and DeSarbo and Rao (1984). DCA provides a reasonably robust approximation to ML Gaussian ordination and requires far less computing time. Similarly, ML Gaussian canonical ordination is technically feasible, but CCA provides a practical and robust approximation to it.

Non-linear methods are appropriate if a reasonable number of species have their optima located within the data set. If the gradient length is reduced to less than about 3 SD, the approximations involved in WA become worse and ultimately (if the gradient length is less than about 1.5 SD) the methods yield poor results because most species are behaving monotonically over the observed range. Thus if the community variation is within a narrow range, the linear ordination methods—PCA and redundancy analysis—are appropriate. If the community variation is over a wider range, non-linear ordination methods—including DCA and CCA—are appropriate.

B. Direct or Indirect?

Direct gradient analysis allows one to study the part (large or smart) of the variation in community composition that can be explained by a particular set of environmental variables. In indirect gradient analysis attention is first focused on the major pattern of variation in community composition; the environmental basis of this pattern is to be established later. If the relevant environmental data are to hand, the direct approach—either fitting separate response surfaces by regression for each major species, or analysing the overall patterns of the species–environment relationship by constrained ordination—is likely to be more effective than the traditional indirect approach. However, indirect gradient analysis does have the advantage that no prior hypothesis is needed about what environmental variables are relevant. One does not need to measure the environmental variables in advance, and one can use informal field knowledge to help interpret the patterns that emerge—hence the emphasis in the literature on ordination as a technique for "hypothesis generation", the implication being that experimental or more explicit statistical approaches can be used for subsequent hypothesis testing. This distinction is not hard and fast, but it does draw attention to the strengths and limitations of indirect gradient analysis.

In Section V.D, we showed in passing how an indirect gradient analysis can be carried out *after* a direct gradient analysis in order to summarize the community variation that remains after known effects have been removed. When the known environmental variables are not the prime object of study, they are called concomitant variables (Davies and Tso, 1982) or covariables. It would be convenient to solve for the residual (unconstrained) axes without having to extract all the constrained axes first. Fortunately, this is straightforward. In the iterative algorithm for PCA and CA, one simply extends Step 3b such that the trial scores are not only made uncorrelated with any previous axis (if present) but are also made uncorrelated with all specified covariables (see Appendix for details). In this way the effects of the covariables are partialled out from the ordination; hence the name "partial ordination". The theory of "partial components analysis" and "partial correspondence analysis", as we call these extensions of PCA and CA, is given by Gabriel (1978, theorem 3) and Ter Braak (1988), respectively. Swaine and Greig-Smith (1980) used partial components analysis to obtain an ordination of within-plot vegetation change in permanent plots. Partial correspondence analysis, or its detrended form, would be more appropriate if the gradients were long.

C. Direct Gradient Analysis: Regression or Constrained Ordination?

Whether to use constrained ordination (multivariate direct gradient analysis) instead of a series of separate regressions (the traditional type of direct gradient analysis) depends on whether or not there is any advantage in analysing all the species simultaneously. Both constrained and unconstrained ordination assume that the species react to the *same* composite gradients of environmental variables, while in regression a separate composite gradient is constructed for each species. Regression can therefore allow more detailed descriptions and more accurate prediction and calibration, if properly carried out (with due regard to its statistical assumptions) and if sufficient data are available. However, ecological data that are collected over a large range of habitat variation require non-linear models, and building good non-linear models by regression is demanding in time and computation. In CCA the composite gradients are linear combinations of environmental variables and the non-linearity enters through a unimodal response model with respect to a few composite gradients, taken care of in CCA by the procedure of weighted averaging. Constrained ordination is thus easier to apply, and requires less data, than regression; it provides a summary of the species–environment relationship, and we find it most useful for the exploratory analysis of large data sets.

Constrained ordination can also be carried out *after* regression, in order to relate the residual variation to other environmental variables. This type of analysis, called "partial constrained ordination", is useful when the explanatory (environmental) variables can be subdivided in two sets, a set of covariables—the effects of which are not the prime object of study—and a further set of environmental variables whose effects are of particular interest.

For example, in the illustration of Section V.C, the study was initiated to investigate differences in vegetation among dune meadows that were exploited under different management regimes (standard farming, biodynamical farming, nature management, among others). Standard CCA showed systematic differences in vegetation among management regimes. A further question is then whether these differences can be fully accounted for by the environmental variables moisture, quantity of manure and thickness of the A1 horizon, whose effects are displayed in Figure 5, or whether the variation that remains after fitting the three environmental variables (three constrained ordination axes) is systematically related to management regimes. This question can be tackled using partial constrained ordination, with the three environmental variables as covariables, and a series of dummy variables (for each of the management regimes) as the variables of interest.

Technically, partial constrained ordination can be carried out by any computer program for constrained ordination. The usual environmental variables are replaced by the residuals obtained by regressing each of the variables of interest on the covariables (see Appendix). Davies and Tso (1982) gave the theory behind partial redundancy analysis; Ter Braak (1988) derived partial canonical correspondence analysis as an approximation to "partial Gaussian canonical ordination".

Partial constrained ordination has the same essential aim as Carleton's (1984) residual ordination, i.e. to determine the variation in the species data that is uniquely attributable to a particular set of environmental variables, taking into account the effects of other (co-) variables; however, Carleton's method is somewhat less powerful, being based on a pre-existing DCA which may already have removed some of the variation of interest. Partial constrained ordination is, by contrast, a true direct gradient analysis technique which seems promising, e.g. for the analysis of permanent plot data (effects of time, with location and/or environmental data as covariables), and a variety of other applications in which effects of particular environmental variables are to be sorted out from the "background" variation imposed by other variables.

VII. CONCLUSIONS

Regression, calibration, ordination and constrained ordination are well-defined statistical problems with close interrelationships. Regression is the tool for investigating the nature of individual species' response to environment, and calibration is the tool for (later) inferring the environment from species composition at an individual site. Both tools come in various degrees of complexity. The simplest are linear and WA regression and calibration. The linear methods are applicable over short ranges of environment, where species' abundance appears to vary monotonically with variation in the environment. The WA methods are applicable over wider ranges of environment; WA regression is a crude method to estimate each species' optimum, and WA calibration just averages the optima of the species that are present. WA works with presence–absence data. If abundances are available, they provide the weights. These WA techniques can be shown to give approximate estimates of the species' optima and environmental values when the species' response surfaces (the relationships between the species' abundance, or probability of occurrence, and the environmental variables) are Gaussian (or for probabilities, Gaussian-logit) in form. Gaussian regression and calibration are also possible, but the WA techniques are simpler and are approximations to the Gaussian methods.

These simple tools are suitable when there are many species of interest and the exact form of the response surface is not critical, and they are very easy to use. If the form of the response surfaces *is* critical, more complex models can be fitted using Generalized Linear Modelling (for regression) and maximum likelihood techniques (for calibration). These more complex tools are becoming important in the theoretical study of species–environment relationships (Austin, 1985) and environmental dynamics (Bartlein *et al.*, 1986). Naturally, they require skilled users who are aware of their statistical assumptions, limitations and pitfalls.

Ordination and constrained ordination can be related to the simpler methods of regression and calibration. Ordination is the tool for exploratory analysis of community data with no prior information about the environment. Constrained ordination is the equivalent tool for the analysis of community variation in relation to environment. Both implicitly assume a common set of environmental variables and a common response model for all of the species. (Without these simplifying assumptions, they could not work; such major simplifications of data can only be achieved at the expense of some realism.) The basic ordination techniques are PCA and CA. PCA constructs axes that are as close as possible to a linear relationship with the species. These axes can be found by a converging sequence of alternating linear regressions and calibrations. Each axis after the first is obtained by partialling out linear relationships with the previous axis. CA is mathematically related to PCA, but has a very different effect. CA axes can be found by a converging sequence of WA regressions and calibrations. In CA, axes after the first are obtained analogously with PCA; in DCA they are obtained by removing all trends, linear or non-linear, with respect to previous axes. CA suffers from the arch effect, which DCA eliminates. DCA is a reasonably robust approximation to Gaussian ordination, in which the axes are constructed so that the species response curves with respect to the axes are Gaussian in form. Gaussian ordination is feasible but not convenient. DCA is much more practical. But there are problems with the detrending, and the method can break down when the connections between sites are too tenuous. Some modifications—including an improved method of detrending—may improve DCA's robustness; alternatively, some forms of nonmetric multidimensional scaling may be more robust (Kenkel and Orlóci, 1986; Minchin, 1987).

Constrained ordination methods have the added constraint that the ordination axes must be linear combinations of environmental variables. This constraint can be implemented as an extra multiple regression step in the general iterative ordination algorithm. PCA then becomes redundancy analysis (a more practical alternative to canonical correlation), Gaussian ordination becomes Gaussian canonical ordination, and CA becomes CCA (Table 2). The constraint makes Gaussian canonical ordination somewhat more stable than its unconstrained equivalent, but still CCA provides a much more practical alternative. All these constrained methods are most powerful if the number of environmental variables is small compared to the number of sites. Then the constraints are much stronger than in normal ordination, and the common problems of ordination (such as the arch effect, the need for detrending and the sensitivity to deviant sites) disappear.

Often, community–environment relationships have been explored by "indirect gradient analysis"—ordination, followed by interpretation of the axes in terms of environmental variables. But if the environmental data are to hand, constrained ordination ("multivariate direct gradient analysis") provides a more powerful means to the same end. Hybrid (direct/indirect) analyses are also possible. In partial ordination and partial constrained ordination, the analysis works on the variation that remains after the effects of particular environmental, spatial or temporal "covariables" have been removed.

The choice between linear and non-linear ordination methods is not a matter of personal preference. Where gradients are short, there are sound statistical reasons to use linear methods. Gaussian methods break down, and edge effects in CA and related techniques become serious; the representation of species as arrows becomes appropriate. As gradient lengths increase, linear methods become ineffective (principally through the "horseshoe effect", which scrambles the order of samples along the first axis as well as creates a meaningless second axis); Gaussian methods become feasible, and CA and related techniques become effective. The representation of species as points, representing their optima, becomes informative. The range 1.5–3 SD for the first axis represents a "window" over which both PCA and CA/DCA, or both redundancy analysis and CCA, can be used to good effect.

ACKNOWLEDGEMENTS

We thank Dr M. P. Austin, Dr P. J. Bartlein, Professor L. C. A. Corsten, J. A. Hoekstra, Dr P. Opdam and Dr H. van Dam for comments on the manuscript. Our collaboration was supported by a Netherlands Science Research Council (ZWO) grant to I.C.P. and a Swedish Natural Science Research Council (NFR) grant to the project "Simulation Modelling of Natural Forest Dynamics". We also thank Dr S. E. Purata V. for supplying unpublished results.

REFERENCES

Alderdice, D.F. (1972) Factor combinations: responses of marine poikilotherms to environmental factors acting in concert. In: *Marine Ecology* (Ed. by O. Kinne), Vol. 1, Part 3, pp. 1659–1722. John Wiley, New York.

Ås (1985) Biological Community patterns in insular environments. *Acta Unit. Ups.* **792**, 1–55.

Atkinson, T.C., Briffa, K.R., Coope, G.R., Joachim, M.J. and Perry, D.W. (1986) Climatic calibration of coleopteran data. In: *Handbook of Holocene Palaeoecology and Palaeohydrology* (Ed. by B.E. Berglund), pp. 851–858. John Wiley, Chichester.

Austin, M.P. (1971) Role of regression analysis in plant ecology. *Proc. Ecol. Soc. Austr.* **6**, 63–75.

Austin, M.P. (1985) Continuum concept, ordination methods, and niche theory. *Ann. Rev. Ecol. Syst.* **16**, 39–61.

Austin, M.P. and Cunningham, R.B. (1981) Observational analysis of environmental gradients. *Proc. Ecol. Soc. Austr.* **11**, 109–119.

Austin, M.P., Cunningham, R.B. and Fleming, P.M. (1984) New approaches to direct gradient analysis using environmental scalars and statistical curve-fitting procedures. *Vegetatio* **55**, 11–27.

Balloch, D., Davies, C.E. and Jones, F.H. (1976) Biological assessment of water quality in three British rivers: the North Esk (Scotland), the Ivel (England) and the Taf (Wales). *Wat. Pollut. Control* **75**, 92–114.

Bartlein, P.J., Webb, T. III and Fleri, E. (1984) Holocene climatic changes in the Northern Midwest: pollen-derived estimates. *Quaff. Res.* **22**, 361–374.

Bartlein, P.J., Prentice, I.C. and Webb, T. III (1986) Climatic response surfaces from pollen data for some eastern North American taxa. *J. Biogeogr.* **13**, 35–57.

Battarbee, R.W. (1984) Diatom analysis and the acidification of lakes. *Phil. Trans. Roy. Soc. London Ser. B* **305**, 451–477.

Bloxom, B. (1978) Constrained multidimensional scaling in N spaces. *Psychometrika* **43**, 397–408.

Böcker, R., Kowarik, I. and Bornkamm, R. (1983) Untersuchungen zur Anwendung der Zeigerwerte nach Ellenberg. *Verh. Ges. Oekol.* **11**, 35–56.

Brown, G.H. (1979) An optimization criterion for linear inverse estimation. *Technometrics* **21**, 575–579.

Brown, P.J. (1982) Multivariate calibration. *J. Roy. Statist. Soc. B* **44**, 287–321.

Carleton, T.J. (1984) Residual ordination analysis: a method for exploring vegetation environment relationships. *Ecology* **65**, 469–477.

Chandler, J.R. (1970) A biological approach to water quality management. *Wat. Pollut. Control* **69**, 415–421.

Charles, D.F. (1985) Relationships between surface sediment diatom assemblages and lakewater characteristics in Adirondack lakes. *Ecology* **66**, 994–1011.

Coombs, C.H. (1964) *A Theory of Data.* John Wiley, New York.

Cox, D.R. and Hinkley, D.V. (1974) *Theoretical Statistics.* Chapman and Hall, London.

Cramer, W. and Hytteborn, H. (1987) The separation of fluctuation and long-term change in vegetation dynamics of a rising sea-shore. *Vegetatio* **69**, 155–167.

Dargie, T.C.D. (1984) On the integrated interpretation of indirect site ordinations: a case study using semi-arid vegetation in southeastern Spain. *Vegetatio* **55**, 37–55.

Davies, P.T. and Tso, M.K.-S. (1982) Procedures for reduced-rank regression. *Appl. Statist.* **31**, 244–255.

Davison, M.L. (1983) *Multidimensional Scaling.* John Wiley, New York.

De Leeuw, J. and Heiser, W. (1980) Multidimensional scaling with restrictions on the configuration. In: *Multivariate Analysis-V* (Ed. by P.R. Krishnaiah), pp. 501–522. North-Holland, Amsterdam.

DeSarbo, W.S. and Rao, V.R. (1984) GENFOLD2: a set of models and algorithms for the general unfolding analysis of preference/dominance data. *J. Class.* **1**, 147–186.

Dobson, A.J. (1983) *Introduction to Statistical Modelling.* Chapman and Hall, London.

Ellenberg, H. (1979) Zeigerwerte der Gefässpflanzen Mitteleuropas. *Scripta Geobotanica* **9**, 1–121.

Fängström, I. and Willén, E. (1987) Clustering and canonical correspondence analysis of phytoplankton and environment variables in Swedish lakes. *Vegetatio* **71**, 87–95.

Feoli, E. and Feoli Chiapella, L. (1980) Evaluation of ordination methods through simulated coenoclines: some comments. *Vegetatio* **42**, 35–41.

Feoli, E. and Orlóci, L. (1979) Analysis of concentration and detection of underlying factors in structured tables. *Vegetatio* **40**, 49–54.

Fresco, L.F.M. (1982) An analysis of species response curves and of competition from field data: some results from heath vegetation. *Vegetatio* **48**, 175–185.

Gabriel, K.R. (1971) The biplot graphic display of matrices with application to principal component analysis. *Biometrika* **58**, 453–467.

Gabriel, K.R. (1978) Least squares approximation of matrices by additive and multiplicative models. *J. Roy. Statist. Soc. B* **40**, 186–196.

Gasse, F. and Tekaia, F. (1983) Transfer functions for estimating paleoecological conditions (pH) from East African diatoms. *Hydrobiologia* **103**, 85–90.

Gauch, H.G. (1982) *Multivariate Analysis in Community Ecology*. Cambridge Univ. Press, Cambridge.

Gauch, H.G. and Whittaker, R.H. (1972) Coenocline simulation. *Ecology* **53**, 446–451.

Gauch, H.G., Chase, G.B. and Whittaker, R.H. (1974) Ordination of vegetation samples by Gaussian species distributions. *Ecology* **55**, 1382–1390.

Gauch, H.G., Whittaker, R.H. and Singer, S.B. (1981) A comparative study of nonmetric ordinations. *J Ecol.* **69**, 135–152.

Gifi, A. (1981) *Nonlinear Multivariate Analysis*. Department of Data Theory, University of Leiden, Leiden.

Gittins, R. (1985) *Canonical Analysis. A Review With Applications in Ecology*. Springer-Verlag, Berlin.

Goff, F.G. and Cottam, G. (1967) Gradient analysis: The use of species and synthetic indices. *Ecology* **48**, 793–806.

Goodall, D.W. and Johnson, R.W. (1982) Non-linear ordination in several dimensions: a maximum likelihood approach. *Vegetatio* **48**, 197–208.

Gourlay, A.R. and Watson, G.A. (1973) *Computational Methods for Matrix Eigen Problems*. John Wiley, New York.

Greenacre, M.J. (1984) *Theory and Applications of Correspondence Analysis*. Academic Press, London.

Heiser, W.J. (1981) *Unfolding Analysis of Proximity Data*. Thesis, University of Leiden, Leiden.

Heiser, W.J. (1987) Joint ordination of species and sites: the unfolding technique. In: *Developments in Numerical Ecology* (Ed. by P. Legendre and L. Legendre), pp. 189–221. Springer-Verlag, Berlin.

Hill, M.O. (1973) Reciprocal averaging: an eigenvector method of ordination. *J. Ecol.* **61**, 237–249.

Hill, M.O. (1974) Correspondence analysis: a neglected multivariate method. *Appl. Statist.* **23**, 340–354.

Hill, M.O. (1977) Use of simple discriminant functions to classify quantitative phytosociological data. In: *First International Symposium an Data Analysis and Informatics* (Ed. by E. Diday, L. Lebart, J.P. Pages and R. Tomassone), Vol. 1, pp. 181–199. INRIA, Chesnay.

Hill, M.O. (1979) *DECORANA—A FORTRAN Program for Detrended Correspondence Analysis and Reciprocal Averaging*. Section of Ecology and Systematics, Cornell University, Ithaca, New York.

Hill, M.O. and Gauch, H.G. (1980) Detrended correspondence analysis: an improved ordination technique. *Vegetatio* **42**, 47–58.

Ihm, P. and Van Groenewoud, H. (1975) A multivariate ordering of vegetation data based on Gaussian type gradient response curves. *J. Ecol.* **63**, 767–777.

Ihm, P. and Van Groenewoud, H. (1984) Correspondence analysis and Gaussian ordination. *COMPSTAT Lectures* **3**, 5–60.

Imbrie, J. and Kipp, N.G. (1971) A new micropaleontological method for quantitative paleoclimatology: application to a late Pleistocene Caribbean core. In: *The Late Cenozoic Glacial Ages* (Ed. by K.K. Turekian), pp. 71–181. Yale University Press, New Haven, CT.

Israëls, A.Z. (1984) Redundancy analysis for qualitative variables. *Psychometrika* **49**, 331–346.

Iwatsubo, S. (1984) The analytical solutions of eigenvalue problem in the case of applying optimal scoring method to some types in data. In: *Data Analysis and Informatics 3* (Ed. by E. Diday), pp. 31–40. North-Holland, Amsterdam.

Jolliffe, I.T. (1986) *Principal Component Analysis*. Springer-Verlag, Berlin.

Jongman, R.H.G., Ter Braak, C.J.F. and Van Tongeren, O.F.R. (1987) *Data Analysis in Community and Landscape Ecology*. Pudoc, Wageningen.

Kalkhoven, J. and Opdam, P. (1984) Classification and ordination of breeding bird data and landscape attributes. In: *Methodology in Landscape Ecological Research and Planning* (Ed. by J. Brandt and P. Agger), Vol. 3, pp. 15–26. Roskilde Universitetsforlag GeoRue, Theme 3, Roskilde.

Kenkel, N.C. and Orlóci, L. (1986) Applying metric and nonmetric multidimensional scaling to ecological studies: some new results. *Ecology* **67**, 919–928.

Kooijman, S.A.L.M. (1977) Species abundance with optimum relations to environmental factors. *Ann. Syst. Res.* **6**, 123–138.

Kooijman, S.A.L.M. and Hengeveld, R. (1979) The description of a non-linear relationship between some carabid beetles and environmental factors. In: *Contemporary Quantitative Ecology and Related Econometrics* (Ed. by G.P. Patil and M.L. Rosenzweig), pp. 635–647. International Co-operative Publishing House, Fairland MD.

Laurec, A., Chardy, P., de la Salle, P. and Rickaert, M. (1979) Use of dual structures in inertia analysis: ecological implications. In: *Multivariate Methods in Ecological Work* (Ed. by L. Orlóci, C.R. Rao and W.M. Stiteler), pp. 127–174. International Co-operative Publishing House, Fairiand MD.

McCullagh, P. and Nelder, J.A. (1983) *Generalized Linear Models*. Chapman and Hall, London.

Macdonald, G.M. and Ritchie, J.C. (1986) Modern pollen spectra from the western interior of Canada and the interpretation of Late Quaternary vegetation development. *New Phytol.* **103**, 245–268.

Meulman, J. and Heiser, W.J. (1984) Constrained multidimensional scaling: more directions than dimensions. In: *COMPSTAT 1984*, pp. 137–142. Physica-Verlag, Vienna.

Minchin, P. (1987) An evaluation of the relative robustness of techniques for ecological ordination. *Vegetatio* **69**, 89–107.

Montgomery, D.C. and Peck, E.A. (1982) *Introduction to Linear Regression Analysis*. John Wiley, New York.

Nishisato, S. (1980) *Analysis of Categorical Data: Dual Scaling and Its Applications.* University of Toronto Press, Toronto.

Oksanen, J. (1983) Ordination of boreal heath-like vegetation with principal component analysis, correspondence analysis and multidimensional scaling. *Vegetation* **52**, 181–189.

Opdam, P.F.M., Kalkhoven, J.T.R. and Phillippona, J. (1984) *Verband tussen Broedvogelgemeenschappen en Begroeiing in een Landschap bij Amerongen.* Pudoc, Wageningen.

Peet, R.K. (1978) Latitudinal variation in southern Rocky Mountain forests. *J. Biogeogr.* **5**, 275–289.

Peet, R.K. and Loucks, O.L. (1977) A gradient analysis of southern Wisconsin forests. *Ecology* **58**, 485–499.

Pickett, S.T.A. (1980) Non-equilibrium coexistence of plants. *Bull. Torrey bot. Club* **107**, 238–248.

Pielou, E.C. (1984) *The Interpretation of Ecological Data.* John Wiley, New York.

Prodon, R. and Lebreton, J.-D. (1981) Breeding avifauna of a Mediterranean succession: the holm oak and cork oak series in the eastern Pyrenees. 1. Analysis and modeling of the structure gradient. *Oikos* **37**, 21–38.

Purata, S.E. (1986) Studies on secondary succession in Mexican tropical rain forest. *Acta Univ. Ups.* Comprehensive Summaries of Uppsala Dissertations from the Faculty of Science 19. Almqvist and Wiksell International, Stockholm.

Rao, C.R. (1964) The use and interpretation of principal components analysis in applied research. *Sankhya A* **26**, 329–358.

Robert, P. and Escoufier, Y. (1976) A unifying tool for linear multivariate statistical methods: the RV-coefficient. *Appl. Statist.* **25**, 257–265.

Sládecek, V. (1973) System of water quality from the biological point of view. *Arch. Hydrobiol. Beiheft* **7**, 1–218.

Smith, P.L. (1979) Splines as a useful and convenient statistical tool. *Am. Stat.* **33**, 57–62.

Swaine, M.D. and Greig-Smith, P. (1980) An application of principal components analysis to vegetation change in permanent plots. *J. Ecol.* **68**, 33–41.

Ter Braak, C.J.F. (1983) Principal components biplots and alpha and beta diversity. *Ecology* **64**, 454–462.

Ter Braak, C.J.F. (1985) Correspondence analysis of incidence and abundance data: properties in terms of a unimodal response model. *Biometrics* **41**, 859–873.

Ter Braak, C.J.F. (1986) Canonical correspondence analysis: a new eigenvector technique for multivariate direct gradient analysis. *Ecology* **67**, 1167–1179.

Ter Braak, C.J.F. (1987a) The analysis of vegetation–environment relationships by canonical correspondence analysis. *Vegetatio* **69**, 69–77.

Ter Braak, C.J.F. (1987b) *CANOCO-a FORTRAN Program for Canonical Community Ordination by [Partial] [Detrended] [Canonical] Correspondence Analysis, Principal Components Analysis and Redundancy Analysis (Version 2.1).* Agriculture Mathematics Group, Wageningen.

Ter Braak, C.J.F. (1988) Partial canonical correspondence analysis. In: *Classification Methods and Related Methods of Data Analysis* (Ed. by H.H. Bock), pp. 551–558. North-Holland, Amsterdam.

Ter Braak, C.J.F. and Barendregt, L.G. (1986) Weighted averaging of species indicator values: its efficiency in environmental calibration. *Math. Biosci.* **78**, 57–72.
Ter Braak, C.J.F. and Looman, C.W.N. (1986) Weighted averaging, logistic regression and the Gaussian response model. *Vegetatio* **65**, 3–11.
Tilman, D. (1982) *Resource Competition and Community Structure.* Princeton University Press, Princeton.
Tso, M.K.-S. (1981) Reduced-rank regression and canonical analysis. *J. Roy. Statist. Soc. B* **43**, 183–189.
van der Aart, P.J.M. and Smeenk-Enserink, N. (1975) Correlations between distribution of hunting spiders (Lycosidae, Ctenidae) and environmental characteristics in a dune area. *Neth. J. Zool.* **25**, 1–45.
van den Wollenberg, A.L. (1977) Redundancy analysis. An alternative for canonical correlation analysis. *Psychometrika* **42**, 207–219.
van Dam, H., Suurmond, G. and Ter Braak, C.J.F. (1981) Impact of acidification on diatoms and chemistry of Dutch moorland pools. *Hydrobiologia* **83**, 425–459.
Webb, T. III and Bryson, R.A. (1972) Late- and postglacial climatic change in the northern Midwest, USA: quantitative estimates derived from fossil spectra by multivariate statistical analysis. *Quat. Res.* **2**, 70–115.
Whittaker, R.H. (1956) Vegetation of the Great Smoky Mountains. *Ecol. Monogr.* **26**, 1–80.
Whittaker, R.H. (1967) Gradient analysis of vegetation. *Biol. Rev.* **49**, 207–264.
Whittaker, R.H., Levin, S.A. and Ropt, R.B. (1973) Niche, habitat and ecotope. *Am. Natur.* **107**, 321–338.
Wiens, J.A. and Rotenberry, J.T. (1981) Habitat associations and community structure of birds in shrubsteppe environments. *Ecol. Monogr.* **51**, 21–41.
Williams, E.J. (1959) *Regression Analysis.* John Wiley, New York.
Wold, H. (1982) Soft modeling. The basic design and some extensions. In: *Systems Under Indirect Observation. Causality–Structure–Prediction* (Ed. by K.G. Jöreskog and H. Wold), Vol. 2, pp. 1–54. North-Holland, Amsterdam.
Zelinka, M. and Marvan, P. (1961) Zür Präzisierung der biologischen Klassifikation der Reinheit fliessender Gewässer. *Arch. Hydrobiol.* **57**, 389–407.

APPENDIX

A general iterative algorithm can be used to carry out the linear and weighted-averaging methods described in this review. The algorithm is essentially the one used in the computer program CANOCO (Ter Braak, 1987b). It operates on response variables, each recording the abundance or presence/absence of a species at various sites, and on two types of explanatory variables: environmental variables and covariables. By environmental variables we mean here explanatory variables of prime interest, in contrast with covariables which are "concomitant" variables whose effect is to be removed. When all three types of variables are present, the algorithm describes how to obtain a *partial constrained ordination*. The other linear and WA techniques are all special cases, obtained by omitting various irrelevant steps.

Let $Y = [y_{ki}]$ $(k = 1, \ldots, m;\ i = 1, \ldots, n)$ be a species-by-site matrix containing the observations of m species at n sites ($y_{ki} \geqq 0$) and let $Z_1 = [z_{1ji}]$ $(l = 0, \ldots, p;\ i = 1, \ldots, n)$ and $Z_2 = [z_{2ji}]$ $(j = 1, \ldots, q;\ i = 1, \ldots, n)$ be covariable-by-site and environmental variable-by-site matrices containing the observations of p covariables and q environmental variables at the same n sites, respectively. The first row of Z_1, with index $l = 0$, is a row of 1's which is included to account for the intercept in Eq. (4). Further, denote the species and site scores on the sth ordination by $\mathbf{u} = [u_k]$ $(k = 1, \ldots, m)$ and $x = [x_i]$ $(i = 1, \ldots, n)$, the canonical coefficients of the environmental variables by $\mathbf{c} = [c_j]$ $(j = 1, \ldots, q)$ and collect the site scores on the $(s-1)$ previous ordination axes as rows of the matrix A. If detrending-by-polynomials is in force (Step A10), then the number of rows of A, s_A say, is greater than $s - 1$. In the algorithm we use the assign statement ":=", for example $a := b$ means "a is assigned the value b". If the left-hand side of the assignment is indexed by a subscript, it is assumed that the assignment is made for all permitted subscript values; the subscript k will refer to species $(k = 1, \ldots, m)$, the subscript i to sites $(i = 1, \ldots, n)$ and the subscript j to environmental variables $(j = 1, \ldots, q)$.

Preliminary Calculations

P1 Calculate species totals $\{y_{k+}\}$, site totals $\{y_{+i}\}$ and the grand total y_{++}. If a linear method is required, set

$$r_k := 1, \quad w_i := 1, \quad w_i^* := \frac{1}{n} \tag{A.1}$$

and if a weighted averaging method is required, set

$$r_k := y_{k+}, \quad w_i := y_{+i}, \quad w_i^* := y_{+i}/y_{++} \tag{A.2}$$

P2 Standardize the environmental variables to zero mean and unit variance. For environmental variable j calculate its mean $\bar{z}$ and variance v

$$\bar{z} := \Sigma_i w_i^* z_{2ji}, \quad v := \Sigma_i w_i^* (z_{2ji} - \bar{z})^2 \tag{A.3}$$

and set $z_{2ji} := (z_{2ji} - \bar{z})/\sqrt{v}$.

P3 Calculate for each environmental variable j the residuals of the multiple regression of the environmental variables on the covariables, i.e.

$$\mathbf{c}_j^* := (Z_1 W Z_1')^{-1} Z_1 W \mathbf{z}_{2j} \tag{A.4}$$

$$\tilde{\mathbf{z}}_{2j} := \mathbf{z}_{2j} - Z_1' \mathbf{c}_j^* \tag{A.5}$$

where $\mathbf{z}_{2j} = (z_{2j}, \ldots, z_{2jn})'$, $W = \text{diag}\,(w_1, \ldots, w_n)$ and $\mathbf{c}_j^*$ is the $(p+1)$-vector of the coefficients of the regression of $\mathbf{z}_{2j}$ on Z_1. Now define $\tilde{Z}_2 = [\tilde{\mathbf{z}}_{2ji}]$ $(j = 1, \ldots, q,\ i = 1, \ldots, n)$.

Iteration Algorithm

Step A0 Start with arbitrary, but unequal site scores $\mathbf{x} = [x_i]$. Set $x_i^0 = x_i$

Step A1 Derive new species scores from the site scores by

$$u_k := \sum_i y_{ki} x_i / r_k \tag{A.6}$$

Step A2 Derive new site scores $\mathbf{x}^* = [x_i^*]$ from the species scores

$$x_i^* := \sum_k y_{ki} u_k / w_i \tag{A.7}$$

Step A3 Make $\mathbf{x}^* = [x_i^*]$ uncorrelated with the covariables by calculating the residuals of the multiple regression of $\mathbf{x}^*$ on Z_1:

$$\mathbf{x}^* := \mathbf{x}^* - Z_1'(Z_1 W Z_1')^{-1} Z_1 W \mathbf{x}^* \tag{A.8}$$

Step A4 If $q \leq s_A$, set $x_i := x_i^*$ and skip Step A5.

Step A5 If q $> s_A$, calculate a multiple regression of $\mathbf{x}^*$ on $\tilde{Z}_2$

$$\mathbf{c} := (\tilde{Z}_2 W \tilde{Z}_2')^{-1} \tilde{Z}_2 W \mathbf{x}^* \tag{A.9}$$

and take as new site scores the fitted values:

$$\mathbf{x} := \tilde{Z}_2' \mathbf{c} \tag{A.10}$$

Step A6 If $s > 1$, make $\mathbf{x} = [x_i]$ uncorrelated with previous axes by calculating the residuals of the multiple regression of $\mathbf{x}$ on A:

$$\mathbf{x} := \mathbf{x} - A'(AWA')^{-1} AW\mathbf{x} \tag{A.11}$$

Step A7 Standardized $\mathbf{x} = [x_i]$ to zero mean and unit variance by

$$\bar{x} := \sum_i w_i^* x_i, \quad \sigma^2 := \sum_i w_i^* (x_i - \bar{x})^2$$

$$x_i := (x_i - \bar{x})/\sigma \tag{A.12}$$

Step A8 Check convergence, i.e. if

$$\sum_i w_i^* (x_i^0 - x_i)^2 < 10^{-10} \tag{A.13}$$

goto Step A9, else set $x_i^0 := -x_i$ and goto Step A1.

Step A9 Set the eigenvalue λ equal to σ in (A.12) and add $\mathbf{x} = [x_i]$ as a new row to the matrix A.

Step A10 If detrending-by-polynomials is required, calculate polynomials of $\mathbf{x}$ up to order 4 and first-order polynomials of $\mathbf{x}$ with the previous ordination axes,

$$x_{2i} := x_i^2, \quad x_{3i} := x_i^3, \quad x_{4i} := x_i^4, \quad x_{(b)i} := -x_i a_{bi} \tag{A.14}$$

where a_{bi} are the site scores of a previous ordination axis $(b = 1, \ldots, s-1)$. Now perform for each of the $(s+2)$-variables in (A.14) the Steps A3–A6 and add the resulting variables as new variables to the matrix A.

Step A11 Set $s := s+1$ and goto Step A0 if required and if further ordination axes can be extracted, else stop.

At convergence, the algorithm gives the solution with the greatest real value of λ to the following transition formulae [where $R = \text{diag}\ (r_1, \ldots, r_m)$ and $W = \text{diag}\ (w_1, \ldots, w_n)$ and where the notation B^0 is used to denote $B'(BWB')^{-1}BW$, the projection operator on the row space of a matrix B in the metric defined by the matrix $[W]$:

$$\mathbf{u} = R^{-1} Y \mathbf{x} \tag{A.15}$$

$$\mathbf{x}^* = (I - \tilde{Z}_1^0) W^{-1} Y' \mathbf{u} \tag{A.16}$$

$$\mathbf{c} = (\tilde{Z}_2 W \tilde{Z}_2')^{-1} \tilde{Z}_2 W \mathbf{x}^* \tag{A.17}$$

$$\lambda \mathbf{x} = (I - A^0) \tilde{Z}_2' \mathbf{c} \tag{A.18}$$

The tilde above Z_2 is there as a reminder that the original environmental variables were replaced by residuals of a regression on Z_1 in (A.5), i.e. in terms of the original variables

$$\tilde{Z}_2' = (I - Z_1^0)Z_2' \qquad \text{(A.19)}$$

Remarks

(1) Note that u_k in the algorithm takes the place of b_k in Section II.
(2) Special cases of the algorithm are: constrained ordination: $p=0$; partial ordination: $q=0$; (unconstrained) ordination: $p=0$, $q=0$; linear calibration and weighted averaging: $p=0$, $q=1$; (partial) multiple regression: $m=1$. The corresponding transition formulae follow from (A.15)–(A.18) with the proviso that, if $q=0$, Z_2 in (A.19) must be replaced by the $n \times n$ identity matrix and generalized matrix inverses are used. Note that, if $p=0$, Z_1 is a $1 \times n$ matrix containing 1's; Z_1 renders the centring of the species data in the linear methods in Section II redundant.
(3) The standardization in P2 removes the arbitrariness in the units of measurement of the environmental variables, and makes the canonical coefficients comparable among each other, but does not influence the values of λ, $\mathbf{u}$ and $\mathbf{x}$ to be obtained in the algorithm.
(4) Step A6 simplifies to Step 3b of the main text if the rows of A are W-orthonormal. The steps A3–A6 form a single projection of $\mathbf{x}^*$ on the column space of $(I - A^0)\tilde{Z}_2'$ if and only if A defines a subspace of the row space of $\tilde{Z}_2$. As each ordination axis defines such a subspace, this is trivially so without detrending. The method of detrending-by-polynomials as defined in Step A10, ensures that A defines also a subspace of $\tilde{Z}_2$ if detrending is in force. The transition formulae (A.15)–(A.18) define an eigenvalue equation of which all eigenvalues are real non-negative (Ter Braak, 1987b).
(5) If a particular scaling of the biplot or the joint plot is wanted, the ordination axes may require linear rescaling. With linear methods one can choose between a Euclidean distance biplot and a covariance biplot, which focus on the approximate Euclidean distances between sites and correlations among species, respectively (Ter Braak, 1983). With weighted averaging methods it is customary to use the site scares $\mathbf{x}^*$ (A.16) and the species scores $\mathbf{u}$ (A.15) to prepare an ordination diagram after a linear rescaling so that the average within-site variance of the species scores is equal to 1 (cf. Section IV.C), as is done in DECORANA (Hill, 1979) and CANOCO (Ter Braak, 1987b).

Originally Published in Volume 23 (this series), pp 187–261, 1992

Inherent Variation in Growth Rate Between Higher Plants: A Search for Physiological Causes and Ecological Consequences

HANS LAMBERS AND HENDRIK POORTER

ADVANCES IN ECOLOGICAL RESEARCH VOL. 34
0065-2504/04 $35.00 DOI 10.1016/S0065-2504(03)34004-8

I. SUMMARY

When grown under optimum conditions, plant species from fertile, productive habitats tend to have inherently higher relative growth rates (RGR) than species from less favourable environments. Under these conditions, fast-growing species produce relatively more leaf area and less root mass, which greatly contributes to their larger carbon gain per unit plant weight. They have a higher rate of photosynthesis per unit leaf dry weight and per unit leaf nitrogen, but not necessarily per unit leaf area, due to their higher leaf area per unit leaf weight. Fast-growing species also have higher respiration rates per unit organ weight, due to demands of a higher RGR and higher rate of nutrient uptake. However, expressed as a fraction of the total amount of carbon fixed per day, they use less in respiration.

Fast-growing species have a greater capacity to acquire nutrients, which is likely to be a consequence, rather than the cause, of their higher RGR. There is no evidence that slow-growing species have a special ability to acquire nutrients from dilute solutions, but they may have special mechanisms to release nutrients when these are sparingly soluble.

We have analysed variation in morphological, physiological, chemical and allocation characteristics underlying variation in RGR, to arrive at an appraisal of its ecological significance. When grown under optimum conditions, fast-growing species contain higher concentrations of organic nitrogen and minerals. The lower specific leaf area (SLA) of slow-growing species is at least partly due to the relatively high concentration of cell-wall material and quantitative secondary compounds, which may protect against detrimental abiotic and biotic factors. As a consequence of a greater investment in protective compounds or structures, the rate of photosynthesis per unit leaf dry weight is less, but leaf longevity is increased.

In short-term experiments with a limiting nutrient availability the RGR of all species is reduced, but potentially fast-growing species still grow faster than inherently slow-growing ones. Therefore, the absence of fast-growing species from infertile environments cannot be explained by their growth rate *per se*. The higher leaf longevity diminishes nutrient losses and is a factor contributing to the success in nutrient-limited habitats. We postulate that natural selection for traits which are advantageous under nutrient-limited conditions has led to the low growth potential of species from infertile and some other unfavourable habitats.

Other examples indicating that selection for traits which allow successful performance under adverse conditions inevitably leads to a lower potential RGR are included. We conclude that it is likely that there are trade-offs between growth potential and performance under adverse conditions, but that current ecophysiological information explaining variation in RGR is too limited to support this contention quantitatively.

II. INTRODUCTION

Plants are distributed over a wide range of habitats varying from tundra to rain forests, from wetlands to deserts and from lowland to alpine regions. Coping with such contrasting, sometimes extreme, environments requires a certain degree of inherent specialization. One of the characteristics in which species of different habitats vary is their growth potential. Plants growing on nutrient-poor soils have a lower growth rate than those on fertile soils. But even when grown under optimum conditions, species which naturally occur on nutrient-poor soils still have a lower growth rate compared to plants characteristic of fertile sites (e.g. Bradshaw

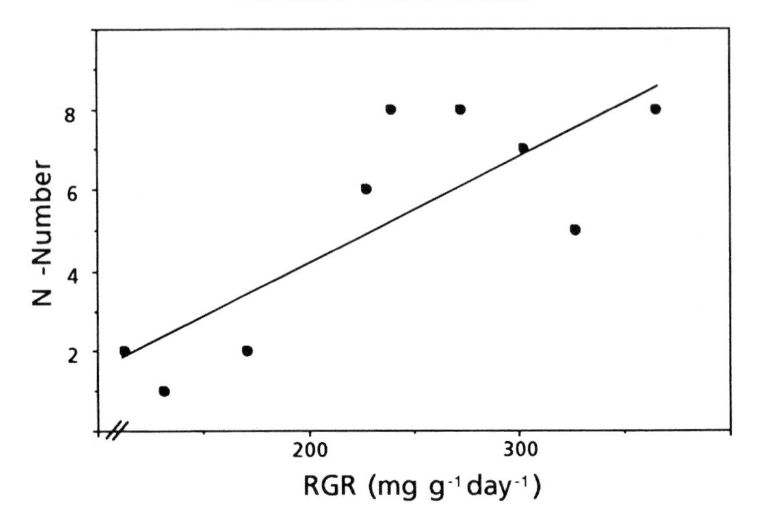

Figure 1 The relationship between the RGR of nine herbaceous C_3 species and the nitrogen index of the species' habitat according to Ellenberg (1979) (high values of the N number correspond to habitats of high nitrogen supply). RGR was determined at an optimum nutrient supply, moderate quantum flux density and fairly low vapour pressure deficit. The species described are, in order of increasing RGR *Corynephorus canescens*, *Festuca ovina*, *Pimpinella saxifraga*, *Phleum pratense*, *Anthriscus sylvestris*, *Poa annua*, *Scrophularia nodosa*, *Rumex crispus* and *Galinsoga parviflora*. (After Poorter and Remkes, 1990.)

et al., 1964; Rorison, 1968; Christie and Moorby, 1975; Grime and Hunt, 1975; Poorter and Remkes, 1990; Figure 1). In addition, species or ecotypes which naturally occur in shaded environments (Pons, 1977; Corré, 1983a), dry habitats (Rozijn and van der Werf, 1986), alpine regions (Woodward, 1979; Atkin and Day, 1990), arctic environments (Warren Wilson, 1966), saline conditions (Ball, 1988), sites which are rich in heavy metals (Wilson, 1988; Verkleij and Prast, 1989), or in other habitats adverse to plant growth all have a lower growth potential than comparable ones from favourable, fertile habitats.

This close association between a species' growth potential and the quality of its natural habitat raises two questions. First, how are the differences in growth rate between species brought about? And, second, what ecological advantage is conferred by a plant's growth potential? These two questions are in fact closely related. A plant is a complex of organs with contrasting functions and subject to conflicting demands. A low or a high potential growth rate may either be the basis or a by-product of adaptation to a certain set of environmental conditions. Hence there may be trade-offs between adaptation to adverse conditions and growth potential. Therefore, the question on the ecological advantage of a potential growth rate cannot be answered until further ecophysiological information is available on the mechanisms explaining variation in growth potential.

Numerous plant characters contribute to a plant's absolute growth rate in its natural habitat, e.g. seed size, germination time, or plant size after overwintering. In this chapter we restrict ourselves to an analysis of the different traits that contribute to a plant's *relative growth rate* and discuss mechanisms which cause variation in any of these traits. We will treat the possible interdependence of various characteristics and try to quantify the importance of each of these in explaining interspecific variation in relative growth rate. Finally, we discuss the ecological implications of interspecific differences in the various traits and in the growth rate itself.

III. GROWTH ANALYSES

Growth analysis is often used as a tool to obtain insight into the functioning of a plant. Different types of analyses exist, depending on what is considered a key factor for growth (cf. Lambers *et al.*, 1989). In the most common approach, leaf area is assumed to be a key factor. The relative growth rate (RGR) (see Table 1 for a list of abbreviations), the rate of increase in plant weight per unit of plant weight already present, is then factorized into two components, the leaf area ratio and the net assimilation rate (Evans, 1972; see Table 2 for a range of published values). The leaf area ratio (LAR) is the amount of leaf area per unit total plant weight. The net assimilation rate (NAR) is defined as the rate of increase in plant weight per unit leaf area. Thus:

$$\mathrm{RGR} = \mathrm{LAR} \times \mathrm{NAR} \qquad (1)$$

LAR and NAR can both be divided into a further set of components. The LAR is the product of the specific leaf area (SLA), the amount of leaf area per unit leaf weight, and the leaf weight ratio (LWR), the fraction of the total plant biomass allocated to leaves. Thus:

$$\mathrm{LAR} = \mathrm{SLA} \times \mathrm{LWR} \qquad (2)$$

Although termed the morphological component, LAR is affected by biomass allocation, chemical composition and leaf anatomy, as will be discussed later.

Table 1 Abbreviations used in this chapter and the preferred units in which they are expressed, listed in alphabetical order

Abbreviation	Meaning	Preferred units
CC	Carbon concentration	mmol C g^{-1}
EXU_a (EXU_w)	Rate of exudation	mg m^{-2} (leaf area) day^{-1} (mg g^{-1} (plant wt) day^{-1})
LAR	Leaf area ratio	m^2 kg^{-1}
LR_a (LR_w)	Rate of leaf respiration	μmol CO_2 m^{-2} (leaf area) s^{-1} (nmol CO_2 g^{-1} (leaf wt) s^{-1})
LWR	Leaf weight ratio	g g^{-1}
NAR	Net assimilation rate	g m^{-2} day^{-1}
NIR	Net nitrogen uptake rate	nmol (g root)$^{-1}$ s^{-1}
PNC	Plant nitrogen concentration	mmol N g^{-1}
PNUE	Photosynthetic nitrogen use efficiency	μmol CO_2 (mol leaf N)$^{-1}$ s^{-1}
RGR	Relative growth rate	mg g^{-1} day^{-1}
RWR	Root weight ratio	g g^{-1}
SLA	Specific leaf area	m^2 kg^{-1}
SR_a (SR_w)	Rate of stem respiration	μmol CO_2 m^{-2} (leaf area) s^{-1} (nmol CO_2 g^{-1} (stem wt) s^{-1})
SRL	Specific root length	m g^{-1}
SWR	Stem weight ratio	g g^{-1}
PS_a (PS_w)	Rate of photosynthesis	μmol CO_2 m^{-2} s^{-1}
RR_a (RR_w)	Rate of root respiration	μmol CO_2 m^{-2} (leaf area) s^{-1} (nmol C g^{-1} (root wt) s^{-1})
VOL_a (VOL_w)	Rate of volatile losses	mg m^{-2} (leaf area) day^{-1}) (mg g^{-1} (plant wt) day^{-1})

The NAR is the net result of dry weight gain and dry weight losses and is largely the balance of the rate of photosynthesis, expressed per unit leaf area (PS), and the rate of leaf respiration (LR), stem respiration (SR) and root respiration (RR), in this case also per unit leaf area. If these physiological processes are expressed in moles of carbon, the net balance of photosynthesis and respiration has to be divided by CC, the carbon concentration of the newly formed material, to obtain the increase in dry weight. The balance is completed by subtracting losses due to volatilization (VOL) and exudation (EXU) per unit time, also expressed on a leaf area basis. Thus:

$$NAR = \frac{(PS_a - LR_a - SR_a - RR_a)}{CC} - EXU_a - VOL_a \quad (3)$$

Table 2 Interspecific variation in growth parameters. All values are expressed per unit dry weight. Species, grown in a controlled environment (glass house, growth room) are indicated with (C) in the specifications, references marked (F) are from plants grown in the field

Parameter	Range	Mean value	Specifications
RGR	31–151	74	(C) 15 tree species (seedlings), Grime and Hunt (1975)
	66–314	159	(C) 93 perennials; Grime and Hunt (1975)
	120–299	176	(C) 22 annuals, Grime and Hunt (1975)
	113–365	224	(C) 24 herbaceous species, Poorter and Remkes (1990)
	19–386	158	(C,F) all species from Table 3
NAR	8–14	10	(C) 24 herbaceous species, Poorter and Remkes (1990)
	2–25	10	(C,F) all species from Table 3
LAR	0.1–4.5	1.5	(F) 35 tropical trees, Ovington and Olson (1970)
	13–36	23	(C) 24 herbaceous species, Poorter and Remkes (1990)
	2–65	18	(C,F) all species from Table 3
SLA	6–37	15	(F) 35 tropical trees, Ovington and Olson (1970)
	25–56	41	(C) 24 herbaceous species, Poorter and Remkes (1990)
	10–131	34	(C,F) all species from Table 3
LWR	0.02–0.34	0.11	(F) 35 tropical trees, Ovington and Olson (1970)
	0.43–0.64	0.54	(C) 24 herbaceous species, Poorter and Remkes (1990)
	0.26–0.81	0.53	(C,F) all species from Table 3
SWR	0.52–0.86	0.70	(F) 35 tropical trees, Ovington and Olson (1970)
	0.07–0.27	0.17	(C) 24 herbaceous species, Poorter and Remkes (1990)
RWR	0.08–0.36	0.20	(F) 35 tropical trees, Ovington and Olson (1970)
	0.22–0.38	0.29	(C) 24 herbaceous species, Poorter and Remkes (1990)

where subscript a indicates that the rates are expressed on a leaf area basis. However, leaf, stem and root respiration are not expected to be directly related to leaf area, but rather to the biomass of the different organs. Equation (3) is therefore extended to include the relations between organ biomass and leaf area:

$$\mathrm{NAR} = \frac{1}{\mathrm{CC}}\left(\mathrm{PS}_a - \mathrm{LR}_w \cdot \frac{1}{\mathrm{SLA}} - \mathrm{SR}_w \cdot \frac{\mathrm{SWR}}{\mathrm{LAR}} - \mathrm{RR}_w \cdot \frac{\mathrm{RWR}}{\mathrm{LAR}}\right) - \mathrm{EXU}_a - \mathrm{VOL}_a \tag{4}$$

where subscript w indicates that rates are expressed per unit dry weight; SWR and RWR are the stem weight ratio and the root weight ratio, the fraction of biomass allocated to stem and roots, respectively. From Eq. (4) it is clear that NAR is not purely a physiological component, as it is often termed, but rather a complex intermingling of a plant's physiology, biomass allocation, chemical composition and leaf area formation. Although NAR is relatively easy to determine, it is not the most appropriate parameter to obtain a clear insight into the relation between physiology and growth. Hence, we rewrite Eqs. (1) and (4) into:

$$\mathrm{RGR} = \frac{(\mathrm{PS}_a \times \mathrm{SLA} \times \mathrm{LWR} - \mathrm{LR}_w \times \mathrm{LWR} - \mathrm{SR}_w \times \mathrm{SWR} - \mathrm{RR}_w \times \mathrm{RWR})}{\mathrm{CC}} - \mathrm{EXU}_w - \mathrm{VOL}_w \qquad (5)$$

where EXU and VOL are expressed per unit total plant weight.

When plants are in a steady state, i.e. when there is a fixed ratio between the increment of nutrients (e.g. nitrogen) and biomass, growth can also be considered in relation to the acquisition of such nutrients;

$$\mathrm{RGR} = \frac{\mathrm{RWR} \times \mathrm{NIR}}{\mathrm{PNC}} \qquad (6)$$

where NIR is the net rate of nitrogen absorption, the rate of nitrogen taken up per unit root weight, and PNC the total plant nitrogen concentration.

Factorizing RGR in its various components does not imply that these components are independent of each other (Hardwick, 1984). Often, an increase in one parameter affects another, either positively or negatively. In the next section we evaluate the importance of LAR and NAR in explaining variation in RGR.

IV. NET ASSIMILATION RATE AND LEAF AREA RATIO

A wealth of information is available on the comparison of growth of two or three species, but few authors have investigated the relation between RGR and growth parameters for a range of species. Potter and Jones (1977) compared nine crop and weed species, Mooney *et al.* (1978) investigated five *Eucalyptus* species and Poorter and Remkes (1990) analysed the growth of 24 wild species common in western Europe. In all of these cases the LAR was the predominant factor explaining the inherent variation in

Table 3 Degree of association between the growth parameters of Eqs. (1) and (2). Means of a compilation of 78 literature references on comparative growth analyses of herbaceous C_3 species. For each reference a linear regression was carved out with the mean of variable *A* for each species as the independent variable, and the mean of variable *B* over the experimental period as the dependent variable. Then the change in the predicted value of variable *B* associated with a 10% change in variable *A* was calculated, starting from the mean values of *A* and *B*. A value close to 10 indicates that a 10% increase in variable *A*, is associated with an almost equal increase in variable *B*, whereas a value close to zero indicates no association. Such an analysis is only fruitful provided the differences in variable *A* are large enough. Therefore, the degree of association was only calculated for pairs of variables in which the smallest and largest value of variable *A* differed at least 10% and 20 mg g^{-1} day^{-1} (RGR), 10% and 2 m^2 kg^{-1} (LAR), 10% and 3 m^2 kg^{-1} (SLA) and 10% (LWR). The values for all references were then averaged.

The literature references are those given in Table 2E of Poorter (1989), supplemented with the C_3 species of Table 2C, the sun species of Table 2D, and Tsunoda (1959), Enyi (1962), Tognoni *et al.* (1967), Khan and Tsunoda (1970a,b), Callaghan and Lewis (1971), Hughes and Cockshull (1971), Eze (1973), Ashenden *et al.* (1975), Smith and Walton (1975), Elias and Chadwick (1979), Grime (1979), Horsman *et al.* (1980), Cook and Evans (1983), Gray and Schlesinger (1983), Spitters and Kramer (1986), Campbell and Grime (1989), Garnier *et al.* (1989), and Muller and Garnier (1990). In each of these analyses root weight determinations were carried out and all species or genotypes were grown under identical conditions.

A	*B*	*n*	DOA	Sign.
RGR	NAR	46	2.4	*
RGR	LAR	46	7.5	***
RGR	SLA	21	7.8	***
RGR	LWR	22	2.0	ns
LAR	NAR	54	−4.3	***
LAR	SLA	27	7.3	***
LAR	LWR	27	3.1	*
SLA	NAR	28	−4.0	*
SLA	LWR	28	−0.7	ns
LWR	NAR	25	−4.8	*

A, variable *A*; *B*, variable *B*; *n*, number of references; DOA: averaged value of the degree of association between variable *A* and variable *B*; Sign.: t-test of the H_0 hypothesis DOA = 00. ns, not significant; *, $P < 0.05$; **, $P < 0.01$; ***, $P < 0.001$.

RGR. Poorter (1989) arrived at the same conclusion after a review of 45 literature sources. An extended compilation is given in Table 3. On average, a 10% increase in RGR is associated with an 7.5% increase in LAR and a 2.4% increase in NAR. Thus, the amount of leaf area a plant realizes with a given total plant weight is an important factor determining the potential growth rate of a plant. Differences in the rate of dry weight gain per unit leaf area are of secondary importance in explaining interspecific variation in RGR.

These generalizations only apply when the same types of plants are compared, e.g. C_3 herbs or trees. Tree species, when compared with herbs, have both a low LAR and a low NAR. C_4 species tend to have a higher RGR due to a higher NAR, when compared with C_3 species. The relatively low RGR of shade-adapted species, grown at a high quantum flux density, is caused by a low NAR, rather than a low LAR (Poorter, 1989).

On average, a 10% increase in LAR is associated with a 4.3% decrease in NAR (Table 3). This is not as expected, because a high LAR decreases the respiratory burden per unit leaf area—see Eq. (4). In a few cases, this discrepancy may have been caused by the lower photosynthesis and NAR, resulting from self-shading in plants with a high LAR and a correspondingly large leaf area. However, in most cases it will be due to less well-defined interactions between the physiology, allocation, anatomy and chemical composition (Konings, 1989). Before discussing these interactions, we will first discuss inherent variation in each of these parameters (cf. Eq. (5)).

V. SPECIFIC LEAF AREA

As outlined in Section IV, variation in RGR is strongly correlated with that in LAR. Differences in LAR can be due to variation in LWR or in SLA. The specific leaf area is defined as the amount of leaf area per unit leaf weight. Its reciprocal, specific leaf weight or specific leaf mass, is also frequently used. Various aspects of inherent and environmentally induced variation in SLA have been reviewed by Dijkstra (1989). Large variations in SLA can be found between different types of plants and species fiom different habitats (Table 2). Evergreens mostly have a low SLA, whereas species with mesomorphic leaves show higher SLAs. Potter and Jones (1977), as well as Mooney *et al.* (1978) and Poorter and Remkes (1990; Figure 2A), found a positive relationship between RGR and SLA. A compilation of the data available from the literature led us to the conclusion that there is a close association between the potential growth rate of a species and its SLA (Table 3). SLA can therefore be considered as the prime factor determining interspecific variation in RGR.

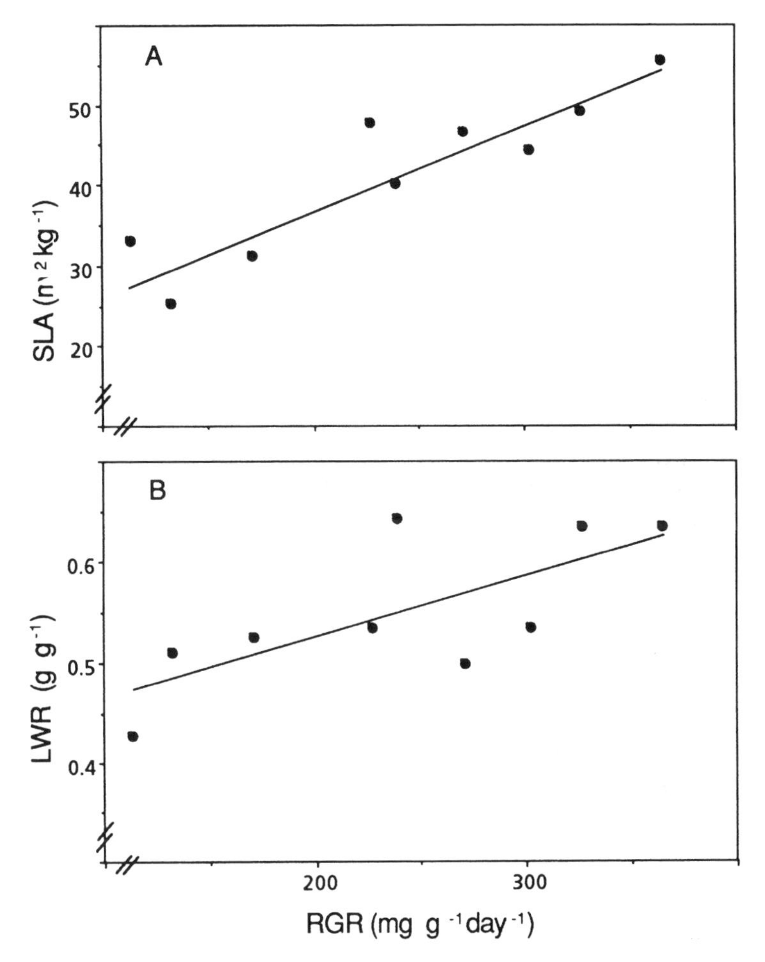

Figure 2 (A) The relationship between specific leaf area (SLA) and RGR and (B) the relationship between the leaf weight ratio (LWR) and RGR for the nine species described in Figure 1. (After Poorter and Remkes, 1990.)

A. Components of SLA

Which traits determine inherent variation in SLA? Starting from a simple leaf prototype, with chlorenchyma, vascular tissue and an epidermis, variation in SLA can be brought about by a change in several leaf characteristics. Firstly, a purely chemical difference between leaves may occur, due to accumulation of, for example, starch or secondary compounds. Starch may account for up to 30–40% of total leaf dry weight (McDonald *et al.*, 1986; Rufty *et al.*, 1988). Slower-growing species tend to invest relatively more in compounds which reduce the plant's palatability, e.g. tannin and lignin (Coley, 1983, 1986; Coley *et al.*, 1985). Accumulation of secondary compounds may be considerable. In some Australian dryland plants, resins make up 10–30% of the total leaf biomass (Dell and McComb, 1978). Lignin and other phenolic compounds form 14–40% of the total leaf dry weight of Californian chaparral shrubs (Merino *et al.*, 1984).

Secondly, a lower SLA may be caused by anatomical differences, e.g. extra layers of palisade parenchyma in sun species as opposed to shade species (Pons, 1977; Björkman, 1981); more support tissue, such as additional sclerenchyma, with an eight-fold difference between two *Agrostis* species (Pammenter *et al.*, 1986; see also Baruch *et al.*, 1985); or smaller cell sizes. A reduction in cell size without altering the total leaf cell volume will drastically increase the cell-wall surface/cell volume ratio and thus decrease SLA. The size of the veinal transport system may also affect SLA. In *Triticum aestivum*, veins contain 10% of the leaf weight (Rawson *et al.*, 1987). Givnish (1986) found the large veins of *Podophyllum peltatum* to comprise 6–20% of the total leaf biomass, depending on leaf size.

Thirdly, variation in SLA can be caused by a difference in investment in leaf hairs, thorns, etc. In two *Espeletia* species, leaf hairs comprise 4 and 20%, respectively, of the total leaf biomass (Baruch and Smith, 1979). Exceptionally, as in *Encelia farinosa*, leaf pubescence may account for up to 60% of total leaf dry weight (Ehleringer and Cook, 1984). However, in most plants neither pubescence nor thorns account for large differences in SLA.

Thus, in general, inherent variation in SLA is not merely caused by a difference in the amount of leaf cells per unit area, but also by variation in leaf anatomy, morphology or chemical composition. These differences in anatomy or morphology will also affect the chemical composition, as each of the above-mentioned anatomical and morphological structures has a distinctive chemical composition (cf. Kimmerer and Potter, 1987). The consequences of such differences in chemical composition for plant growth are discussed in Section VIII.

B. Plasticity in SLA

Plants grown under a low quantum flux density generally have a higher SLA and thinner leaves (Young and Smith, 1980), which is associated with fewer mesophyll cell layers (Pons, 1977; Björkman, 1981) and less non-structural carbohydrate (Waring *et al.*, 1985). To a small extent the higher SLA is also associated with lower concentrations of phenolic compounds, including lignin (Waring *et al.*, 1985; Mole and Waterman, 1988). There is hardly any evidence of differences in plasticity with respect to quantum flux density between species with different growth rates (Pons, 1977; Corré, 1983a,b; Grime *et al.*, 1989). However, the fast-growing *Holcus lanatus* increases SLA more at a low quantum flux density than the slow-growing *Deschampsia flexuosa* (Poorter, 1991). A similar difference was found in a comparison of *Veronica montana*, a (presumably slow-growing) woodland perennial, and *V. persica,* a (presumably fast-growing) annual weed. When grown under a leaf canopy as opposed to unshaded conditions, the SLA of *V. persica* changed considerably more than that of *V. montana* (Fitter and Ashmore, 1974).

Plants grown at low nutrient availability either show no change (Corré, 1983c; Sage and Pearcy, 1987; van der Werf *et al.*, 1992a) or a decrease in SLA (Sage and Pearcy, 1987; van der Werf *et al.*, 1992a). A decrease is at least partly due to accumulation of non-structural carbohydrates (Lambers *et al.*, 1981a; Waring *et al.*, 1985) or secondary compounds like lignin or other phenolics (Gershenzon, 1984; Waring *et al.*, 1985). No differences in plasticity have been found between fast- and slow-growing species (Corré, 1983c; van der Werf *et al.*, 1992a). However, data on the comparison of species grown under a range of conditions are scarce and differences in potential RGR between species sometimes small. A better-founded evaluation therefore awaits further experiments.

VI. BIOMASS ALLOCATION

Biomass allocation can be defined in terms of leaf, stem and root weight ratio, the fraction of total plant biomass allocated to leaves, stems and roots, respectively. A more frequently used parameter, the shoot: root ratio or its inverse, does not acknowledge the distinct functions of leaves and stems and is avoided here. Various aspects of inherent differences in biomass partitioning between leaves and roots have been discussed by Konings (1989). A low availability of nitrogen, phosphorus and water enhances allocation to roots, whereas a low quantum flux density promotes allocation to the leaves. The mechanism behind this "functional equilibrium" (Brouwer, 1963, 1983) is still poorly understood (Lambers, 1983). Genotypic differences in biomass partitioning between leaves, stem and roots have been correlated with differences in the level of gibberellins (*Zea mays*, Rood *et al.*, 1990a; *Brassica rapa*, Rood *et al.*, 1990b; *Lycopersicon esculentum*, Koornneef *et al.*, 1990; cf. Section XII.A) and abscisic acid (*Zea mays*, Saab *et al.*, 1990; *Lycopersicon esculentum*,

O.W. Nagel and H. Konings, personal communication) and with the plant's sensitivity to endogenous gibberellin (*Lycopersicon esculentum*, Jupe *et al.*, 1988).

A. Biomass Allocation at an Optimum Nutrient Supply

Some authors have found a negative correlation between LWR and RGR (Hunt *et al.*, 1987, plants grown at low quantum flux density; Shipley and Peters, 1990, at high quantum flux density), others a positive one (Ingestad, 1981; Poorter and Remkes, 1990, at intermediate quantum flux density, Figure 2B). Differences in growth conditions may have a decisive impact on the final result (Poorter and Lambers, 1991). Irrespective of the way the plants are grown, current information shows that LWR is less important than SLA in explaining inherent variation in RGR (Table 3).

Interestingly, monocotyledonous herbaceous species invest relatively more biomass in roots and less in leaves, compared to dicotyledonous ones with the same inherent RGR (Garnier, 1991). Moreover, the general trend of increasing RGR with increasing investment in leaf weight (Poorter and Remkes, 1990) only holds for dicotyledonous species and not for grasses (Garnier, 1991).

B. Plasticity in Biomass Allocation

In general, plants grown at a low quantum flux density show a shift in the allocation of biomass from roots and stem to leaves. This shift is generally more pronounced in faster-growing species (Pons, 1977; Werner *et al.*, 1982; Grime *et al.*, 1989; but see Corré, 1983b).

A decrease in nutrient availability often decreases LWR and increases RWR, particularly in fast-growing species (Christie and Moorby, 1975; Tilman and Cowan, 1989; Shipley and Peters, 1990). However, as noted previously, an interaction between nutrient supply and plant size may seriously affect the observed relationship between allocation and RGR, especially at a low nutrient supply (Ingestad, 1962, cited in Corré, 1983c). Taking into account only those references in which this artefact was most certainly avoided, a generally higher plasticity in allocation is still observed for fast-growing species (Christie and Moorby, 1975; Robinson and Rorison, 1988; Campbell and Grime, 1989; van der Werf *et al.*, 1992a; but see Bradshaw *et al.*, 1964; Crick and Grime, 1987).

VII. GROWTH, MORPHOLOGY AND NUTRIENT ACQUISITION OF ROOTS

The simple growth equation: RGR = NAR × LAR, suggests that any investment in biomass other than leaf area reduces the plant's RGR. Such an approach tends to consider the roots merely as a carbohydrate-consuming organ and does not give credit to their role in the acquisition of nutrients and water or their function in transport, storage and anchorage. In this section we will concentrate on the root's role in the acquisition of ions. Growth can then best be approached from an alternative point of view, where RGR is defined in terms of the root weight ratio (RWR), the net rate of nitrogen acquisition (NIR) and the plant's nitrogen concentration (PNC)—Eq. (6). Equation (6) suggests that a high RGR can be achieved by a large investment in root biomass, by a high rate of nitrogen uptake per unit root weight (specific ion uptake rate), or by a combination of these. However, a large investment in root weight may in fact reduce RGR because investment in leaves is reduced.

A. Root Growth and Nutrient Acquisition at an Optimum Nutrient Supply

Root systems of fast- and slow-growing species differ in their architecture. Slow-growing *Festuca* species from nutrient-poor habitats tend to have "herringbone" morphologies, i.e. next to one main axis, the root systems have only primary laterals (Fitter *et al.*, 1988). Results of a simulation model indicate that a herringbone morphology allows the most effective exploration and exploitation of mobile resources (Fitter, 1987). On the other hand, such a morphology is less efficient for long-distance transport, because the total transport path is longer, and hence requires a greater investment in root biomass. Fast-growing grassland species have a more random or nearly dichotomous root morphology.

Slow-growing grass species from nutrient-poor habitats generally have a higher specific root length (SRL, the root length per unit root weight) and relatively more fine roots (Berendse and Elberse, 1989; Boot, 1989; Boot and Mensink, 1990). However, SRL tends to vary with age in an unpredictable manner (Fitter, 1985) and some studies show a higher SRL for fast-growing grasses (Robinson and Rorison, 1985). No correlation between SRL and RGR was found in a comparison of 24 monocotyledons and dicotyledons species, grown at an optimum nutrient supply (Poorter and Remkes, 1990). A higher SRL is likely to contribute to the acquisition of ions which diffuse slowly in the soil (Clarkson, 1985), but also of more mobile ones when plants have to compete for these. Similarly, root hairs contribute to the acquisition of relatively immobile nutrients. There is a large variation between plant species with respect to root hair density and root hair length, but variation in these root characteristics does not seem to be related to the maximum growth rate of the species or its performance in nutrient-poor environments (Robinson and Rorison, 1987; Boot, 1989). The "proteoid" roots which are found in some dicotyledonous species are discussed in Section VII.C.

At an optimum nutrient supply, inherently fast-growing species have a somewhat lower RWR than slow-growing species from the same life form (Section VI). Since fast-growing species also have a higher nitrate and organic nitrogen concentration (Section VIII.A), it follows that the fast-growing species must have higher nitrogen absorption rates (cf. Eq. (6); Chapin, 1980).

Nassery (1970) compared the growth and phosphate uptake of the fast-growing *Urtica dioica* with that of the slow-growing. *Deschampsia flexuosa*. At an optimum nutrient supply, the fast-growing species had the highest nutrient uptake rate per unit root weight. This conclusion was corroborated by Christie and Moorby (1975), Chapin and Bieleski (1982), Chapin *et al.* (1986a), Garnier *et al.* (1989) and Poorter *et al.* (1991); Figure 3A). This is not to say that fast-growing species grow faster because their rate of nitrate or phosphate uptake is higher. Rather, the rapid uptake may be a result of their higher growth rate. At an optimum supply, the rate of ion uptake is, at least partly, determined by "demand", which results in a strong negative feedback when the growth rate is low (Clarkson, 1986; Rodgers and Barneix, 1988). An increased demand diminishes the negative feedback, thus enhancing net uptake (Jackson *et al.*, 1976; Doddema and Otten, 1979; Lambers *et al.*, 1982). When pregrown at a low phosphate concentration, the rate of phosphate uptake at a saturating concentration was 35% higher for the fast-growing *Urtica dioica* than for the slow-growing *Deschampsia flexuosa*. Pregrown at an optimum phosphate supply, this difference is about 300% (Nassery, 1970). Similar results have been obtained for other species, both with phosphate (Harrison and Helliwell, 1979; Clarkson and Scattergood, 1982) and a range of other ions (Lee, 1982; Glass, 1983). It is therefore very likely that a low ability to incorporate large amounts of absorbed nutrients into organic matter, rather than their low absorption capacity, controls the growth rate of plants from nutrient-poor sites at optimum nutrient supply.

Is there evidence for an inherently higher affinity (lower K_m) of the uptake system of slow-growing species from nutrient-poor sites? The phosphate uptake system of slow-growing *Carex* species has a high affinity for phosphate, in comparison with that of crop species (Atwell *et al.*, 1980). Data of Muller and Garnier (1990) on growth at low nitrate

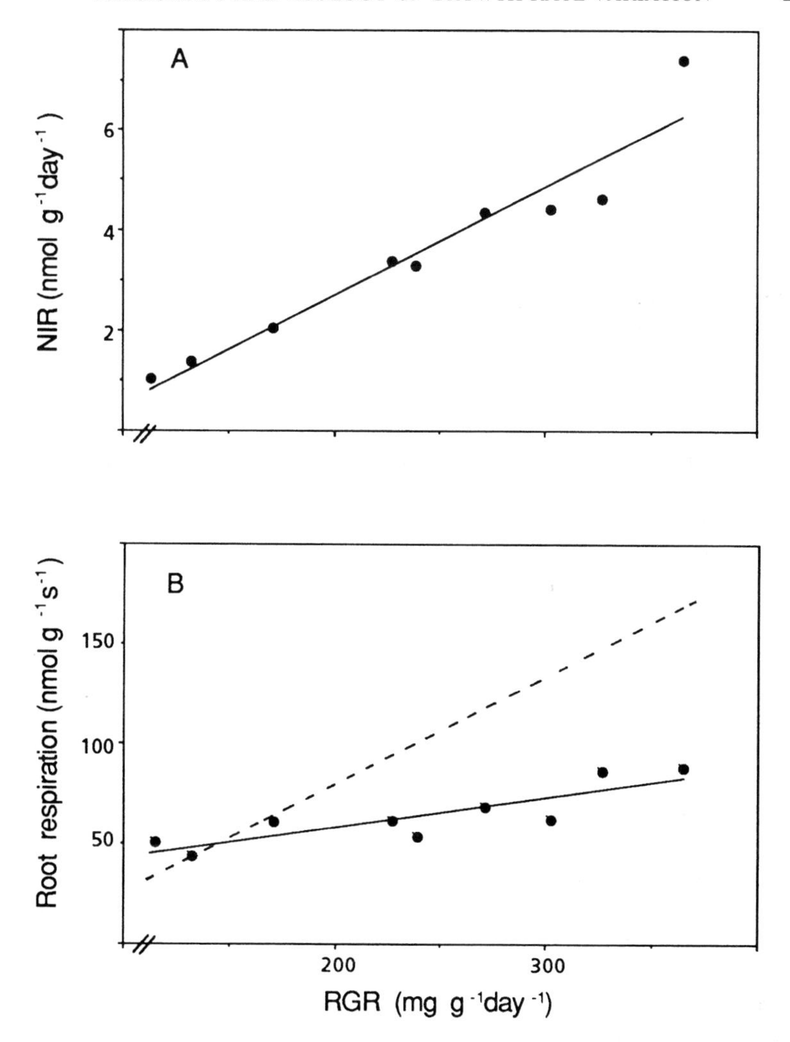

Figure 3 (A) The net influx of nitrate into roots (NIR) and (B) the rate of root respiration of fast-growing and slow-growing species, described in Figure 1. The dashed line in (B) gives the calculated rate of root respiration assuming specific costs for maintenance, growth and ion uptake to be the same as those for a slow-growing *Carex* species. (After Poorter *et al.*, 1990; Poorter *et al.*, 1991.)

concentrations, contrary to those of Freijsen and Otten (1984), suggest a lower K_m for nitrate of the uptake system in slower-growing species. However, in neither of these studies was the K_m determined. So far the experimental data do not support the hypothesis that the nitrate uptake system of slow-growing species from nutrient-poor habitats has a higher affinity for nitrate (van de Dijk, 1980; Bloom, 1985; Oscarson *et al.*, 1989). However, not many systematic comparisons of kinetic parameters of nitrate uptake in contrasting species are available from the literature and experiments have been carried out under different experimental conditions, so that a final conclusion cannot yet be reached.

B. The Plasticity of Parameters Related to Root Growth and Nutrient Acquisition

As discussed in Section VI, fast-growing species tend to be more plastic with respect to biomass allocation. At a growth-limiting nitrogen supply, inherently fast-growing species have a higher RWR than slow-growing ones of the same life-form. Both a low phosphate supply (Powell, 1974; Christie and Moorby, 1975) and a low nitrogen supply (Robinson and Rorison, 1985; Boot, 1989) mostly increase SRL. Slow-growing and fast-growing species respond in a similar manner (Fitter, 1985).

Root hairs are of vital importance for the acquisition of ions which diffuse slowly in soil, e.g. phosphate (Clarkson, 1985), and stimulation of root hair formation at a low nutrient supply has been reported frequently (Föhse and Jungk, 1983). The inherently slow-growing species *Deschampsia flexuosa* has a remarkable plasticity with respect to root hair formation, in comparison with fast-growing grass species (Robinson and Rorison, 1987; Boot and Mensink, 1990). Both fast-growing and slow-growing grasses respond to nutrient shortage with increased root hair length, but the tendency appears to be greatest in slow-growing ones (Boot and Mensink, 1990; Liljeroth *et al.*, 1990). The greater plasticity in density and length of root hairs in response to nutrient supply in inherently slow-growing species is likely to contribute to their successful performance in phosphate-poor environments.

Roots of crop species (Drew *et al.*, 1973; de Jager, 1982), fast-growing herbaceous wild plants (de Jager and Posno, 1979), trees (Philipson and Coutts, 1977) and desert perennials (von Willert *et al.*, 1992) have an amazing capacity to proliferate growth in nutrient-rich or moist patches in the root environment. Crick and Grime (1987) compared morphological plasticity in the slow-growing *Scirpus sylvaticus* and the faster-growing *Agrostis stolonifera*. *A. stolonifera* had a greater capacity to proliferate its fine roots in nutrient-rich patches. Therefore, it can dynamically exploit a fertile root environment and successfully compete with neighbouring plants, whereas the relatively large, but unresponsive, root system of the slow-growing *S. sylvaticus* is more advantageous when nutrients are strongly limiting and become available in temporally unpredictable pulses. A similar conclusion is drawn from data on two cold-desert bunchgrass species (Jackson and Caldwell, 1989) differing in RGR (Eissenstat and Caldwell, 1987). However, using a somewhat different technique from the one used by Crick and Grime (1987), Grime *et al.* (1991) concluded that the capacity to proliferate roots locally may be similar in slow- and fast-growing species. Our current information is insufficient to draw final conclusions on variation in the root's capacity to locally proliferate fine roots between fast-growing and slow-growing species.

Plants do not only respond to nutrient-rich patches with an increased root production, but also with an increased capacity for nutrient uptake (Drew and Saker, 1978; Lambers *et al.*, 1982). Jackson *et al.* (1990) compared the effect of a local increase in phosphate supply on two cold-desert *Agropyron* species, referred to above. Both species responded with an 80% increase in phosphate absorption capacity, and no differences between the two species were found with respect to the physiological plasticity of phosphate uptake.

Apart from spatial variation in nutrient availability, there may be variation in time. Campbell and Grime (1989) found that a potentially fast-growing species grows faster than an inherently slow-growing species when nutrient pulses are long, whereas the opposite is true under a regime of short pulses. In this case the frequency of pulses was constant, but the duration of the pulse varied, and thus the total amount of nutrients supplied. In an experiment with two *Plantago major* subspecies, where the frequency of pulses was increased, but total nutrient supply was constant, the faster-growing subspecies achieved an increasingly higher RGR than the slow-growing one (Poorter and Lambers, 1986).

It is concluded that fast-growing species are characterized by a high degree of plasticity in root morphology, such as the adjustment of their RWR and perhaps also the local proliferation of roots in nutrient-rich patches. Such a high degree of morphological plasticity is likely to be an integral part of the mechanism of resource acquisition in productive environments (Crick and Grime, 1987). Theoretical models predict that in infertile soils such a strategy leads to net losses of mineral nutrients from the plant (Sibly and Grime, 1986). Under poor conditions, alternative mechanisms requiring a lower investment of mineral nutrients but leading to greater losses of carbon from the roots, might have greater survival value. Here rapid proliferation of fine roots might incur net nutrient losses, as the nutrient costs of investing new roots in a generally poor soil might be higher than its returns. There is as yet no evidence for a difference in plasticity of nutrient acquisition between fast-growing and slow-growing species.

C. Other Root Characteristics Related to Nutrient Acquisition

Like leaves, roots are subject to a continuous turnover. Aerts (1989) reports a turnover rate of 0.6–1.7 g roots (g roots)$^{-1}$ year^{-1}. A high turnover rate of roots, like that of leaves, incurs a net loss of nutrients; as discussed in Section XV. We are not aware of systematic comparisons between fast-growing and slow-growing species with respect to root turnover.

Specialized root structures termed "proteoid" roots, are formed on members of the Proteaceae (Lamont, 1982), many of which are slow-growing species from very nutrient-poor soils (Barrow, 1977). Proteoid roots consist of sections of dense "bottle-brush-like" clusters of short (5–10 mm) rootlets, covered with a dense mat of root hairs. Such structures induce the release of various nutrients from sparingly soluble sources (Section XI.A). Proteoid roots, or functionally similar structures, are also known from fast-growing crop species (Gardner *et al.*, 1982a; Hoffland *et al.*, 1989a).

Another special structure, which allows plants to grow in phosphate-poor environments, is the mycorrhiza, an association between a fungus and roots. Both fast- and slow-growing species have the capacity to form such an association, predominantly under phosphate-poor conditions. Some species are inherently non-mycorrhizal, but this characteristic is also not associated with the inherent growth rate of the species (cf. Tester *et al.*, 1987).

D. Conclusions

Variation in RGR between species is certainly associated with variation in root attributes. Slow-growing species tend to have a "herring bone" architecture, rather than the more random structure of fast-growing species, and also a higher specific root length. Fast-growing species tend to invest relatively less biomass in roots when grown at an optimum supply of nutrients, but have a greater capacity to adjust their investment in root biomass as well as associated structures than slow-growing species. The nutrient uptake rate per unit root dry weight of fast-growing species is higher than that of slow-growing species, but this may well be an effect of rapid growth, rather than its cause.

VIII. CHEMICAL COMPOSITION

Plant dry matter is composed of a number of major compounds, which can be grouped into the following seven categories: lipids, lignin, organic

N-compounds, (hemi)cellulose, non-structural sugars, organic acids and minerals. Apart from "primary" compounds, there is a wealth of "secondary" compounds, defined by the absence of a clearly defined role in the metabolic processes of the plant (Baas, 1989; Waterman and McKey, 1989). Lignin is often included in the category of "secondary" compounds and this will also be done here.

A. Primary Compounds

When grown at an optimum nutrient supply, inherently fast-growing species have a higher total and organic nitrogen concentration per unit plant dry weight than slow-growing ones (Poorter *et al.*, 1990; Figure 4A). Their higher organic nitrogen concentration is due partly to a greater biomass investment in leaves, which tend to have a higher nitrogen concentration than other vegetative plant organs, and partly to a higher nitrogen concentration in all vegetative organs *per se* (Poorter *et al.*, 1990). Soluble protein constitutes a larger fraction of the total leaf nitrogen concentration in a fast-growing *Plantago major* subspecies than in a slower-growing one (Dijkstra and Lambers, 1989a).

Fast-growing (sub)species generally contain more minerals and organic acids per unit dry weight than slow-growing ones, when grown at an optimum nutrient supply (Chapin and Bieleski, 1982; Dijkstra and Lambers, 1989b; Poorter and Bergkotte, 1992). Some comparative studies show that phosphorus, particularly inorganic phosphate, accumulates to a greater extent in fast-growing (sub)species (Nassery, 1970; Chapin and Bieleski, 1982; Dijkstra and Lambers, 1989b). Others show the opposite (Christie and Moorby, 1975; Chapin *et al.*, 1982). Accumulation of nitrate appears to be characteristic of fast-growing, "nitrophilous" species (Dittrich, 1931; Smirnoff and Stewart, 1985; Poorter and Bergkotte, 1992). Such accumulation is most pronounced at a high nitrate supply (Stulen *et al.*, 1981) and at a relatively low quantum flux density, when nitrate replaces soluble carbohydrates and carboxylates as an osmotic solute (Stienstra, 1986; Veen and Kleinendorst, 1986; Blom-Zandstra and Lampe, 1985).

It is sometimes claimed (cf. Chapin, 1988) that slow-growing species show more "luxury consumption" than fast-growing ones. Apart from the fact that this is not compatible with most of the comparative data cited in this section, it still remains to be demonstrated whether luxury consumption exists at all. We propose to define "luxury consumption", as the absorption beyond a rate which leads to more growth, rather than as "vacuolar storage during the period of active growth" (Chapin, 1988). Accumulation of nitrate does not imply "luxury consumption", since nitrate

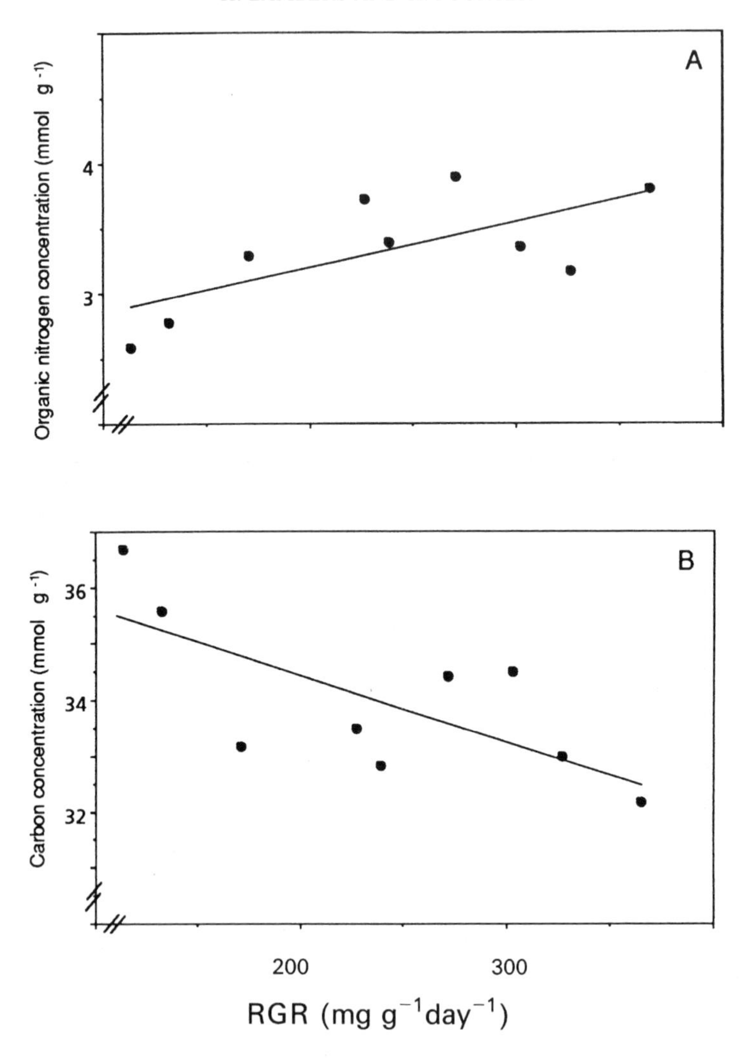

Figure 4 Aspects of the chemical composition of fast-growing and slow-growing species, described in Figure 1. (A) Organic nitrogen concentration of plant dry matter. (B) Carbon concentration of plant dry matter. (C) Dry matter content (dry weight × 100/fresh weight) of plant biomass. (D) Construction costs of plant biomass, calculated as outlined in Penning de Vries *et al.* (1974) and Lambers and Rychter (1989). (After Poorter *et al.*, 1990; Poorter and Bergkotte, 1992.)

may replace organic solutes, which leads to more carbon being available for metabolic processes. For example, faster-growing genotypes of *Lactuca sativa* accumulate nitrate, rather than organic solutes (Blom-Zandstra *et al.*, 1988). For phosphate, the situation may be different, but

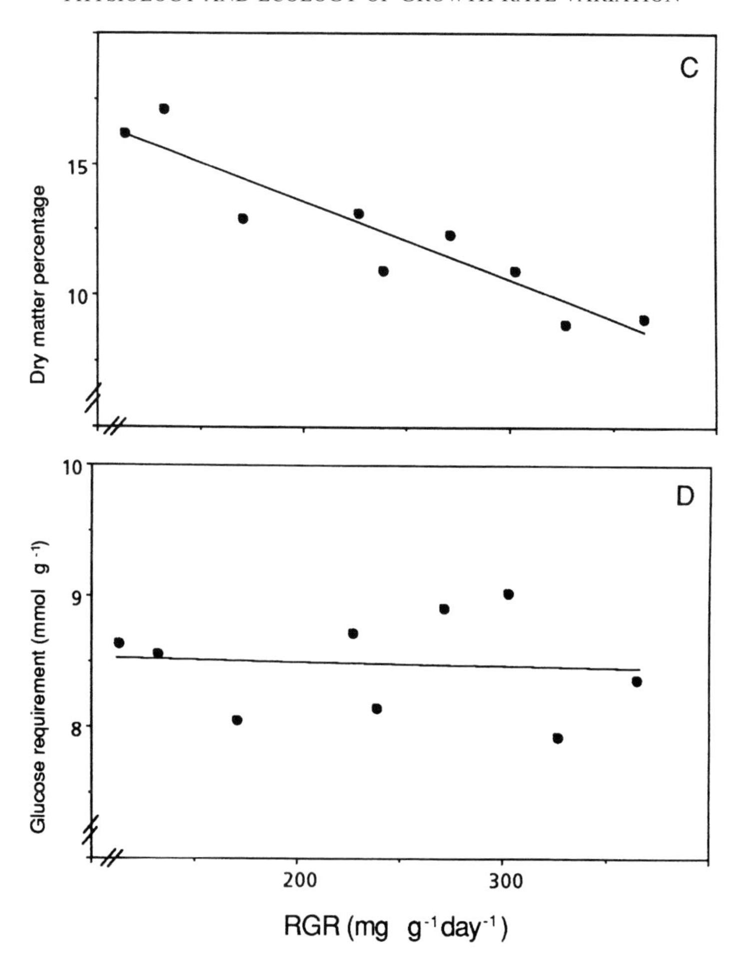

Figure 4 Continued.

data in the literature are conflicting (e.g. Chapin and Bieleski, 1982; Chapin *et al.*, 1982).

Slow-growing (sub)species contain more cell-wall components (lignin and (hemi)cellulose) than fast-growing ones, when grown at optimum nutrient supply (Dijkstra and Lambers, 1989a; Poorter and Bergkotte, 1992). Using pyrolysis-mass-spectrometry, Niemann *et al.* (1992) showed that fast-growing species contain relatively more compounds associated with the cytoplasm and slow-growing ones more of those associated with cell walls. At a relatively fixed cell size, accumulation of cell-wall components and solutes will also alter the water content (g H_2O g^{-1} DW) of the different organs. Indeed, both leaves, stems and roots of slow-growing (sub)species contain rather low amounts of water per unit dry weight (Dijkstra and Lambers, 1989b; Poorter and Bergkotte, 1992; Figure 4C). In the *Lactuca sativa* genotypes mentioned above, replacement of nitrate for organic acids partly explains genotypec differences in leaf dry matter content (Reinink *et al.*, 1987; Blom-Zandstra *et al.*, 1988).

B. Secondary Compounds

Plants contain a suite of "secondary" compounds, which serve a range of distinct ecological functions, including allelopathy, the deterrence of herbivores, attraction of pollinators and attraction of organisms predating on herbivores (Harborne, 1982; Chou and Kuo, 1986; Baas, 1989; Dicke and Sabelis, 1989). Much attention has been paid to the role of secondary plant compounds in reducing herbivory. In this context they are often classified as "quantitative" vs. "qualitative" defence compounds (Harborne, 1982; Waterman and McKey, 1989). Typically, quantitative secondary compounds are composed of C, H and O only, have low turnover rates and act as digestibility reducers when present in large amounts. Qualitative secondary compounds tend to be specific toxins which occur in low concentrations and may be subject to rapid turnover. Some of these toxins contain nitrogen (cyanogenic glycosides, analogues of amino acids, alkaloids), but many do not (cardenolides, glucosinolates, saponins).

Slow-growing species or genotypes accumulate more quantitative secondary plant compounds than fast-growing ones (Coley, 1986; Waterman and McKey, 1989). The concentration of lignin (Coley, 1983), condensed tannin (Waterman and McKey, 1989), volatile terpenoids (Morrow and Fox, 1980) or diplacol (Merino *et al.*, 1984) may comprise 10–30% of a plant's leaf dry weight.

There is no clear-cut correlation between a plant's inherent growth rate and the concentration of qualitative secondary compounds in its tissues. Perhaps the only generalization that can be made is that fast-growing species, if they have any antiherbivore chemicals at all, accumulate only qualitative compounds. There is a vast array of qualitative secondary compounds, which generally account for less than 1% of the dry weight (Baas, 1989). However, their distribution has not been systematically investigated for a range of species grown under standard conditions.

Slow-growing species may also accumulate qualitative compounds, often in a manner which complements accumulation of quantitative secondary compounds. A special case is that of young leaves. Most herbivores preferentially feed on leaves with a high protein and water content, a low leaf toughness and a low concentration of antiherbivore compounds (Coley, 1983; Kimmerer and Potter, 1987; Waterman and McKey, 1989). In young leaves of slow-growing plants, leaf toughness and the concentration of digestibility-reducing compounds is often low, whereas the nutrient and water content is relatively high compared to older leaves. Hence, young leaves appear to be attractive for herbivores, but they also accumulate toxic compounds, e.g. anthocyanins in tropical trees (Coley and Aide, 1989) or saponins in *Ilex opaca* (Potter and Kimmerer, 1989). Also mature leaves of *Ilex opaca* accumulate saponins, namely in their mesophyll cells, which are not protected by digestibility-reducing compounds like lignin, crystals and tannin. These quantitative secondary compounds accumulate instead in other cells of the same leaves (Kimmerer and Potter, 1987).

One obvious risk of the accumulation of specific toxins is that herbivorous organisms coevolve and become insensitive to the defence (Harborne, 1982). In the case of *Ilex opaca*, mentioned above, specialist leaf miners consume only the mesophyll tissue, apparently able to neutralize the toxic saponins in these cells. The chances for tolerance to evolve against quantitative antiherbivore compounds are considerably smaller. However, some degree of tolerance against palatability-reducing compounds has been demonstrated (Bernays *et al.*, 1989) and some herbivores even incorporate these quantitative compounds, thus possibly gaining protection (Taper and Case, 1987).

C. Defence under Suboptimal Conditions

Environmental conditions, such as nutrient supply and water stress, may restrict plant growth more than expected from their effect on photosynthesis. Under such conditions non-structural carbohydrates accumulate, leading to an excess of carbon in the plant. When plants have such an excess of carbon, accumulation of carbon-based defences is expected. Similarly, at a high availability of nutrients and a relatively low quantum flux density, accumulation of nitrogen-based secondary compounds is predicted. Confirmation of this carbon/nutrient balance theory (Bryant *et al.*, 1983) has been found in studies where plants, grown at a low nutrient availability, show an increase in the concentration of condensed tannins, total phenols and/or phenol glycosides (Waring *et al.*, 1985; Bryant *et al.*, 1987; Nicolai, 1988; Margna *et al.*, 1989; but see Denslow *et al.*, 1987). Similarly, Johnson *et al.* (1987) observed a positive correlation between N-supply and the concentration of alkaloids.

D. Effects of Chemical Defence on Growth Potential

What are the costs associated with the accumulation of antiherbivore compounds? The amount of glucose needed to produce 1 g of a toxic compound may be high (cf. Baas, 1989). However, as the concentration of toxins in plant tissues is rather low, their accumulation hardly affects the cost of synthesizing plant biomass.

The glucose needed to produce 1 g of digestibility-reducing compounds is generally lower than that of qualitative secondary compounds (Baas, 1989; Lambers and Rychter, 1989). In the case of tannin, approximately the same amount of glucose is required as to construct cellulose or starch. However, because these compounds may accumulate in large quantities in plant tissue, they may comprise a large part of the plant's carbon resources. This carbon cannot be used for the construction of the photosynthetic apparatus, so that the photosynthetic return per unit weight of a well-protected leaf is less than that of a leaf which allocates less carbon to quantitative defence compounds.

The costs of the accumulation of secondary compounds exceed the specific costs of synthesis. The enzyme apparatus to produce these compounds must also be maintained and there is a turnover of different compounds (e.g. monoterpenes: Burbott and Loomis, 1969; alkaloids: Waller and Nowacki, 1978; cyanogenic compounds: Adewusi, 1990). Moreover, toxic compounds must be stored in special compartments or structures in which they cannot harm the plant's metabolism. Examples include oil glands containing essential oils in *Eucalyptus* species (Welch, 1920), leaf hairs containing carvone in *Mentha spicata* (Gershenzon *et al.*, 1989) and the separation of cyanogenic glycosides from the enzymes releasing HCN in *Phaseolus lunatus* (Frehner and Conn, 1987).

An alternative approach to calculate the costs of accumulating secondary compounds relates their concentration to the growth of a plant. Coley (1986) found a negative correlation between the rate of leaf production and the leaf tannin concentration of *Cecropia peltata* (Figure 5). However, the negative correlation may be a reflection of correlating, but in themselves unrelated, plant characteristics. Accumulation of secondary compounds may even be the phenotypic result of slow growth, as discussed above.

E. The Construction Costs of Plant Material

At an optimum nutrient supply, fast-growing species have a lower carbon concentration in the various organs than slow-growing ones

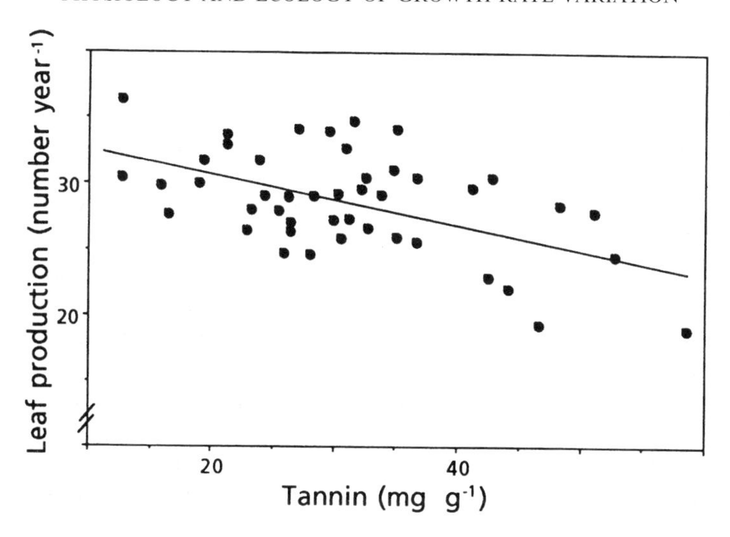

Figure 5 Leaf production and tannin concentration in *Cecropia peltata*. (After Coley, 1986.)

(Poorter and Bergkotte, 1992; Figure 4B). This is a consequence of a difference in chemical composition. The different components of the plant's biomass vary in carbon concentration ranging from high in lipids to zero for minerals (cf. Table 4). Therefore, a plant with a high proportion of biomass invested in compounds with a high proportion of carbon, like lipids and protein, has to fix more carbon to construct one unit of plant weight than a plant that consists mainly of (hemi)cellulose, organic acids and minerals (cf. Eq. (5), Section III). The difference in carbon concentration of fast- and slow-growing species is about 10%, whereas that in RGR is over 300% (Figure 4B). Thus, variation in carbon concentration has only a rather small effect on variation in RGR.

Construction of the different compounds not only requires glucose for C-skeletons, but also for the generation of ATP and NAD(P)H (Section X). Generally, compounds with a high carbon concentration are more reduced and require more glucose for their synthesis (Penning de Vries *et al.*, 1974; Table 4). Hence, glucose costs for the synthesis of biomass can be derived from the carbon concentration of biomass, provided a correrction is made for the mineral concentration (Vertregt and Penning de Vries, 1987). Alternatively, these costs can be calculated from the concentration of the various primary and secondary compounds of plant biomass (Penning de Vries *et al.*, 1974; Lambers and Rychter, 1989). Although variation in glucose requirement exists, there is not much

Table 4 The carbon concentration (mmol/g) of a number of primary and secondary compounds present in plant biomass, as well as the requirement of glucose (mmol g^{-1}) and oxygen (mmol g^{-1}), and the carbon dioxide release (mmol g^{-1}) during synthesis of these compounds from glucose and nitrate. The values for oxygen requirement and carbon dioxide release are used to calculate the expected respiratory quotient (RQ) during synthesis of these compounds. (The principles of the calculations are outlined in Penning de Vries *et al.* (1983) and Lambers and Rychter (1989), but different values have been used, where appropriate.)

Component	Carbon concentration	Glucose costs	O_2-requirement	CO_2-release	RQ
Volatile terpenoids					
Limonene	61.3	18.8	29.6	59.2	2.0
Lipids	53.8	16.8	11.0	36.5	3.3
Protein	36.8	13.8	13.5	37.9	2.8
Phenolics					
Lignin	46.3	11.8	5.9	13.1	2.2
Ellagitannin	32.8	8.6	12.1	22.7	1.9
Structural carbohydrates					
Hemicellulose	32.0	7.1	3.6	3.6	1.0
Cellulose	30.8	6.5	2.1	2.1	1.0
Non-structural carbohydrates					
Starch	30.8	6.5	2.1	2.1	1.0
Sucrose	29.3	6.1	1.5	1.5	1.0
Organic acids					
Citric acid	26.1	4.3	5.2	0.0	< 1
Malic acid	24.8	3.7	0.0	−7.5	< 1
Minerals	0.0	—	—	—	—

difference between slow-growing evergreens, slow-growing deciduous plants and faster-growing species (Chapin, 1989; Lambers and Rychter, 1989). In a comparison of a range of herbaceous species, the costs for construction of plant biomass was very similar for fast-growing and slow-growing species (Poorter and Bergkotte, 1992; Figure 4D). There are two reasons for this relative constancy of construction costs. Firstly, the production of protein, which is present in larger amounts in fast-growing species, requires a similar amount of glucose to that of quantitative secondary compounds, characteristic of slow-growing species (Table 4; cf. Sections VIII.A and VIII.B). And, secondly, the higher concentration of costly proteins coincides with an increased concentration of cheap compounds, such as organic acids and minerals (cf. Section VIII.A).

F. Conclusions

Fast-growing species are characterized by a high organic nitrogen and mineral concentration, whereas slow-growing species accumulate relatively more quantitative secondary compounds, which play a role in reducing herbivory. The cost of constructing leaves with these contrasting chemical composition differs only marginally, but the photosynthetic return per unit weight of the leaves of fast-growing plants will be much higher.

IX. PHOTOSYNTHESIS

A. Species-specific Variation in the Rate of Photosynthesis

Fast-growing crop species (Evans, 1983; Makino *et al.*, 1988) and their accompanying weeds (Sage and Pearcy, 1987) tend to have higher maximum rates of photosynthesis (expressed per unit leaf area) than evergreen trees and shrubs (Field *et al.*, 1983; Langenheim *et al.*, 1984). Similarly, sun species have a higher rate of light-saturated photosynthesis per unit area than slower-growing shade species, when the plants are grown at an optimum quantum flux density (e.g. Pons, 1977; Björkman, 1981; Seemann *et al.*, 1987). Fast-growing tree and shrub species have higher rates of photosynthesis per unit leaf area than slower-growing ones (Mooney *et al.*, 1978, 1983; Field *et al.*, 1983; Oberbauer *et al.*, 1985). Some of these differences may be phenotypic, rather than inherent for a species, reflecting a poor nutrient or water supply in the natural habitat.

Generally, fast-growing species tend to have higher rates of photosynthesis than slow-growing ones, at least when photosynthesis is expressed per unit leaf weight (Gottlieb, 1978; Dijkstra and Lambers, 1989a; Poorter *et al.*, 1990). The difference may persist when expressed per unit leaf area (Schulze and Chapin, 1987; Evans, 1989a), as long as species of vastly different life forms are compared. In comparisons of species of similar life forms, e.g. herbaceous species (Dijkstra and Lambers, 1989a; Poorter *et al.*, 1990), fast-growing and slow-growing species have very similar rates of photosynthesis per unit leaf area. Hence, variation in the rate of photosynthesis per unit leaf area does not offer an explanation for differences in RGR between species of similar life form.

Very little research has been done to elucidate differences in photosynthesis between inherently slow-growing and fast-growing species. Information providing a framework for further analysis of inherent differences in photosynthesis is discussed below and is confined in the main to C_3 species and not C_4 or CAM plants.

Variation in photosynthetic capacity may reflect differences in organic nitrogen concentration. This capacity is related to a leaf's nitrogen concentration, because the major part of all organic nitrogen in the mesophyll cells of a C_3 plant is found in the chloroplasts (Evans, 1989a). Indeed, leaves of fast-growing species have a higher nitrogen concentration (Section VIII). However, at the same nitrogen concentration in the leaf, there is still a wide variation in light-saturated rates of photosynthesis between species (Evans, 1989a).

B. Photosynthetic Nitrogen Use Efficiency

The rate of photosynthesis per unit leaf nitrogen, the photosynthetic nitrogen use efficiency (PNUE), is higher for fast-growing herbaceous species than for slow-growing ones with the same life form, at least when measured at the moderate quantum flux density at which the plants were grown (Poorter *et al.*, 1990). Therefore, we will discuss plant traits which may explain inherent variation in PNUE.

With increasing nitrogen concentration per unit leaf area, photosynthesis is saturated at an increasingly higher quantum flux density. If measured at a quantum flux density which saturates photosynthesis at a low, but not at a high nitrogen concentration in the leaf, a curvilinear photosynthesis–leaf–nitrogen relationship is inevitable (Evans, 1989a). Such a situation was found in a comparison of *Plantago major* subspecies (Dijkstra, 1989). The slow-growing subspecies, with a low PNUE when determined at the relatively low quantum flux density at which plants were grown, has a higher chlorophyll concentration per unit leaf area (Dijkstra and Lambers, 1989a). At the relatively low quantum flux density, this will cause shading of the chloroplasts near the lower leaf surface. However, measured at light- and CO_2-saturation the PNUE was the same for both subspecies. These results agree with those on six fast- and slow-growing monocotyledonous species. When the quantum flux density during the measurements of photosynthesis was increased from that at which the plants were grown to light saturation, both the slope of the CO_2-response curve and the CO_2-saturated rate of photosynthesis increased significantly more for the slow-growing species than for the fast-growing ones (A. van der Werf, personal communication). Since many leaves often function at a quantum flux density well below the saturation level of photosynthesis, comparison of PNUE values determined at a low quantum flux density is valid, providing the quantum flux density is the same for all plants compared (Schulze, 1982; Karlsson, 1991).

Measuring photosynthesis at light saturation in field-grown plants, Field and Mooney (1986) found the highest PNUE values for annuals, intermediate values for drought-deciduous shrubs, and the lowest values for evergreen trees and shrubs. Though some of these differences may have been phenotypic, there is likely to be a strong inherent component as well (cf. Evans, 1989a). What could be the basis of the relatively law PNUE of slow-growing species?

1. Partitioning of Nitrogen between Chloroplasts and other Cell Components

Variation in PNUE might reflect a difference in investment of nitrogen in photosynthetic and non-photosynthetic leaf components. In C_3 species with a high PNUE, approximately 75% of the nitrogen in mesophyll cells is located in the chloroplasts (Evans, 1989a). It is likely that non-photosynthetic and photosynthetic cells require a similar amount of nitrogen not associated with photosynthesis. Part of this nitrogen is associated with primary cell walls, which are claimed to contain up to 20% structural proteins by weight (Jones and Robinson, 1989). These proteins ("extensin") are rich in hydroxyproline and appear to be associated with resistance to microbial attack and some forms of abiotic stress (Esquerré-Tugayé *et al.*, 1979; Lamport and Catt, 1981). Apart from "extensin", other (hydroxy)proline-rich proteins occur in plant cell walls and some of these are probably also associated with plant defence reactions (Lamport, 1980; Kleis-San Francisco and Tierney, 1990). Cell wall components such as (hemi)cellulose and lignin represent a considerably greater fraction of the leaf dry weight in slow-growing herbaceous species than of that in fast-growing ones, whereas the reverse is true for the organic nitrogen fraction (Poorter and Bergkotte, 1992). We do not know if the amount of protein per unit weight of cell walls of fast- and slow-growing species is the same. If so, then the fraction of organic nitrogen that is tied up in cell walls is certainly much greater in slow-growing species. This indicates that a low PNUE may be partly associated with greater investment in cell-wall components.

Leaves with a very thick or multiple epidermis (Esau, 1977), crystal cells (Kimmerer and Potter, 1987), collenchyma and sclerenchyma elements (Konings *et al.*, 1989), or cells with specific functions (e.g. water storage; Schmidt and Kaiser, 1987), must invest part of the nitrogen in their leaves in these structures. Thus, leaves which contain some of these additional elements are bound to have a lower PNUE.

Similarly, accumulation of relatively large quantities of nitrogen-containing molecules as compatible solutes (e.g. proline and glycinebetaine;

Wyn Jones and Gorham, 1983), storage proteins (Franceschi *et al.*, 1983; Staswick, 1988), peptides that sequester heavy metals (Lolkema *et al.*, 1984; Robinson and Jackson, 1986), protective compounds (e.g. polyamines; Galston and Sawhney, 1990; Kuehn *et al.*, 1990), antifungal polypeptides (e.g. thionins; Reimann-Philipp *et al.*, 1989; Apel *et al.*, 1990), or toxic antiherbivore compounds (e.g. cyanogenic glycosides: Kakes, 1987; cyanolipids: Poulton 1990; alkaloids: Hartmann *et al.*, 1989) also decreases PNUE.

We conclude that a low PNUE may be a consequence of a large investment of nitrogen in cell walls, specialized cells or compounds that are not associated with photosynthesis.

2. *Suboptimal Partitioning of Nitrogen within the Chloroplast*

When grown at optimum nitrogen supply, slow-growing herbaceous species have higher chlorophyll concentrations per unit leaf area and unit nitrogen (Poorter *et al.*, 1990). Slow-growing herbs have double the concentration of chlorophyll found in fast-growing species (0.6 vs. 0.3 mmol m^{-2}). This requires the extra investment of at least 15 mmol of nitrogen per square metre of leaf area, which amounts to 12% or more of a leaf's total nitrogen concentration in the slow-growing herbs, but only increases the leaf's absorptance by 7% (Evans, 1989b). Although extra investment in chlorophyll increases photosynthetic performance under shade conditions (Evans, 1989b), there may well be some excess in capacity of the light-harvesting machinery. But this offers only a partial explanation for the low PNUE of slow-growing species.

3. *Activation of Rubisco*

Activation of Rubisco requires carbamylation of the enzyme (Salvucci, 1989). The degree of carbamylation depends on the quantum flux density. In some, but not all higher plant species, Rubisco is also regulated by a naturally occurring tight-binding inhibitor: 2-carboxyarabinitol 1-phosphate (Servaites, 1990). The difference in photosynthesis per unit Rubisco at high quantum flux density cannot be explained by variation in the degree of enzyme activation by either of the above mechanisms (Seemann, 1989). Intrinsic differences in the enzyme from different species, rather than regulation of its activity by activation, are the likely cause of variation in the specific activity of Rubisco.

4. *Variation in Rubisco Specific Activity*

Variation in PNUE has been further analysed in a comparison of *Alocasia macrorrhiza*, a tropical understorey plant, with two crop species (Seemann *et al.*, 1987; Seemann, 1989). *Alocasia* has a considerably lower photosynthetic capacity per unit leaf nitrogen or Rubisco protein. The relatively low PNUE of *Alocasia* is partly a consequence of a relatively low specific activity (carboxylating activity per unit enzyme) of its Rubisco. A low specific activity is not restricted to shade-tolerant or inherently slow-growing species, but has also been found in comparisons of fast-growing crop species (Seemann and Berry, 1982, cited in Evans, 1989a; Makino *et al.*, 1988). So far we do not know the biochemical basis of variation in specific activity of Rubisco or if there is any systematic link between specific activity and ecological traits.

5. *Feedback Inhibition of Photosynthesis*

Comparing the rate of photosynthesis at normal and high internal partial pressure (p_i) of CO_2, provides insight into the extent of feedback control of photosynthesis (Sharkey *et al.*, 1986; Sage and Sharkey, 1987). Such control may play an important role when the products of photosynthesis, i.e. sucrose and its pbosphorylated precursors, accumulate, due to for example a relatively low temperature (Sage and Sharkey, 1987) or a limited activity of the sink (Plaut *et al.*, 1987). Under these conditions the photosynthetic apparatus is only partly used, so that the PNUE is less than maximal. We do not know if such a situation of feedback inhibition is the norm in slow-growing species, but if so it could offer an explanation for their relatively low PNUE.

6. *Effects of the CO_2 Concentration inside the Leaf*

Photosynthesis increases with increasing internal partial pressure of CO_2 inside the leaf (p_i), up to a maximum. Differences in p_i between species or populations have been reported (Mooney and Chu, 1983) and might offer a partial explanation for a low PNUE in leaves with a low SLA and a high nitrogen concentration per unit leaf area, as found for slow-growing species (Poorter *et al.*, 1990). Parkhurst *et al.* (1988) found a significant CO_2 pressure gradient inside the leaf, more so for hypostomatous than for amphistomatous leaves (Parkhurst and Mott, 1990). Moreover, the CO_2 pressure at the site of carboxylation (p_c) might be significantly lower than p_i. Hence, a greater internal CO_2 diffusion limitation also partly explains a low PNUE. If generally valid, species with a low PNUE should show less discrimination against $^{13}CO_2$ than those with a higher PNUE. The degree of discrimination reflects the ratio of the CO_2 concentration inside the leaf and that in the atmosphere (p_a) (cf. Farquhar *et al.*, 1982).

Our own data on a range of herbaceous species grown under the same conditions do not show a correlation between carbon isotope discrimination and RGR (Figure 6B). In a comparison of species from high altitudes with related species from low altitude, the high-altitude species show less discrimination, which correlates with their lower ratio of p_i/p_a (Körner and Diemer, 1987; Körner *et al.*, 1988). Leaf carbon isotope discrimination of annuals (Smedley *et al.*, 1991) and short-lived perennials (Ehleringer and Cooper, 1988) is less than that of longer-lived (and presumably slower-growing) perennials, also when grown under the same environmental field conditions. Our plants (Figure 6) were grown under well-watered conditions and at a relatively low vapour pressure deficit, which may have led to a relatively high p_i (cf. Schulze *et al.*, 1987; Brugnoli and Lauteri, 1991). We cannot fully exclude that variation in p_i between species occurs when plants are supplied with a limiting amount of water or at a higher vapour-pressure deficit and that this correlates with that in PNUE. However, the variation in PNUE in comparisons of widely different species (Evans, 1989a) is certainly too great to be fully accounted for by variation in p_i. Moreover, the variation in PNUE as found in a comparison of herbaceous species cannot be accounted for by variation in p_i (Figure 6A,B).

7. *Conclusions*

To summarize the above, slow-growing species with a relatively low PNUE may invest relatively large quantities of nitrogen in components not associated with photosynthesis. They may also have a suboptimal distribution of nitrogen between elements of the photosynthetic apparatus or a Rubisco enzyme with a low catalytic capacity. A larger degree of feedback inhibition or a relatively low CO_2 concentration at the site of carboxylation might also play a role. There is as yet no convincing evidence for any of these possible explanations.

C. Is There a Compromise between Photosynthetic Nitrogen Use Efficiency and Water Use Efficiency?

The leaf's stomatal conductance tends to be regulated in each a way that a compromise is reached between gain of CO_2 and loss of H_2O (Farquhar and

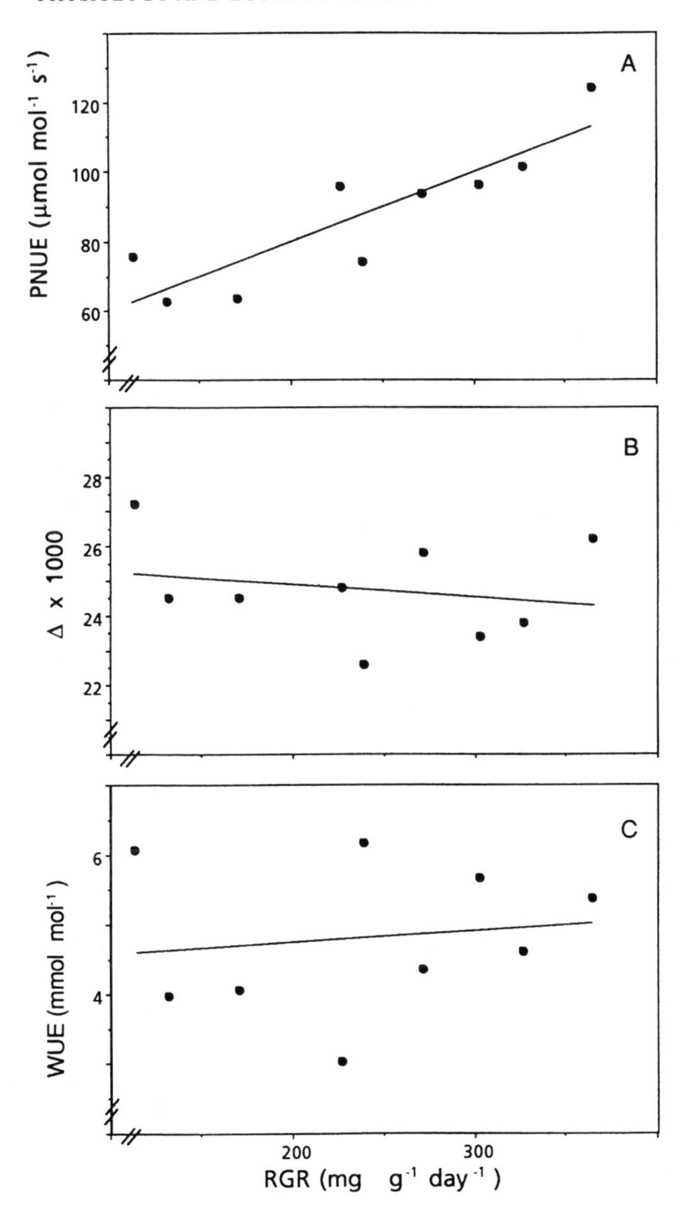

Figure 6 Gas exchange-related characteristics of fast-growing and slow-growing species, described in Figure 1. (A) The rate of photosynthesis expressed on a leaf nitrogen basis. (B) Carbon isotope discrimination. (C) The water use efficiency. Gas exchange characteristics were determined under conditions used for growing the plants. (After Poorter *et al.* (1990) and unpublished data of H. Poorter and G. D. Farquhar; further information, based on 24 species, is unpublished.)

Sharkey, 1982). Increased conductance would lead to roughly proportionally greater transpiration, but marginally greater photosynthesis. However, different environmental conditions may require a different compromise (Cowan, 1977). Species which naturally occur in environments where the water supply is low and/or the evaporative demand is high might have a lower stomatal conductance and a lower intercellular pressure of CO_2 (p_i). This tends to increase their water use efficiency of photosynthesis (WUE), but, all other parameters being equal, leads to a lower PNUE, as discussed above. Hence, a low PNUE might be a reflection of a high WUE. In a comparison of five Californian evergreen species, Field *et al.* (1983) indeed found that species with a lower PNUE tended to have a higher WUE. The water-efficient species typically occurred in the driest habitats. Pavlik (1983) did not find a different WUE for two dune grasses differing in their PNUE.

If there were a general trend in a wider comparison, for species with a low PNUE to have a high WUE, such species are expected to show less discrimination against $^{13}CO_2$ (Farquhar *et al.*, 1982; Hubick and Farquhar, 1989). However, five Californian evergreens with a low PNUE showed exactly the same discrimination as the average C_3 plant (Field and Mooney, 1986), which does not lend support to the contention that a low PNUE generally reflects a high WUE. Our own data on the PNUE and WUE of nine herbaceous species (Figures 6A,C) do not support the hypothesis either. Therefore, if such a compromise between PNUE and WUE exists, it can at best be relevant in comparisons of species with similar life form grown at a relatively high quantum flux density and with a high evaporative demand.

D. Photosynthesis under Suboptimal Conditions

The photosynthetic capacity decreases with decreasing nitrogen concentration in the leaf, in a (curvi)linear fashion (Evans, 1989a). In the slow-growing *Carex diandra*, p_i increases with decreasing steady state nitrogen supply (H. Lambers and R. Welschen, unpublished). Such an increase with decreasing leaf nitrogen has also been found for other species, including fast-growing ones (Morgan, 1986; Sage and Pearcy, 1987; Ghashghaie and Saugier, 1989; but see Sage and Pearcy, 1987). This increase in p_i may contribute to curvilinearity in the photosynthesis–leaf nitrogen relationship, at least when photosynthesis is measured at an ambient CO_2 concentration. An alternative explanation is that the CO_2 gradient between the intercellular spaces and the site of carboxylation and/or the gradient in p_i within the leaf increases with increasing leaf nitrogen concentration (Evans, 1989a). A relatively lower investment in chlorophyll-complexing thylakoid proteins at a low nitrogen supply, as long as this does not significantly affect the leaf's absorptanoe, also tends to contribute to curvilinearity (Section IX.B). Curvilinearity of the photosynthesis–leaf nitrogen relationship does not appear to be due to inactivation of Rubisco at a high leaf nitrogen concentration (Evans and Terashima, 1988). Despite these reasons which tend to produce curvilinearity, such a relationship its not invariably found, due to the proportionally greater investment of nitrogen in Rubisco with increasing leaf nitrogen concentration (Evans, 1989a).

Curvilinearity of the photosynthetic capacity vs. leaf nitrogen relationship can be interpreted as a relatively inefficient use of nitrogen for photosynthesis at a high nitrogen concentration in the leaf and as such might be expected to occur particularly in slow-growing species adapted to nutrient-poor soils. Indeed, at high N-supply the PNUE of slow-growing species is lower than at high N-supply. A similar tendency, though not as strong, has also been observed for fast-growing species, so that the PNUE of plants grown at a low N-supply is rather similar for fast- and slow-growing species (Boot *et al.*, 1992; A. van der Werf, personal communication). Perhaps the low PNUE of slow-growing species at a high N-supply merely indicates that the rate of biomass accumulation and cell expansion of these species is saturated by the supply of N before that of N-uptake and N-assimilation.

When plants are grown at a low quantum flux density, their leaves have a lower photosynthetic capacity (e.g. Pons, 1977; Young and Smith, 1980). Allocation of nitrogen to the light-harvesting machinery is increased at the expense of that to Rubisco and other enzymes of the Calvin cycle (Björkman, 1981). This change in pattern of nitrogen distribution over the various components of the photosynthetic apparatus leads to optimization of the use of nitrogen for photosynthesis. There is some evidence that slow-growing, shade-tolerant species, have a greater capacity to adjust their nitrogen partitioning in this manner than fast-growing ones (Evans, 1989b).

E. Conclusions

From the above discussion it transpires that slow-growing species have a relatively low PNUE, when grown at optimum nitrogen supply. As yet, there is no satisfactory explanation for this difference. At the low N-supply which the slow-growing species encounter in their natural habitat, the PNUE of fast- and slow-growing species is rather similar. A further appreciation of the ecological advantage, if any, associated with a low PNUE, clearly awaits more information on the physiological, biochemical or anatomical background of the variation in PNUE and possible negative correlations of PNUE with WUE.

X. RESPIRATION

Respiration provides the driving force for three major energy-requiring processes: maintenance, growth and ion uptake. Maintenance respiration is mainly associated with turnover of various cellular components and the conservation of solute gradients across membranes. Growth respiration is used to supply ATP and NADH, needed to convert glucose into the different chemical compounds. In roots, respiratory energy is also needed for the absorption of nutrients from the environment.

A. Species-specific Variation in the Rate of Respiration

Fast-growing species have a higher rate of nitrate uptake (Figure 3A) and a higher RGR. Therefore, it is not surprising that they have a higher rate of shoot and root respiration (Dijkstra and Lambers, 1989a; Poorter *et al.*, 1990, 1991; van der Werf *et al.*, 1992b; Figure 3B). However, the rate of root respiration of fast-growing species is not as high as would be expected from their higher RGR and NIR. Using data of van der Werf *et al.* (1988), who determined specific costs for root growth, maintenance and ion uptake for two slow-growing *Carex* species, the calculated rate of root respiration of fast-growing species is approximately four times higher than that of slow-growing ones (Poorter *et al.*, 1991; Figure 3B, broken line). This value is at variance with that determined experimentally, showing that the rate of root respiration of fast-growing species is only 50% higher than that of slow-growing ones (Figure 3B, regression line through data points). Why do the fast-growing species respire at a rate which is so much lower than expected from their RGR and NIR?

1. Variation in Respiratory Efficiency

The relatively low rate of respiration of fast-growing species might be due to a relatively more efficient respiration. Apart from the cytochrome pathway, which yields three molecules of ATP per oxygen atom reduced, plants have an alternative, non-phosphorylating respiratory pathway. Engagement of this path, rather than the cytochrome path, yields only one third as much ATP (Lambers, 1985). Do fast-growing species employ this alternative pathway to a lesser extent than the slow-growing ones and therefore produce more ATP for the same amount of glucose respired and oxygen reduced? Despite a wide variation in the participation of the alternative, nonphosphorylating electron transport path in respiration, varying from 0 to 44% of total root respiration (van der Werf et al., 1989) and from 4 to 58% for leaves (Collier and Cummins, 1989; Atkin and Day, 1990), there is no evidence that this respiratory path contributes less to respiration in fast-growing species or genotypes (Dijkstra and Lambers, 1989b; Atkin and Day, 1990; Poorter *et al.*, 1991; van der Werf *et al.*, 1992b). In fact, Collier and Cummins (1989) found that respiration rates of the leaves of ruderals collected in the field were higher and that their alternative path was engaged to a greater extent than that of understorey species. Next to a greater alternative path activity, the ruderals also had a greater capacity for respiration via this path. This might contribute to their functioning in fluctuating environments, in line with results on root respiration of other species (Lambers *et al.*, 1981b).

The rate of ATP production per oxygen consumed might also be higher in fast-growing species if their mitochondria operate under substrate limitation or further away from "state 4"conditions. (State 4 is the condition where the respiration is strongly restricted by the availability of ADP, as opposed to state 3, where ADP is available in saturating amounts.) If so, they do not produce ATP whilst still reducing oxygen (cf. Whitehouse *et al.*, 1989). In state 3 conditions, they operate more efficiently, producing maximally three ATP per oxygen atom. If the fast-growing species would respire closer to state 3, their ADP: O ratio *in vivo* might be higher. Straightforward methods exist to test this hypothesis (Day and Lambers, 1983), but so far no comparative data are available on fast- and slow-growing species.

2. *Variation in Specific Costs of Energy-Requiring Processes*

The current data provide no indication that fast-growing species produce ATP more efficiently than slow-growing ones. Therefore, it is very likely that the relatively low respiration rate of fast-growing species is caused by lower specific respiratory costs for energy-requiring processes, such as maintenance, growth or ion acquisition, at least under optimum growth conditions.

We have very little information on the biochemistry and physiology of "maintenance processes". Protein turnover and the maintenance of solute gradients are considered major components (Penning de Vries, 1975), but quantitative information on either of these processes in roots is scarce (van der Werf *et al.*, 1992c). Assuming the same turnover rate per unit protein, we expect the maintenance respiration of fast-growing species, which have a higher protein concentration (Section VIII.A), to be higher than that of slow-growing species, but this is not borne out by experimental results. Other processes which require respiratory energy and which do not contribute to growth or ion uptake will also increase the maintenance costs. For example, the relatively large turnover of carbohydrate pools in slow-growing species as compared to fast-growing ones (Farrar, 1989), might partly explain the high respiratory costs of roots in slow-growing species. However, maintenance respiration is only a small portion of total respiration, at least in young plants grown at an optimum nutrient supply (van der Werf *et al.*, 1989; Poorter *et al.*, 1991). Thus variation in the specific costs for maintenance is unlikely to affect substantially the rate of root respiration.

In the case of growth respiration, the construction costs per unit biomass may vary with the chemical composition of the plant. Table 4 provides information on the gas exchange which could be expected to occur with the synthesis of a range of primary and secondary compounds. The calculated oxygen uptake and carbon dioxide release include gas exchange associated with ATP production, (de)carboxylating reactions,. as well as the use of oxygen in reactions catalysed by such enzymes as mixed function oxygenases. We calculated that the oxygen requirement for the synthesis of roots of fast-growing species is higher than that for slow-growing ones (Poorter *et al.*, 1991). Hence, differences in chemical composition cannot explain why respiration rates of slow-growing species are only marginally lower than those of species growing three times as fast and absorbing ions at over four times the rate of the slow-growing ones.

If the specific costs for maintenance and growth cannot explain the relatively high respiration rates of slow-growing species, then the most likely explanation for the observed differences in root respiration between slow-growing and fast-growing species is variation in the specific cost for nutrient acquisition. Slow-growing species, which are often associated with unproductive environments (Section II), are likely to be geared towards nutrient uptake from a very dilute solution in comparison with the fast-growing species from relatively nutrient-rich habitats. This may require more energy, either because it involves extrusion and re-uptake of compounds which release ions from complexes in the soil (Section XI.A) or because the ratio of proton entry and ion absorption is higher, compared to the system of fast-growing species (cf. Clarkson, 1986; McClure *et al.*, 1990). For *Hordeum vulgare*, two transport systems for nitrate uptake have been described. One of these systems is an energy-requiring one, operating at very low external concentrations. The other one does not require metabolic energy and functions at an external nitrate concentration sufficiently high, compared to the cytoplasmic concentration, to allow passive diffusion of nitrate across the plasma membrane (Glass *et al.*, 1990; Siddiqi *et al.*, 1990). Perhaps fast-growing species maintain a relatively low nitrate concentration in the cytoplasm of their roots cells, due to rapid reduction of nitrate in the cytosol, transport into their, presumably relatively large, vacuoles (cf. Section XII.A), or efficient export to the shoot, made possible by the relatively high rate of transpiration per unit root weight (Poorter *et al.*, 1990). If so, the system which does not require metabolic energy may predominate in the roots of fast-growing species, whereas the energetically more expensive one is more important in the roots of slow-growing ones. This is not to say that fast-growing species lack a similar system. Rather, it may be constitutive in slow-growing species and inducible under nutrient-deficient conditions in fast-growing ones, when the electrochemical gradient across the plasma membrane does not allow passive uptake of nitrate (van der Werf *et al.*, 1992b, cf. Section VII.A). An alternative and more attractive explanation for presumably higher costs for nitrate uptake is that the ratio between ion influx and efflux is lower in slow-growing species (cf. Pearson *et al.*, 1981; Deane-Drummond and Glass, 1983; Oscarson *et al.*, 1987).

B. Respiration at Suboptimal Nitrogen Supply or Quantum Flux Density

With a decreasing nitrogen supply, there is a decline in the nitrogen concentration and the rate of respiration per unit mass, both in roots (Lambers *et al.*, 1981a; Duarte *et al.*, 1988; Granato *et al.*, 1989; van der Werf *et al.*, 1992b) and in leaves (Waring *et al.*, 1985; Boot *et al.*, 1992). The respiration rate is less, because of reduced energy requirements for biosynthetic processes, ion transport and loading of sucrose in the phloem. Upon a decrease of the nitrogen supply, both the nitrogen concentration and the rate of leaf respiration of the faster-growing *Agrostis vinealis* decline to a greater extent than that of the slow-growing *Corynephorus canescens* (Boot *et al.*, 1992). Similar results have been obtained in a comparison of the fast-growing *Holcus lanatus* and the slow-growing *Deschampsia flexuosa* (C.A.D.M. van de Vijver, R.G.A. Boot and H. Poorter, unpublished). At optimum nitrogen supply, *Agrostis* has a higher leaf nitrogen concentration (cf. Section VIII.A), whereas it is reduced to the same level as in *Corynephorus* when the nitrogen supply is reduced. There is some evidence that specific costs for ion transport and/or maintenance increase at a limiting nitrogen supply, particularly in fast-growing species (A. van der Werf *et al.*, 1992b).

When plants are transferred to a low quantum flux density, the rate of root respiration declines (Kuiper and Smid, 1985; H.H. Prins, R. Hetem and H. Poorter, unpublished), as expected from the lower rate of root growth under such conditions. Upon prolonged exposure to a low quantum flux density, there is only a small, or no difference in root respiration between plants grown at high vs. low quantum flux density (Lambers and Posthumus, 1980; A. van der Werf and P. Poot, unpublished). This is attributed to the increase in LWR upon prolonged exposure to a low quantum flux density, which increases the root's energy requirement for uptake of ions destined for the shoot. Also, the rate of leaf respiration is less for plants grown at a low quantum flux density (Pons, 1977; Waring *et al.*, 1985). Presumably the lower rates of respiration reflect the lower energy requirement for biosynthetic and transport processes. We do not know of any comparative data on fast-growing and slow-growing species.

C. Conclusions

At an optimum nutrient supply, the specific rate of root respiration of fast-growing species is lower than expected from their high RGR and high NIR. A satisfactory explanation for this relatively low respiration rate cannot be provided yet, but it may well be due to a relatively low energy requirement for ion acquisition. At a nutrient supply or quantum flux density which is suboptimal for growth, the rate of respiration in both leaves and roots is less than for plants growing under optimum conditions. This is at least partly explained by reduced rates of energy-requiring processes. Upon a change in nitrogen supply, fast-growing species adjust their respiration rate to a greater extent than slow-growing ones, possibly with concomitant changes in specific costs for energy-requiring processes.

XI. EXUDATION AND VOLATILE LOSSES

Plants lose photosynthates through exudation and volatilization as well as during respiration. Exudation may occur both above- and below-ground, whereas volatilization predominantly occurs above-ground.

A. The Quantitative and Qualitative Importance of Exudation

It is well established that roots exude a range of organic compounds, including sugars (McCully and Canny, 1985), organic acids (Gardner *et al.*, 1983) and amino acids (McDougall, 1970), especially when phosphate is in short supply (Graham *et al.*, 1981; Lipton *et al.*, 1987). Estimates of the loss of exudates from roots vary widely, largely depending on the methods used to quantify this process and perhaps on the species under investigation (Cheshire and Mundie, 1990). Soluble exudates constitute less than 0.5% of the carbon present in the plant (Cheshire and Mundie, 1990). Their production ranges from 10 to 100 mg per gram root dry weight produced; root cap plus mucigel may provide a further 20–50 mg (Newman, 1985; Gregory and Atwell, 1991). At the very most 5% of the photosynthates are lost through. "rhizodeposition", i.e. the loss of organic matter via both processes (Lambers, 1987), with the exception of plants with proteoid roots in which very high values are found (see below). If losses due to continual cell death are also included, losses due to rhizodeposidition (*sensu latu*) may amount to 10% of all photosynthates produced (Helal and Sauerbeck, 1986; Lynch and Whipps, 1990).

Exudates are of distinct importance for the acquisition of sparingly available nutrients and for interactions with symbionts. Highly efficient chelators (phytosiderophores) are excreted by roots of Gramineae and these allow the roots to absorb Fe, Zn, Mn and Cu from poorly soluble sources in calcareous soils (Römheld and Marschner, 1986; Marschner *et al.*, 1989; Zhang *et al.*, 1989). At least part of the excreted phytosiderophores are absorbed again by the roots as a metal–siderophore complex (Römheld and Marschner, 1990). Non-gramineous species also release chelating compounds, generally of a phenolic nature, but these are less efficient than the true phytosiderophores (Römheld, 1987). Many species, particularly those with proteoid roots, release citric acid (Gardner *et al.*, 1983; Hoffland *et al.*, 1989a; Hoffland *et al.*, 1989b) which may amount to as much as 23% of the biomass at plant harvest (Dinkelaker *et al.*, 1989). The excretion of citric acid greatly enhances the root's capacity to use insoluble phosphate (Hoffland *et al.*, 1990). Proteoid roots also have an increased capacity for reduction of iron and manganese in the rhizosphere and, consequently, to mobilize sparingly soluble Fe or Al phosphates (Gardner *et al.*, 1982a, b). Excretion of citric acid may also be significant in releasing cations from humic substances (Albuzzio and Ferrari, 1989). We conclude that the release of chelating substances, such as citric acid and possibly other organic acids is important in acquiring nutrients from both calcareous substrates and acidic soils where cations are bound to a humic complex. Many slow-growing species are associated with such calcareous or acidic soils (Grime and Hunt, 1975). Hence the release of root exudates is likely to confer an advantage in such soils and should not merely be considered as a loss of carbon.

Some micro-organisms exude organic compounds that precipitate heavy metals outside the cells. There is no evidence that this mechanism is important in ecotypes of higher plants that tolerate high concentrations of heavy metals in the root environment (Verkleij and Schat, 1990).

Release of organic substances from roots can also play a pivotal role in symbiotic associations. For example, flavonoids, released from the roots of Leguminosae induce the nodulation genes of *Rhizobium*, the primary step in the nodulation process (Richardson *et al.*, 1988; Hartwig *et al.*, 1990). Plants capable of a mycorrhizal symbiosis release organic compounds to the rhizosphere when they contain very little phosphate in the roots, presumably due to the fact that their membranes contain less than an optimum amount of phospholipids (Ratnayaka *et al.*, 1978).

Some root exudates, predominantly of a phenolic nature, play a role in allelopathic interactions between plants. Although the exact nature of the compounds released, their biochemical effect on neighbouring plants and their ecological significance are often not known, the existence of allelopathic interactions, including those based on exuded compounds, is beyond doubt (Putnam and Tang, 1986; Kuiters, 1990). Root exudation can also affect, both negatively and positively, the rates of a number of soil biological processes, such as denitrification (Woldendorp, 1963), nitrification (Haider *et al.*, 1987; Vitousek *et al.*, 1989) and mineralization (Sparling *et al.*, 1982). Different plant species affect these soil biological processes to varying degrees (Janzen and Radder, 1989; Berendse *et al.*, 1989; van Veen *et al.*, 1989), but the exact nature of the effects is generally not fully known. There is some evidence that losses through exudation are quantitatively more important in a slower-growing *Hordeum vulgare* variety than in a faster-growing one (Liljeroth *et al.*, 1990).

Losses of carbon through exudation also occur above-ground. We have very little information on their quantitative importance and inherent variation of this process (Tukey, 1970).

B. The Quantitative and Qualitative Importance of Volatile Losses

Although it is probably fair to state that volatile losses are generally not of great quantitative significance, they are of fairly wide importance, e.g. in allelopathic interactions and herbivory (Harborne, 1982; Rhoades, 1985; Dicke and Sabelis, 1989).

Emissions of isoprene in three fern species accounted for 0.02 to 2.6% of the carbon fixed during photosynthesis, increasing with photon flux density and temperature and varying between species (Tingey *et al.*, 1987). These values are in the same range as those found for tree leaves

(Sanadze, 1969; Tingey *et al.*, 1981). They are up to 2.5 times lower than those based on field measurements for a number of tree species, which may reflect species differences in emission or effects of high quantum flux density (Flyckt *et al.*, 1980, cited in Tingey *et al.*, 1981). In extreme cases, such as in *Ledum groenlandicum* during part of the year (Prudhomme, 1983) and *Populus tremuloides* at high temperatures (35–45°C) (Monson and Fall, 1989), up to 8% of recently fixed carbon may be lost as volatiles.

Losses of specific volatiles, though quantitatively minor (less than 0.001% of the photosynthates produced daily), are responsible for attraction of predators upon attack of leaves by herbivores, e.g. spider mites, and thus reduce herbivore damage (Dicke and Sabelis, 1989).

C. Conclusions

Losses through exudation *sensu lato* can significantly reduce a plant's growth rate. Carbon loss through exudation and volatilization is most certainly of ecological importance in a nutrient-poor environment and in interactions of a plant species with other organisms. Although not supported by hard evidence, we hypothesize that exudation is more important in slow-growing species from nutrient-poor sites. Exudation might allow such species to acquire nutrients which are otherwise unavailable.

XII. OTHER DIFFERENCES BETWEEN FAST- AND SLOW-GROWING SPECIES

Apart from the above-mentioned traits, which directly affect the growth of a plant, some other aspects of fast-growing and slow-growing species and mutants thereof have been investigated. In recent years fascinating information has become available on the role of a specific class of phytohormones in the control of a plant's growth rate—the gibberellins.

A. Hormonal Aspects

Fast-growing genotypes contain more gibberellin than slower-growing ones (Rood *et al.*, 1983; Rood *et al.*, 1990a; Rood *et al.*, 1990b; Rood *et al.*, 1990c; Dijkstra *et al.*, 1990; H. Konings and M. Berrevoets, personal communication). Rapid growth of hybrids ("heterosis") has been associated with higher levels of gibberellin in both herbaceous species and trees (Rood and Pharis, 1987; Bate *et al.*, 1988). Interestingly, in a *Zea mays* hybrid, both the superior growth and the higher level of gibberellin, compared to those of its parents, are restricted to favourable conditions and not displayed during growth at low temperature (Rood and Pharis, 1987). Gibberellins control leaf size (*Brassica rapa*, Zanewich *et al.*, 1990; *Thlaspi arvense*, Metzger and Hassebrock, 1990; *Lycopersicon esculentum*, H. Konings and M. Berrevoets, personal communication) and it seems likely that the variation in leaf size between fast- and slow-growing species is at least partly associated with differences in the concentration of endogenous gibberellins.

Mutants of *Lycopersicon esculentum* with reduced levels of gibberellin, have a range of characteristics similar to slow-growing species, e.g. a higher RWR, but lower LAR and SLA, relatively more dry matter per unit fresh weight, and a low rate of photosynthesis per unit leaf dry weight (H. Konings and M. Berrevoets, personal communication). Treatment with gibberellin reduces genotypic differences in RGR, indicating that this hormone plays a role in intraspecific variation in growth potential, probably via its effect on leaf area development and biomass partitioning (Dijkstra and Kuiper, 1989; Dijkstra *et al.*, 1990; Rood *et al.*, 1990a, b, c; Zanewich *et al.*, 1990; H. Konings and M. Berrevoets, personal communication).

The detailed mechanism of the gibberellin effects on growth is largely unknown. However, it is well documented that this phytohormone affects stem growth via both cell elongation and cell division (Métraux, 1987; Jupe *et al.*, 1988; Rood *et al.*, 1990b). Effects of gibberellin on cell enlargement could account for a number of the chemical differences between fast- and slow-growing species (cf. Section VIII.A). The larger surface-to-volume ratio of plants with smaller cells is expected to be associated with a relatively large investement in cell-wall components and hence a high dry matter percentage. Smaller cells are bound to have relatively small vacuoles, which would explain the relatively low capacity of slow-growing species to accumulate organic acids, nitrate and other minerals. The effects of gibberellin on cell division could account for differences in biomass partitioning (cf. Section VI.A). A low level of gibberellins prevents rapid incorporation of photosynthates into leaf and stem biomass, so that a relatively large proportion is translocated to and incorporated into the roots.

Clearly, further work is needed on the mechanism of gibberellin action on cell growth and on the level of this phytohormone in different species. This may well lead to physiological explanations for inherent variation in RGR and provide insight into evolutionary mechanisms causing such variation.

B. Miscellaneous Traits

Fast-growing grass species tend to have wider leaves than slow-growing ones; leaves of dicotyledonous fast-growing species also tend to be larger than those of slow-growing ones (Christie and Moorby, 1975; Ceulemans, 1989; Körner and Pelaez Menendez-Riedl, 1989; H. Poorter, unpublished). Leaves of fast-growing poplar hybrids are larger than those of either of their parents, *Populus trichocarpa* and *P. deltoides*. Leaf cells of *P. trichocarpa* are larger than those of *P. deltoides,* whereas *P. deltoides* has more cells per leaf. The greater leaf size of the hybrids can be explained by inheritance of a larger cell number from *P. deltoides* and larger cells from *P. trichocarpa* (Ceulemans, 1989). In general, variation in leaf size appears to be due predominantly to fewer cells per leaf, rather than to cell size (Körner and Pelaez Menendez-Riedl, 1989).

The diurnal pattern of leaf growth may vary between species and hybrids thereof. Leaves of *Populus trichocarpa* mainly grow in the light, with little growth in the dark, those of *P. deltoides* grow during the dark period, with little stimulation in the light, whereas their fast-growing interspecific hybrids grow during both night and dark (Ceulemans, 1989).

For a wide range of species and for families of *Poa annua*, a slightly negative correlation between nuclear DNA content and RGR has been reported (Grime *et al.*, 1988). A negative correlation between RGR and seed size has been suggested (Fenner, 1978; Gross, 1984). However, no such relationship has been detected in a much larger data set (Thompson, 1987). Seeds of fast-growing species germinate more rapidly than those of slow-growing ones (Grime *et al.*, 1988).

XIII. AN INTEGRATION OF VARIOUS PHYSIOLOGICAL AND MORPHOLOGICAL ASPECTS

In the previous sections several aspects of the physiology, morphology, allocation and biochemical composition have been discussed in relation to the potential growth rate of plant species. We now address the question of what proportion each parameter contributes to the observed differences in growth rate, using (Eq. (5)) as a framework.

A. Carbon Budget

For the nine species presented in Figures 1–4 and 6, RGR varies more than three-fold. The carbon concentration (Section VIII) is lower for fast-growing species, which is partly due to their higher mineral concentration. The lower carbon concentration contributes to a higher RGR, but is only of minor importance in explaining the observed variation in RGR. Quantifications of exudation and volatile losses are scarce but these processes are unlikely to determine a large part of the variation in RGR either (Section IX). Hence, the major differences are due to variation in photosynthetic gain and respiratory losses of carbon. Indeed, carbon gain of the fastest-growing of these 9 species is about 3.1 times that of the slowest-growing one (Figure 7). This is caused by a much higher SLA (Section V) and a slightly higher LWR (Section VI), rather than by a higher rate of photosynthesis per unit leaf area (Section VIII). Species also differ in the way they utilize the fixed carbon. Although fast-growing species have higher shoot and root respiration rates, the proportion of fixed carbon used in respiration is less. This is due largely to a higher rate of photosynthesis per unit plant weight and, as far as root respiration is concerned, also to the lower RWR of fast-growing species. It is to be expected that the lower RWR contributes to the higher RGR, provided the smaller root size is compensated for by a higher specific activity. Indeed, fast-growing species do have a much higher rate of net ion uptake (Section VII) and water absorption (Poorter *et al.*, 1990) than slow-growing ones. Compared to slow-growing species, their rate of root respiration is not as high as would be expected from their four times higher rate of nitrate uptake and their three times higher relative growth rate. This relatively low respiration rate contributes to the rapid growth of fast-growing species.

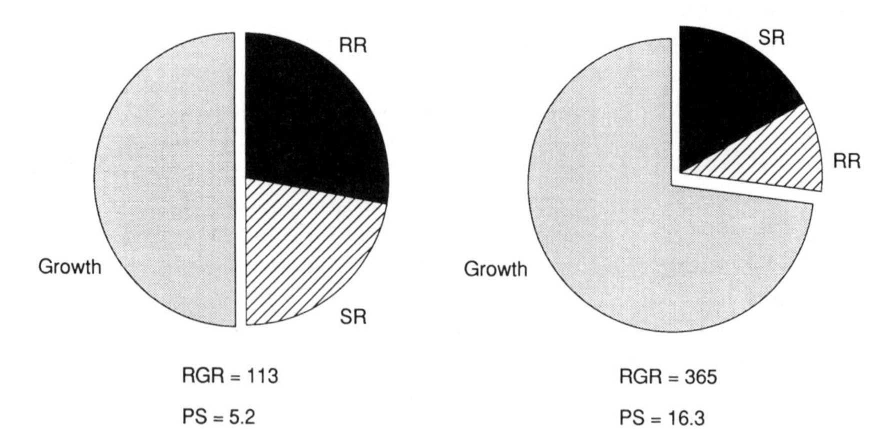

Figure 7 The carbon budget of a slow-growing species (*Corynephorus canescens*, left) and a fast-growing species (*Galinsoga parviflora*, right). In the upper line the RGR of these species is given (expressed in mg g^{-1} day^{-1}). The second line gives the daily gross CO_2 fixation (mmol (g plant)$^{-1}$). (After Poorter *et al.*, 1990.)

B. Interrelations

Up till now we have treated the different physiological and morphological aspects as being independent of each other. This is a simplified view, as there are numerous interrelations. For example, a shift in biomass allocation from leaves to roots implies a decrease in the photosynthetically active area and an increase in the respiratory burden. Consequently, a decrease in the rate of carbon fixation and growth of such a plant is expected. At the same time, a high allocation to roots may ensure better access to nutrients and water, which may result in an increased plant nitrogen concentration, a higher photosynthetic capacity per unit leaf area and also a higher stomatal conductance. Moreover, there may be a decrease in self-shading. Consequently, the rate of photosynthesis per unit leaf area will increase. Also, root biomass is generally energetically cheaper to construct (Poorter and Bergkotte, 1992), so that growth rate will not decrease to the same extent as LWR decreases.

Another simplification we have made so far is to compare diferent types of plants over a limited time course. Investment of biomass in compounds that reduce herbivory or increase a plant's stress tolerance then inevitably leads to a decrease in the rate of photosynthesis per unit plant weight (Section IX.D). However, due to these investments, the life expectancy of a leaf increases, the net result being a possibly similar or even higher rate of photosynthesis integrated over the entire life span of a leaf (Schulze, 1982).

Strong correlations between different plant traits do not necessarily imply causal relations. For example, the correlation of RGR with SLA is probably partly fortuitous. Based on the conventional growth analysis (Eqs. (1)–(5)), we conclude that variation in SLA is the main cause for inherent variation in RGR. Simplified, it means that an increase in leaf area relative to leaf weight increases the growth rate. From a mechanistic point of view this statement is not entirely correct. Although photosynthesis indeed is an area-related process, the actual quantum capture and CO_2-fixation require, light–harvesting complexes, the coupling factor, Rubisco and other photosynthetically active components (Section IX). As this photosynthetic machinery incorporates the major part of a leaf's organic nitrogen, PNUE and leaf organic nitrogen may be used as a good approximation of the efficiency and size of the photosynthetic machinery (Section IX). Fast-growing species have both a higher PNUE and a higher organic nitrogen concentration. These two parameters appear to be closely correlated with SLA and, combined, show the same 2.2-fold variation for the nine species presented here, as found for SLA. This correlation partly reflects a causal relation, in so far as a high SLA coincides with a low concentration of chlorophyll per unit leaf area (Poorter *et al.*, 1990), which reduces internal shading and therefore increases PNUE (Section IX.B). However, is seems likely that the strong positive correlation between RGR and SLA is also partly fortuitous, in that SLA correlates merely in a non-causal manner with a number of other factors which really determine RGR. Clearly, the parameters of Eq. (5) cannot provide the full answer to the causes of inherent variation in RGR.

XIV. SPECIES-SPECIFIC PERFORMANCE UNDER SUBOPTIMAL CONDITIONS

Up till now we have paid most attention to plants grown under conditions favourable for plant growth. But how do fast- and slow-growing species perform under suboptimal conditions?

When grown at a low nutrient concentration in the environment, the RGR of potentially fast-growing species is reduced more than that of slow-growing ones (e.g. Christie and Moorby, 1975; Robinson and Rorison, 1987; Boot and Mensink, 1991). However, the inherently fast-growing species are still growing faster than slow-growing ones. This seems to be the general trend (Mahmoud and Grime, 1976; Chapin, 1983; Chapin *et al.*, 1983; Berendse and Elberse, 1989; C.A.D.M. van de Vijver, R.G.A. Boot and H. Poorter, unpublished; but see Hommels *et al.*, 1990; Muller and Garnier, 1990). This raises the question whether similar results would have been obtained in a situation where a fast-growing species competes with a slow-growing one under nutrient stress. This question will be addressed in Section XV.A.

The higher RGR of the inherently fast-growing *Holcus lanatus*, in comparison with the slow-growing *Deschampsia flexuosa*, at a low nutrient supply is explained by differences in LAR (C.A.D.M. van de Vijver, R.G.A. Boot and H. Poorter, unpublished). This is probably true for most comparisons (Christie and Moorby, 1975; van Andel and Biere, 1989). However, there are very few comparative data showing the cause of the higher RGR of fast-growing species at low nutrient supply.

When grown at a low quantum flux density, fast-growing sun species have a similar RGR to slow-growing shade species (Pons, 1977). In extreme cases, such as that of the fast-growing tropical pioneer tree *Cecropia obtusifolia*, the RGR declines to a very low rate when plants are grown in the understorey, rather than large gaps. This is due to the decrease in NAR (Popma and Bongers, 1988). The successful performance of most shade species in shaded habitats is likely to be due to a different response of germination, stem elongation and other processes, to light quality, rather than to quantum flux density (cf. Fitter and Ashmore, 1974; Pons, 1977).

XV. THE ECOLOGICAL CONSEQUENCES OF VARIATION IN POTENTIAL GROWTH RATE

A. What Ecological Advantage can be Conferred by a Plant's Growth Potential?

The ecological advantage of a high RGR seems straightforward: fast growth results in the rapid occupation of a large space, which is advantageous in a situation of competition for limiting resources. A high RGR may also facilitate rapid completion of the life cycle of a plant, which is essential for ruderals. But what is the survival value of slow growth? Grime and Hunt (1975) and Chapin (1980, 1988) mention several possibilities:

(*i*) Slow-growing species make modest demands and will therefore less likely exhaust the available resources, e.g. nutrients. However, this does not seem to be an evolutionary stable strategy, as a neigbouring individual with a faster nutrient uptake could absorb most nutrients (cf. Schulze and Chapin, 1987). Moreover, these modest demands cannot explain slow growth under, e.g. alpine or saline conditions.

(*ii*) Slow-growing species function closer to their optimum than fast-growing species in an adverse environment. However, the "ecological" optimum of a plant species often differs from its "physiological" optimum. As the physiological optimum of slow-growing species more or less equals that of fast-growing species (Grime and Hunt, 1975), and all plants, especially fast-growing species, have a great ability to adapt to different environmental conditions (Bradshaw, 1965; Grime *et al.*, 1986; van der Werf, 1992a), we do not expect fast-growing species to be at a disadvantage in such cases. In fact, both in growth analyses (Section XIV) and in short-term competition experiments (Mahmoud and Grime, 1976; Berendse and Elberse, 1989) carried out at a limiting nutrient supply, potentially fast-growing species grow faster and have a greater competitive ability than slow-growing ones.

(*iii*) Slow-growing species incorporate less photosynthates and nutrients into structural biomass and may thus form reserves for later growth, enabling them to maintain physiological integrity during periods which severely restrict growth, e.g. low nutrient availability. However, under adverse conditions growth is restricted before photosynthesis, causing sugars to accumulate (Chapin *et al.*, 1986b; McDonald *et al.*, 1986). Hence, it is unlikely that survival during periods of nutrient shortage depends on storage of photosynthates. The presence of stored nutrients may indeed buffer fluctuations in nutrient supply in the field. However, perhaps with the exception of phosphate, there is no convincing evidence that slow-growing species accumulate nutrients to a greater extent (Section VIII.A). None the less, slow-growing species may deplete their smaller resources less rapidly, due to their lower RGR. So far, data on the occurrence of fluctuations in nutrient supply and the plant's response to them are too scarce and conflicting (Grime *et al.*, 1986; Poorter and Lambers, 1986; Campbell and Grime, 1989) to allow the conclusion that an inherently low RGR has survival value in this context.

(*iv*) In a low-resource environment a high growth rate cannot be realized, so a high potential RGR is a selectively neutral trait. However, as noted before, potentially fast-growing species are still growing faster than potentially slow-growing species. Fast-growing species will then attain a larger size (van Andel and Biere, 1989), which has been shown to be advantageous in terms of competition and fitness (Black, 1958; Stanton, 1984). So, although a very high RGR is not attainable, a slightly higher RGR may still be of advantage.

We conclude that a low potential growth rate *per se* does not confer ecological advantage. Why then do slow-growing species occur more frequently in unfavourable habitats than fast-growing ones? An alternative explanation for the observed differences in potential growth rate is that not RGR itself, but rather one of the components linked with RGR, has been the target of selection (Lambers and Dijkstra, 1987; cf. Grime, 1979).

B. Selection for Traits Associated with a Low SLA

The most likely trait selected for is SLA, as variation in this trait is closely correlated with that in RGR (Section V; cf. Poorter, 1989; Poorter and Remkes, 1990).

In a situation where water or nutrients are limiting, conservation of the scarce resource is at least as important as its capture. Hence, plants under water stress should decrease their transpiration (von Willert *et al.*, 1992). But also for nutrients it has been shown that unproductive species are more successful due to less leaf turnover, so that nutrient losses are restricted (Monk, 1966; Berendse *et al.*, 1987; Karlsson and Nordell, 1987; Aerts and Berendse, 1989). How can turnover be decreased? This depends on the environmental factor which affects leaf longevity.

Herbivory can be reduced by increasing leaf toughness (Coley, 1983; Grubb, 1986), accumulating palatability-reducing compounds (Coley, 1987; Waterman and McKey, 1989) and investment in leaf hairs (Woodman and Fernandes, 1991) or thorns. The abrasive effects of high wind speeds can be reduced by investment in fibre (Woodward, 1983; Pammenter *et al.*, 1986). Trampling resistence may be be the result of a large amount of cell wall material per cell (Dijkstra, 1989). Transpiration can be decreased and water use efficiency can be increased by the construction of leaf hairs or epicuticular waxes (Ehleringer, 1983; Richards *et al.*, 1986; Ehleringer and Cook, 1990). Epicuticular waxes may also serve a function by decreasing damage by ultraviolet light, preventing contact between rain water and the interior of a leaf and so restricting leaching of nutrients out of a leaf (Mulroy, 1979). Furthermore, waxes may confer disease resistance (Carver *et al.*, 1991) and diminish deleterious effects of salt spray (McNeilly *et al.*, 1987). Each of these additional investments increases the leaf's longevity, but decreases SLA with a concomitant decrease in the rate of photosynthesis per unit leaf weight. Consequently, all of these inherent adaptations to adverse conditions diminish the plant's growth potential, but positively influence its fitness.

Is there any indication that plants without these types of adjustment could survive in unfavourable habitats? This would require introduction of plants that only differ in one specific trait in different environments. However, such isogenic genotypes are not available, and variation in one trait could be expected to affect related traits (Section VIII.B). The best ecological information available is that from introduction of foreign species, e.g. the introduction into Venezuela of two African C_4 species. The introduced species with a high SLA have outcompeted a native C_4 species, which possesses a low SLA, in relatively fertile places, but not in more infertile habitats (Baruch *et al.*, 1985). On subantarctic islands the introduced grass *Agrostis stolonifera*, with a high SLA, is able to survive in the wind-sheltered places but is not found outside these shelters, whereas *Agrostis magellanica*, characterized by a lower SLA due to more sclerenchyma, occurs in the wind-swept parts of these islands (Pammenter *et al.*, 1986). Similarly, *Stephanomeria malheurensis*, a species with a relatively low SLA which occurs in the same environment as its progenitor *S. exigua* ssp. *coronaria* with a higher SLA, is restricted to sites where it may encounter greater stress. The number of individuals of *S. exigua* ssp. *coronaria* by far exceeds, that of *S. malheurensis*, though their RGR is very similar (Gottlieb, 1978).

Here again, a high SLA appears associated with competive ability and a low SLA with persistence.

C. Selection for Other Traits Underlying RGR

It is likely that other traits underlying RGR have also been the target of natural selection. For example, the relatively high RWR of slow-growing species is a cause of their low RGR (Section VII.A). However, RWR is a rather plastic trait, especially in fast-growing species, so that correlations between plant performance and RWR may reflect phenotypic, rather than inherent variation.

In so far as a high RWR reflects a greater ability to compete for soilderived resources (Baan Hofman and Ennik, 1980, 1982; Aerts *et al.*, 1991), it may be of advantage in specific environments. Also, if indeed slow-growing species exude specific compounds which effectively release nutrients from a nutrient-poor soil (Section XI.A), this trait which reduces RGR may be selected for at nutrient-poor sites. Exudates, lute volatiles, may also have allelopathic effects and aid a plant to thrive in an environment amongst competitors.

A low nitrogen concentration in the leaf, though reducing the photosynthetic capacity, may add to the leaf's longevity by decreasing herbivory (Section VIII.B), and thus the photosynthetic yield during the leaf's entire life span may be higher when the leaf nitrogen concentration is relatively low.

These patterns of investment, which inexorably reduce the plant's growth potential, tend to add to a species' success in nutrient-poor or and environments where losses due to herbivory severely reduce the plant's fitness. Thus, they imply a trade-off between growth potential and adaptation to specific habitat features that limit growth.

D. Consequences of a High Growth Potential for Plant Performance in Specific Environments

Fast-growing species often grow in competitive situations in a vegetation which develops a high leaf area index during the season. Optimization of the use of nitrogen for photosynthesis and maximization of production then requires the discharge of the oldest, shaded leaves and retranslocation of a part of the N to the youngest, more exposed leaves (Hirose and Werger, 1987; Hirose *et al.*, 1988; Pons *et al.*, 1989). Under such circumstances, where leaves function only a relatively short period, protection against adverse conditions, requiring a large investment in quantitative secondary compounds, does not lead to an increased photosynthetic return.

Consequently, in fairly dense vegetations where light capture is essential and leaf turnover is high, there is a selection pressure for leaves with a high SLA (Schulze, 1982; Poorter, 1989).

Also when grown at a limiting nitrogen supply, when the leaf area index is relatively low, fast-growing species tend to have higher rates of leaf turnover (Williamson, 1976). Here it inevitably leads to greater losses (Berendse and Elberse, 1989) and possibly ultimately to the disappearance of the fast-growing species from such environments. The rapid leaf turnover of fast-growing species in a dense canopy might depend on the same, as yet poorly understood, mechanism of N-translocation operating at a limiting nitrogen supply (cf. Horgan and Wareing, 1980; Simpson *et al.*, 1982; Kuiper *et al.*, 1988). If correct, then the poor performance of fast-growing plants in a nutrient-poor environment is the consequence of their adaptation to nutrient-rich situations.

E. A Low Growth Potential and Plant Performance in Adverse Environments, Other than Nutrient-poor Habitats

So far, we have concentrated on nutrient-poor habitats, when referring to adverse soil conditions. Although there is quite convincing information that traits associated with a low RGR confer selective advantage in some unproductive environments, it is by no means certain that this is invariably so.

Sites that are rich in heavy metals tend to be inhabited by slow-growing ecotypes (Wilson, 1988; Verkleij and Prast, 1989). However, considering the mechanisms involved in heavy-metal tolerance (Verkleij and Schat, 1990), it appears unlikely that such stress tolerance inexorably reduces an ecotype's RGR. Moreover, some cadmium-tolerant ecotypes of *Silene vulgaris* (synonymous for *S. cucubalus*) have very similar RGRs to a sensitive ecotype (Verkleij and Prast, 1989; Verkleij *et al.*, 1990). Possibly, some heavy-metal tolerant ecotypes have evolved in habitats which are not only rich in heavy metals, but nutrient poor as well, so that their inherently low RGR is not causally related to their stress tolerance. Similarly, there is some evidence that salt-tolerant ecotypes of *Beta vulgaris* do not necessarily have a lower RGR than sensitive ones (J. Rozema, personal communication), again suggesting that stress-tolerance and a low RGR are not correlated all that tightly.

F. Conclusions

We conclude that there are trade-offs between investment in structures that lead to a high growth potential and in structures associated with conservation of nutrients and biomass, when accumulation of large amounts of secondary plant compounds are involved. In so far as nutrient losses are associated with rapid leaf turnover there may well be consequences of efficient functioning in one environment for the performance in another. A final conclusion awaits further information on the regulation of the nitrogen concentration in leaves of fast- and slow-growing species, both as dependent on the nitrogen supply and on the light climate in the canopy.

It is likely that there are trade-offs between tolerance of adverse conditions, other than nutrient-poor conditions, and growth potential, but there is no convincing evidence that stress-tolerance and a low RGR are invariably causally related. More information on the mechanisms underlying RGR and stress-tolerance is required before further generalizations can be made.

XVI. CONCLUDING REMARKS AND PERSPECTIVES

Generalizing the above leads to suites of traits of a "typical fast-growing" and a "typical slow-growing" plant species (Table 5). Most of these traits refer to slow-growing species from nutrient-poor sites. Species from other adverse habitats may also have a lower RGR (Section II and XVE), but much less comparative data are available. The difference with species from more favourable conditions is probably less pronounced and information on special traits of such slow-growing species is scanty.

A typically fast-growing species occurs in productive habitats and, under optimum growth conditions, invests heavily in leaf area (high SLA, high LAR), possibly as a result of higher gibberellin production. Although their rate of photosynthesis per unit leaf area is not necessarily higher, fast-growing species have higher rates of photosynthesis per unit leaf weight: The higher photosynthesis is partly caused by a higher leaf organic nitrogen concentration, partly by variation in the efficiency with which nitrogen is used for photosynthesis (PNUE). One of the likely causes of the low PNUE of slow-growing species is the investment of nitrogen in compounds and structures associated with the protection of their leaves against both biotic and abiotic adverse conditions, but this is unlikely to offer the full explanation.

Fast-growing species have higher rates of shoot and root respiration, expressed per unit shoot and root weight, respectively. But their root

Table 5 Typical characteristics of fast-growing and slow-growing C_3 species, summarizing information presented in the text. Unless stated otherwise, the differences refer to plants grown at optimum nutrient supply. A ? indicates that there are indications but no hard data available in the literature

Characteristic	Fast-growing species	Slow-growing species
Habitat		
Nutrient supply	high	low
Productivity	high	low
Morphology and allocation		
Leaf area ratio	high	low
Specific leaf area	high	low
Leaf weight ratio	higher	lower
Root weight ratio	lower	higher
Investment of nitrogen in leaves (% of total plant N)	high	low
Physiology		
Photosynthesis/leaf area (when species of similar life form are compared)	equal	equal
Photosynthesis leaf weight	high	low
Shoot respiration/shoot weight	higher	lower
Root respiration/root weight	higher	lower
Photosynthetic nitrogen use efficiency	higher	lower
Respiratory losses (% of total C fixed)	low	high
Exudation rate/root weight	low?	high?
Ion uptake rate	high	low
Gibberellin content	high?	low?
Chemical composition		
Nitrogen concentration	high	low
Concentration of minerals	high	low
Water content	high	low
Carbon concentration	low	high
Concentration of quantitative secondary compounds	low	high
Concentration of qualitative secondary compounds	variable	variable
Plasticity with respect to nutrient supply		
of SLA	equal	equal
of allocation	high	low
of photosynthesis	high	somewhat lower
Other aspects		
Leaf turnover	high	low
Root turnover	?	?

respiration is not as much higher as to be expected from their much higher rate of growth and nutrient uptake. We hypothesize that slow-growing species constitutively have a rather costly uptake system which is more effective under nutrient-poor conditions.

Fast-growing species have a greater ability to adjust their biomass allocation when exposed to nutrient-poor conditions. There is no evidence that the greater plasticity of allocation of biomas and nitrogen of fast-growing species *per se* confers any disadvantage under nutrient-limited conditions. However, the leaf longevity of fast-growing species is also shorter and further reduced under nutrient-poor conditions, leading to relatively large losses of nutrients. This inefficient use of nutrients is considered one of the main reasons for the lack of success of inherently fast-growing species in nutrient-poor environments. Trade-offs between investment in photosynthetic machinery and the degree to which a plant is defended against herbivory, leaf damage due to strong winds, trampling, drought, salt and/or diseases are likely to have occurred, mainly by adaptations which decrease SLA.

A number of topics relating to characteristic differences between fast-growing and slow-growing species still need continued attention. These include the physiological mechanisms determining inherent variation in specific leaf area, biomass allocation, photosynthetic nitrogen use efficiency, root respiration, the importance of gibberellins and the life-span of the different plant organs. There is also a lack of information on the quantitative importance of losses through exudates and volatiles and their association with nutrient acquisition and interactions between a plant and other organisms. A cost-benefit analysis for symbiotic associations, as dependent on environmental conditions, is needed to evaluate their significance in the acquisition of nutrients. Further evidence is also required to substantiate our belief that biomass allocation of fast-growing species is more plastic with respect to factors other than nutrient supply. If proven correct, information on the regulation of such plasticity and on the ecological significance thereof is needed. Finally, the physiological and ecological costs and benefits of the various inherent adaptations to adverse environments warrant further research.

We have attempted to provide a general background on inherent variation in growth rate and to identify major research areas which need further investigation. Such investigations require a combined approach from ecologists, physiologists, biochemists, phytochemists and theoretical biologists. They are bound to yield information which is of great importance for our understanding of the functioning of plants, both in their natural environment and in a crop situation.

ACKNOWLEDGEMENTS

We would like to thank all colleagues who generously allowed us to use some of their unpublished data, and the following colleagues for their constructive criticism on (parts of) earlier drafts of this manuscript: Frank Berendse, Arjen Biere, René Boot, Marion Cambridge, Heinjo During, Eric Garnier, Henk Konings, Dick Pegtel, Thijs Pons, Jacques Roy, Adrie van der Werf, Marinus Werger and Chin Wong. We thank Marion Cambridge for her linguistic advice.

REFERENCES

Adewusi, S.R.A. (1990) Turnover of dhurrin in green sorghum seedlings. *Plant Physiol.* **94**, 1219–1224.

Aerts, R. (1989) Nitrogen use efficiency in relation to nitrogen availability and plant community composition. In: *Causes and Consequences of Variation in Growth Rate and Productivity of Higher Plants* (Ed. by H. Lambers, M.L. Cambridge, H. Konings and T.L. Pons), pp. 285–297. SPB Academic Publishing, The Hague.

Aerts, R. and Berendse, F. (1989) Aboveground nutrient turnover and net primary production of an evergreen and a deciduous species in a heathland ecosystem. *J. Ecol.* **77**, 342–356.

Aerts, R., Boot, R.G.A. and van der Aart, P.J.M. (1991) The relation between above- and belowground biomass allocation patterns and competitive ability. *Oecologia* **87**, 551–559.

Albuzzio, A. and Ferrari, G. (1989) Modulation of the molecular size of humic substances by organic acids of the root exudates. *Plant Soil* **113**, 237–241.

van Andel, J. and Biere, A. (1989) Ecological significance of variability in growth rate and plant productivity. In: *Causes and Consequences of Variation in Growth Rate and Productivity of Higher Plants* (Ed. by H. Lambers, M.L. Cambridge, H. Konings and T.L. Pons), pp. 257–267. SPB Academic Publishing, The Hague.

Apel, K., Bohlmann, H. and Reimann-Philipp, U. (1990) Leaf thionins, a novel class of putative defence factors. *Physiol. Plant.* **80**, 315–321.

Ashenden, T.W., Stewart, W.S. and Williams, W. (1975) Growth responses of sand dune populations of *Dactylis glomerata* L. to different levels of water stress. *J. Ecol.* **63**, 97–107.

Atkin, O.K. and Day, D.A. (1990) A comparison of the respiratory processes and growth rates of selected Australian alpine and related lowland plant species. *Aust. J. Plant Physiol.* **17**, 517–526.

Atwell, B.J., Veerkamp, M.T., Stuiver, C.E.E. and Kuiper, P.J.C. (1980) The uptake of phosphate by *Carex* species from oligotrophic to eutrophic swamp habitats. *Physiol. Plant.* **49**, 487–494.

Baan Hofman, T. and Ennik, G.C. (1980) Investigation into plant characters affecting the competitive ability of perennial ryegrass (*Lolium perenne* L.). *Neth. J. agric. Sci.* **28**, 97–109.

Baan Hofman, T. and Ennik, G.C. (1982) The effect of root mass of perennial ryegrass (*Lolium perenne* L.) on the competitive ability with respect to couchgrass (*Elytrigia repens* (L.) Desv.). *Neth. J. agric. Sci.* **30**, 275–283.

Baas, W.J. (1989) Secondary plant compounds, their ecological significance and consequences for the carbon budget. Introduction of the carbon/nutrient cycle theory. In: *Causes and Consequences of Variation in Growth Rate and Productivity of Higher Plants* (Ed. by H. Lambers, M.L. Cambridge, H. Konings and T.L. Pons), pp. 313–340. SPB Academic Publishing, The Hague.

Ball, M.C. (1988) Salinity tolerance in the mangroves *Aegiceras corniculatum* and *Avicennia marina*. I. Water use in relation to growth, carbon partitioning and salt balance. *Aust. J. Plant Physiol.* **15**, 447–464.

Barrow, N.J. (1977) Phosphorus uptake and utilization by tree seedlings. *Aust. J. Bot.* **25**, 571–584.

Baruch, Z. and Smith, A.P. (1979) Morphological and physiological correlates of niche breadth in two species of *Espeletia* (Compositae) in the Venezuelan Andes. *Oecologia* **38**, 71–82.

Baruch, Z., Ludlow, M.M. and Davis, R. (1985) Photosynthetic responses of native and introduced C_4 grasses from Venezuelan savannas. *Oecologia* **67**, 388–393.

Bate, N.J., Rood, S.R. and Blake, T.J. (1988) Gibberellins and heterosis in poplar. *Can. J. Bot.* **66**, 1148–1152.

Berendse, F. and Elberse, W.T. (1989) Competition and nutrient losses from the plant. In: *Causes and Consequences of Variation in Growth Rate and Productivity of Higher Plants* (Ed. by H. Lambers, M.L. Cambridge, H. Konings and T.L. Pons), pp. 269–284. SPB Academic Publishing, The Hague.

Berendse, F., Oudhof, H. and Bol, J. (1987) A comparative study on nutrient cycling in wet heathland ecosystems. I. Litter production and nutrient losses from the plant. *Oecologia* **74**, 174–184.

Berendse, F., Bobbink, R. and Rouwenhorst, G. (1989) A comparative study on nutrient cycling in wet heathland ecosystems. II. Litter decomposition and nutrient mineralization. *Oecologia* **78**, 338–348.

Bernays, E.A., Cooper Driver, G. and Bilgener, M. (1989) Herbivores and plant tannins. In: *Advances in Ecological Research* (Ed. by M. Begon, A.H. Fitter, E.D. Ford and A. Macfadyen), Vol. 19, pp. 263–302. Academic Press, London.

Björkman, O. (1981) Responses to different quantum flux densities. In: *Encyclopedia of Plant Physiology* (Ed. by O.L. Lange, P.S. Nobel, C.B. Osmond and H. Ziegler), Vol. 12A, pp. 57–107. Springer-Verlag, Berlin.

Black, J.N. (1958) Competition between plants of different initial seed sizes in swards of subterranean clover (*Trifolium subterranean* L.) with particular reference to leaf area and the light microclimate. *Aust. J. agric. Res.* **9**, 299–318.

Blom-Zandstra, M. and Lampe, J.E.M. (1985) The role of nitrate in the osmoregulation of lettuce (*Lactuca saliva* L.) grown at different light intensities. *J. exp. Bot.* **36**, 1043–1052.

Blom-Zandstra, M., Lampe, J.E.M. and Ammerlaan, H.M. (1988) C and N utilization of two lettuce genotypes during growth under non-varying light conditions and after changing the light intensity. *Physiol. Plant.* **74**, 147–153.

Bloom, A. (1985) Wild and cultivated barleys show similar affinities for mineral nitrogen. *Oecologia* **65**, 555–557.

Boot, R.G.A. (1989) The significance of size and morphology of, root systems for nutrient acquisition and competition. In: *Causes and Consequences of Variation in Growth Rate and Productivity of Higher Plants* (Ed. by H. Lambers, M.L. Cambridge, H. Konings and T.L. Pons), pp. 299–311. SPB Academic Publishing, The Hague.

Boot, R.G.A. (1990) *Plant Growth as Affected by Nitrogen Supply*. PhD thesis, University of Utrecht.

Boot, R.G.A. and Mensink, M. (1994) The effect of nitrogen supply on the size and morphology of root systems of perennial grasses from contrasting habitats. *Plant Soil* **129**, 291–299.

Boot, R.G.A. and Mensink, M. (1991) The influence of nitrogen availability on growth parameters of fast- and slow-growing perennial grasses. In: *Plant Root Growth. An Ecological Perspective, Special Publication of the British Ecological Society* (Ed. by D. Atkinson), pp. 161–168. Blackwell Scientific Publications, London.

Boot, R.G.A., Schildwacht, P.M. and Lambers, H. (1992) Partitioning of nitrogen and biomass at a range of N-addition rates and their consequences for growth and gas exchange in two perennial grasses from inland dunes. *Physiol. Plant.* (in press).

Bradshaw, A.D. (1965) Evolutionary significance of phenotypic plasticity in plants. *Adv. Gen.* **13**, 115–155.

Bradshaw, A.D., Chadwick, M.J., Jowett, D. and Snaydon, R.W.D. (1964) Experimental investigations into the mineral nutrition of several grass species. IV. Nitrogen level. *J. Ecol.* **52**, 665–676.

Brouwer, R. (1963) Some aspects of the equilibrium between overground and underground plant parts. *Mededelingen van het Instituut voor Biologisch en Scheikundig Onderzoek van Landbouwgewassen* **213**, 31–39.

Brouwer, R. (1983) Functional equilibrium: sense or nonsense? *Neth. J. agric. Sci.* **31**, 335–348.

Brugnoli, E. and Lauteri, M. (1991) Effects of salinity on stomatal conductance, photosynthetic capacity, and carbon isotope discrimination of salt-tolerant (*Gossypium hirsutum* L.) and salt-sensitive (*Phaseolus vulgaris* L.) C_3 non-halophytes. *Plant Physiol.* **95**, 628–635.

Bryant, J.P., Chapin, F.S. III and Klein, D.R. (1983) Carbon/nutrient balance of boreal plants in relation to vertebrate herbivory. *Oikos* **40**, 357–368.

Bryant, J.P., Clausen, T.P., Reichardt, P.B., McCarthy, M.C. and Werner, R.A. (1987) Effect of nitrogen fertilization upon the secondary chemistry and nutritional value of quaking aspen (*Populus tremuloides* Michx.) leaves for the large aspen tortrix (*Choristoneura conflictana* (Walker)). *Oecologia* **73**, 513–517.

Burbott, A.J. and Loomis, W.D. (1969) Evidence for metabolic turnover of monoterpenes in peppermint. *Plant Physiol.* **44**, 173–179.

Callaghan, T.V. and Lewis, M.C. (1971) The growth of *Phleum alpinum* L. in contrasting habitats at a sub-antarctic station. *New Phytol.* **70**, 1143–1154.

Campbell, B.D. and Grime, J.P. (1989) A comparative study of plant responsiveness to the duration of episodes of mineral nutrient enrichment. *New Phytol.* **112**, 261–267.

Carver, T.L.W., Thomas, B.J., Ingerson-Morris, S.M. and Roderick, H.W. (1991) The role of the abaxial leaf surface waxes of *Lolium* spp in resistance to *Erysiphe graminis*. *Plant Pathol.* **39**, 573–583.

Ceulemans, R. (1989) Genetic variation in functional and structural productivity components in *Populus*. In: *Causes and Consequences of Variation in Growth Rate and Productivity of Higher Plants* (Ed. by H. Lambers, M.L. Cambridge, H. Konings and T.L. Pons), pp. 69–85. SPB Academic Publishing, The Hague.

Chapin, F.S. III (1980) The mineral nutrition of wild plants. *Ann. Rev. Ecol. System.* **11**, 233–260.

Chapin, F.S. III (1983) Adaptation of selected trees and grasses to low availability of phosphorus. *Plant Soil* **72**, 283–287.

Chapin, F.S. III (1988) Ecological aspects of plant nutrition. *Adv. Min. Nutr.* **3**, 161–191.

Chapin, F.S. III (1989) The costs of tundra plant structures: Evaluation of concepts and currencies. *Am. Nat.* **133**, 1–19.

Chapin, F.S. III and Bieleski, R.L. (1982) Mild phosphorus stress in barley and a related low-phosphorus-adapted barleygrass: Phosphorus fractions and phosphate absorption in relation to growth. *Physiol. Plant.* **54**, 309–317.

Chapin, F.S. III, Follet, J.M. and O'Connor, K.F. (1982) Growth, phosphate absorption, and phosphorus chemical fractions in two *Chionochloa* species. *J. Ecol.* **70**, 305–321.

Chapin, F.S. III, Tyron, P.R. and van Cleve, K. (1983) Influence of phosphorus supply on growth and biomass allocation of Alaskan taiga tree seedlings. *Can. J. Forestry Res.* **13**, 1092–1098.

Chapin, F.S. III, van Cleve, K. and Tyron, P.R. (1986) Relationship of ion absorption to growth rate in taiga trees. *Oecologia* **69**, 238–242.

Chapin, F.S. III, Shaver, G.R. and Kedrowski, R.A. (1986b) Environmental controls over carbon, nitrogen and phosphorus fractions in *Eriophorum* in Alaskan tussock tundra. *J. Ecol.* **74**, 167–195.

Cheshire, M.V. and Mundie, C.M. (1990) Organic matter contributed to soil by plant roots during the growth and decomposition of maize. *Plant Soil* **121**, 107–114.

Chou, C.-H and Kuo, Y.-L. (1986) Alleopathic research of subtropical vegetation in Taiwan. Alleopathic exclusion of understorey by *Leucaena leucophylla* (Lam.) de Wit. *J. chem. Ecol.* **12**, 1431–1448.

Christie, E.K. and Moorby, J. (1975) Physiological responses of semi-arid grasses. I. The influence of phosphorus supply on growth and phosphorus absorption. *Aust. J. agric. Res.* **26**, 423–436.

Clarkson, D.T. (1985) Factors affecting mineral nutrient acquisition by plants. *Ann. Rev. Plant Physiol.* **26**, 77–115.

Clarkson, D.T. (1986) Regulation of the absorption and release of nitrate by plant cells: A review of current ideas and methodology. In: *Fundamental, Ecological and Agricultural Aspects of Nitrogen Metabolism in Higher Plants* (Ed. by H. Lambers, J.J. Neeteson and I. Stulen), pp. 3–27. Martinus Nijhof/Dr W. Junk, The Hague.

Clarkson, D.T. and Scattergood, C.B. (1982) Growth and phosphate transport in barley and tomato plants during the development of, and recovery from, phosphate-stress. *J. exp. Bot.* **33**, 865–875.

Coley, P.D. (1983) Herbivory and defensive characteristics of tree species in a lowland tropical forest. *Ecol. Mon.* **53**, 209–233.

Coley, P.D. (1986) Costs and benefits of defense by tannins in a neotropical tree. *Oecologia* **70**, 238–241.

Coley, P.D. (1987) Interspecific variation in plant anti-herbivore properties: The role of habitat quality and rate of disturbance. *New Phytol.* **106** (suppl), 251–263.

Coley, P.D. and Aide, T.M. (1989) Red coloration of tropical young leaves: a possible antifungal defence? *J. trop. Ecol.* **5**, 293–300.

Coley, P.D., Bryant, J.P. and Chapin, F.S. III. (1985) Resource availability and plant antiherbivory defense. *Science* **230**, 895–899.

Collier, D. and Cummins, W.R. (1989) A field study on the respiration rates in the leaves of temperate plants. *Can. J. Bot.* **67**, 3478–3481.

Cook, M.G. and Evans, L.T. (1983) Some physiological aspects of the domestication and improvement of rice (*Oryza* ssp.). *Field Crops Res.* **6**, 219–238.

Corré, W.J. (1983a) Growth and morphogenesis of sun and shade plants. I. The influence of light intensity. *Acta Bot. Neerlandica* **32**, 49–62.

Corré, W.J. (1983b) Growth and morphogenesis of sun and shade plants. II. The influence of light quality. *Acta Bot. Neerlandica* **32**, 185–202.

Corré, W.J. (1983c) Growth and morphogenesis of sun and shade plants. III. The combined effects of light intensity and nutrient supply. *Acta Bot. Neerlandica* **32**, 277–294.

Cowan, I.R (1977) Water use in higher plants. In: *Water, Planets, Plants and People* (Ed. by A.K. McIntyre), pp. 71–107. Australian Academy of Science, Canberra.

Crick, J.C. and Grime, J.P. (1987) Morphological plasticity and mineral nutrients capture in two herbaceous species of contrasting ecology. *New Phytol.* **107**, 403–414.

Day, D.A. and Lambers, H. (1983) The regulation of glycolysis and electron transport in roots. *Physiol. Plant.* **58**, 155–160.

Deane-Drummond, C.E. and Glass, A.D.M. (1983) Short term studies of nitrate uptake into barley plants using ion-specific electrodes and $^{36}ClO_3^-$ I. Control of net uptake by NO_3^- efflux. *Plant Physiol.* **73**, 100–104.

Dell, B. and McComb, A.J. (1978) Plant resins: Their formation, secretion and possible functions. *Adv. bot. Res.* **6**, 277–316.

Denslow, J.S., Vitousek, P.M. and Schultz, J.C. (1987) Bioassays of nutrient limitation in a tropical rain forest soil. *Oecologia* **74**, 370–376.

Dicke, M. and Sabelis, M.W. (1989) Does it pay to advertise for body guards? In: *Causes and Consequences of Variation in Growth Rate and Productivity of Higher Plants* (Ed. by H. Lambers, M.L. Cambridge, H. Konings and T.L. Pons), pp. 341–358. SPB Academic Publishing, The Hague.

van de Dijk, S.J. (1980) Two ecologically distinct subspecies of *Hypochaeris radicata* L. II. Growth response to nitrate and ammonium, growth strategy and formative aspects. *Plant Soil* **57**, 111–122.

Dijkstra, P. (1989) Cause and effect of differences in specific leaf area. In: *Causes and Consequences of Variation in Growth Rate and Productivity of Higher Plants* (Ed. by H. Lambers, M.L. Cambridge, H. Konings and T.L. Pons), pp. 125–140. SPB Academic Publishing, The Hague.

Dijkstra, P. and Kuiper, P.J.C. (1989) Effects of exogenously applied growth regulators on shoot growth of inbred lines of *Plantago major* differing in relative growth rate: differential response to gibberellic acid and (2-chloroethyl-) trimethylammonium chloride. *Physiol. Plant.* **77**, 512–518.

Dijkstra, P. and Lambers, H. (1989a) Analysis of specific leaf area and photosynthesis of two inbred lines of *Plantago major* differing in relative growth rate. *New Phytol.* **113**, 283–290.

Dijkstra, P. and Lambers, H. (1989b) A physiological analysis of genetic variation in relative growth rate within *Plantago major*. *Func. Ecol.* **3**, 577–585.

Dijkstra, P., ter Reegen, H. and Kuiper, P.J.C. (1990) Relation between relative growth rate, endogenous gibberellins, and the response to applied gibberellic acid for *Plantago major*. *Physiol. Plant.* **79**, 629–634.

Dinkelaker, B., Römheld, V. and Marschner, H. (1989) Citric acid excretion and precipitation of calcium citrate in the rhizosphere of white lupin. *Plant Cell Environ.* **12**, 285–292.

Dittrich, W. (1931) Zur Physiologic des Nitratumsatzes in höheren Pflanzen (unter besonderer Berücksichtigung der Nitratspeicherung). *Planta* **12**, 69–119.

Doddema, H. and Otten, H. (1979) Uptake of nitrate by mutants of *Arabidopsis thaliana*, disturbed in uptake or reduction of nitrate. III. Regulation. *Physiol. Plant.* **45**, 339–346.

Drew, M.C. and Saker, L.R. (1978) Nutrient supply and the growth of the seminal root system in barley. III. Compensatory increase in growth of lateral roots, and in rates of phosphate uptake, in response to a localized supply of phosphate. *J. exp. Bot.* **29**, 435–451.

Drew, M.C., Saker, L.R. and Ashley, T.W. (1973) Nutrient supply and the growth of the seminal root system in barley. I. The effect of nitrate concentration on the growth of axes and laterals. *J. exp. Bot.* **24**, 1189–1202.

Duarte, P., Oscarson, P., Tillberg, J.-E. and Larsson, C.-M. (1988) Nitrogen and carbon utilization in shoots and roots of nitrogen-limited *Pisum*. *Plant Soil* **111**, 241–244.

Ehleringer, J. (1983) Characterization of a glabrate *Encelia farinosa* mutant: morphology, ecophysiology, and field observations. *Oecologia* **57**, 303–310.

Ehleringer, J.R. and Cook, C.S. (1984) Photosynthesis in *Encelia farinosa* Gray in response to decreasing leaf water potential. *Plant Physiol.* **75**, 688–693.

Ehleringer, J.R. and Cook, C.S. (1990) Characteristics of *Encelia* species differing in leaf reflectance and transpiration rate under common garden conditions. *Oecologia* **82**, 484–489.

Ehleringer, J.R. and Cooper, T.A. (1988) Correlations between carbon isotope ratio and microhabitat in desert plants. *Oecologia* **76**, 562–566.

Eissenstat, D.M. and Caldwell, M.M. (1987) Characterization of successful competitors: an evaluation of potential growth rate in two cold desert tussock grasses. *Oecologia* **71**, 167–173.

Elias, C.O. and Chadwick, M.J. (1979) Growth characteristics of grass and legume cultivars and their potential for land reclamation. *J. appl. Ecol.* **16**, 537–544.

Ellenberg, H.E. (1979) Zeigerwerte der Gefässpflanze Mitteleuropas. *Scripta Geobotanica* **9**.

Enyi, B.A.C. (1962) Comparative growth-rates of upland and swamp rice varieties. *Ann. Bot.* **26**, 467–487.

Esau, K. (1977) *Anatomy of Seed Plants*. John Wiley, New York.

Esquerré-Tugavé, M.T., Lafitte, C., Mazau, D., Toppan, A. and Touzé, A. (1979) Cell surfaces in plant microorganism interactions II. Evidence for the accumulation of hydroxyproline-rich glycoproteins in the cell wall of diseased plants as a defense mechanism. *Plant Physiol.* **64**, 320–326.

Evans, G.C. (1972) *The Quantitative Analysis of Plant Growth*. Blackwell Scientific Publications, Oxford.

Evans, J.R. (1983) Nitrogen and photosynthesis in the flag leaf of wheat (*Triticum aestivum* L.). *Plant Physiol.* **72**, 297–302.

Evans, J.R. (1989a) Photosynthesis and nitrogen relationships in leaves of C_3 plants. *Oecologia* **78**, 9–19.

Evans, J.R. (1989b) Photosynthesis—the dependence on nitrogen partitioning. In: *Causes and Consequences of Variation in Growth Rate and Productivity of Higher Plants* (Ed. by H. Lambers, M.L. Cambridge, H. Konings and T.L. Pons), pp. 159–174. SPB Academic Publishing, The Hague.

Evans, J.R. and Terashima, I. (1988) Photosynthetic characteristics of spinach leaves grown with different nitrogen treatments. *Plant Cell Physiol.* **29**, 157–165.

Eze, J.M.O. (1973) The vegetative growth of *Helianthus annuus* and *Phaseolus vulgaris* as affected by seasonal factors in Freetown, Sierra Leone. *Ann. Bot.* **37**, 315–329.

Farquhar, G.D. and Sharkey, T.D. (1982) Stomata and photosynthesis. *Ann. Rev. Plant Physiol.* **33**, 317–345.

Farquhar, G.D., O'Leary, M.H. and Berry, J.A. (1982) On the relationship between carbon isotope discrimination and the intercellular carbon dioxide concentration in leaves. *Aust. J. Plant Physiol.* **11**, 539–552.

Farrar, J.F. (1989) The carbon balance of fast-growing and slow-growing species. In: *Causes and Consequences of Variation in Growth Rate and Productivity of Higher Plants* (Ed. by H. Lambers, M.L. Cambridge, H. Konings and T.L. Pons), pp. 241–256. SPB Academic Publishing, The Hague.

Fenner, M. (1978) A comparison of the abilities of colonizers and closed-turf species to establish from seed in artificial swards. *J. Ecol.* **66**, 953–963.

Fenner, M. (1983) Relationships between seed weight, ash content and seedling growth in twenty-four species of Compositae. *New Phytol.* **95**, 697–706.

Field, C. and Mooney, H.A. (1986) The photosynthesis–nitrogen relationship in wild plants. In: *On the Economy of Plant Form and Function* (Ed. by T.J. Givnish), pp. 25–55. Cambridge University Press, Cambridge.

Field, C., Merino, J. and Mooney, H.A. (1983) Compromises between water-use efficiency and nitrogen-use efficiency in five species of California evergreens. *Oecologia* **60**, 384–389.

Fitter, A.H. (1985) Functional significance of root morphology and root system architecture. In: *Ecological Interactions in Soil, British Ecological Society Special Publication* (Ed. by A.H. Fitter, D. Atkinson, D.J. Read and M.B. Usher), pp. 87–106. Blackwell Scientific Publications, Oxford.

Fitter, A.H. (1987) An architectural approach to the comparative ecology of plant root systems. *New Phytol.* **106**, 61–77.

Fitter, A.H. and Ashmore, C.J. (1974) Response of two *Veronica* species to a simulated woodland light climate. *New Phytol.* **73**, 997–1001.

Fitter, A.H., Nichols, R. and Harvey, M.L. (1988) Root system architecture in relation to life history and nutrient supply. *Funct. Ecol.* **2**, 345–351.

Föhse, D. and Jungk, A. (1983) Influence of phosphate and nitrate supply on root hair formation of rape, spinach and tomato plants. *Plant Soil* **74**, 359–368.

Franceschi, V.R., Wittenbach, V.A. and Giaquinta, R.T. (1983) Paraveinal mesophyll of soybean leaves in relation to assimilate transfer and compartmentation. III. Immunohistochemical localization of specific glycopeptides in the vacuole after depodding. *Plant Physiol.* **72**, 586–589.

Frehner, M. and Conn, E.E. (1987) The linamarin β-glucosidase in Costa Rican wild lima beans (*Phaseolus lunatus* L.) is apoplastic. *Plant Physiol.* **84**, 1296–1300.

Freijsen, A.H.J. and Otten, H. (1984) The effect of nitrate concentration in a flowing solution system on growth and nitrate uptake of two *Plantago* species. *Plant Soil* **77**, 159–169.

Galston, A.W. and Sawhney, R.K. (1990) Polyamines in plant physiology. *Plant Physiol.* **94**, 406–410.

Gardner, W.K., Parbery, D.G. and Barber, D.A. (1982a) The acquisition of phosphorus by *Lupinus albus* L. I. Some characteristics of the soil/root interface. *Plant Soil* **68**, 19–32.

Gardner, W.K., Parbery, D.G. and Barber, D.A. (1982b) The acquisition of phosphorus by *Lupinus albus* L. II. The effects of varying phosphors supply and soil type on some characteristics of the soil/root interface. *Plant Soil* **68**, 33–41.

Gardner, W.K., Parbery, D.G. and Barber, D.A. (1983) The acquisition of phosphorus by *Lupinus albus* L. III. The probable mechanism by which phosphorus movement in the soil/root interface is enhanced. *Plant Soil* **70**, 107–124.

Garnier, E. (1991) Above and below-ground resource capture in herbaceous plants: Relationships with growth and biomass allocation. *Trends Ecol. Evol.* **6**, 126–131.

Garnier, E., Koch, G.W., Roy, J. and Mooney, H.A. (1989) Responses of wild plants to nitrate availability. Relationships between growth rate and nitrate uptake parameters, a case study with two *Bromus* species, and a survey. *Oecologia* **79**, 542–550.

Gershenzon, J. (1984) Changes in the levels of plant secondary metabolites under water and nutrient stress. In: *Phytochemical Adaptations to Stress* (Ed. by B.N. Timmermann, C. Steelink and F.A. Loewus). *Recent Advances in Phytochemistry* Vol. 18, pp. 273–320, Plenum, New York.

Gershenzon, J., Maffei, M. and Croteau, R. (1989) Biochemical and histochemical localization of monoterpene biosynthesis in the glandular trichomes of spearmint (*Mentha spicata*). *Plant Physiol.* **89**, 1351–1357.

Ghashghaie, J. and Saugier, B. (1989) Effects of nitrogen deficiency on leaf photosynthetic response of tall fescue to water deficit. *Plant Cell Environ.* **12**, 261–271.

Givnish, T.J. (1986) Biomechanical constraints on crown geometry in forest herbs. In: *On the Economy of Plant Form and Function* (Ed. by T.J. Givnish), pp. 525–583. Cambridge University Press, Cambridge.

Glass, A.D.M. (1983) Regulation of ion transport. *Ann. Rev. Plant Physiol.* **34**, 311–326.

Glass, A.D.M., Siddigi, M.Y., Ruth, T.J. and Rufry, T.W. Jr. (1990) Studies of the uptake of nitrate in barley. II. Energetics. *Plant Physiol.* **93**, 1585–1589.

Gottlieb, L.D. (1978) Allocation, growth rates and gas exchange in seedlings of *Stephanomeria exigua* ssp. *coronaria* and its recent derivative *S. malheurensis*. *Am. J. Bot.* **65**, 970–977.

Graham, J.H., Leonard, R.T. and Menge, J.A. (1981) Membrane-mediated decrease in root exudation responsible for phosphorus inhibition of vescicular-arbuscular mycorrhyza formation. *Plant Physiol.* **68**, 548–552.

Granato, T.C., Raper, C.D. Jr. and Wilkerson, G.G. (1989) Respiration rate in maize roots is related to concentration of reduced nitrogen and proliferation of lateral roots. *Physiol. Plant* **76**, 419–424.

Gray, J.T. and Schlesinger, W.H. (1983) Nutrient use by evergreen and deciduous shrubs in Southern California. II. Experimental investigations in the relationship between growth, nitrogen uptake and nitrogen availability. *J. Ecol.* **71**, 43–56.

Gregory, P.J. and Atwell, B.J. (1991) The fate of carbon in pulse-labelled crops of barley and wheat. *Plant and Soil* **126**, 205–213.

Grime, J.P. (1979) *Plant Strategies and Vegetation Processes*. Wiley, Chichester.

Grime, J.P. and Hunt, R. (1975) Relative growth rate: Its range and adaptive significance in a local flora. *J. Ecol.* **63**, 393–422.

Grime, J.P., Crick, J.C. and Rincon, J.E. (1986) The ecological significance of plasticity. In: *Plasticity in Plants* (Ed. by D.H. Jennings and A.J. Trewawas), pp. 5–30. The Company of Biologists, Cambridge.

Grime, J.P., Hodgson, J.G. and Hunt, R. (1988) *Comparative Plant Ecology*. Unwin Hyman, London.

Grime, J.P., Hall, W., Hunt, R., Neal, A.M., Ross-Frases, W. and Sutton, F. (1989) A new development of the temperature-gradient tunnel. *Ana. Bot.* **64**, 279–287.

Grime, J.P., Campbell, B.D., Mackey, J.M.L. and Crick, J.C. (1991) Root plasticity, nitrogen capture and competitive ability. In: *Plant Root Growth. An Ecological Perspective. Special Publication of the British Ecological Society* (Ed. by D. Atkinson).

Gross, K.L. (1984) Effects of seed size and growth form on seedling establishment of six monocarpic perennial plants. *J. Ecol.* **72**, 369–387.

Grubb, P.J. (1986) Sclerophylls, pachyphylls apd pycnophylls: The nature and significance of hard leaf surfaces. In: *Insects and Plant Surface* (Ed. by B. Juniper and R. Southwood), pp. 137–150. Edward Arnold, London.

Haider, K., Mosier, A. and Heinemeyer, O. (1987) The effect of growing plants on denitrification at high soil nitrate concentrations. *Soil Sci. Soc. Am. J.* **51**, 97–102.

Harborne, J.B. (1982) *Introduction to Ecological Biochemistry*. Academic Press, London.

Hardwick, R.C. (1984) Some recent developments in growth analysis—A review. *Ann. Bot.* **54**, 807–812.

Harrison, A.F. and Helliwell, D.R (1979) A bioassay for comparing phosphorus availability in soils. *J. appl. Ecol.* **16**, 497–505.

Hartmann, T., Ehmke, A., Eilert, U., von Bortsel, K. and Theurig, C. (1989) Sites of synthesis, translocation and accumulation of pyrrolizidine alkaloid N-oxides in *Senecio vulgaris* L. *Planta* **177**, 98–107.

Hartwig, U.A., Maxwell, C.A., Joseph, C.M. and Phillips, D.A. (1990) Chrysoeriol and luteolin released from alfafa seeds induce nod genes in *Rhizobium meliloti*. *Plant Physiol.* **92**, 116–122.

Helal, H.M. and Sauerbeck, D. (1986) Effect of plant roots on carbon metabolism of soil microbial biomass. *Z. Planzenernähr. Bodenk* **149**, 181–188.

Hirose, T. and Werger, M.J.A. (1987) Nitrogen use efficiency in instantaneous and daily photosynthesis of leaves in the canopy of a *Solidago altissima* stand. *Physiol. Plant.* **70**, 215–222.

Hirose, T., Werger, M.J.A., Pons, T.L. and van Rhenen, J.W.A. (1988) Canopy structure and leaf nitrogen distribution in a stand of *Lysimachia vulgaris* L. as influenced by stand density. *Oecologia* **77**, 145–150.

Hirose, T., Werger, M.J.A. and van Rhenen, J.W.A. (1989) Canopy development and leaf nitrogen distribution in a stand of *Carex acutiformis*. *Ecology* **70**, 1610–1618.

Hoffland, E., Findenegg, G.R. and Nelemans, J.A. (1989a) Solubilization of rock phosphate by rape. I. Evaluation of the role of the nutrient uptake pattern. *Plant and Soil* **113**, 155–160.

Hoffland, E., Findenegg, G.R. and Nelemans, J.A. (1989b) Solubilization of rock phosphate by rape. II. Local root exudation of organic acids as a response to P-starvation. *Plant and Soil* **113**, 161–165.

Hoffland, E., Findenegg, G.R., Leffelaar, P.A. and Nelemans, J.A. (1990) Use of a simulation model to quantify the amount of phosphate released from rock phosphate by rape. *Transactions of the 14th International Congress of Soil Science (Kyoto)* **II**, 170–175.

Hommels, C.H., Kuiper, P.J.C. and Telkamp, G.P. (1990) Study on nutrient stress tolerance of nine *Taraxacum* microspecies with a contrasting mineral ecology by cultivation in a low nutrient flowing solution. *Physiol. Plant.* **79**, 389–399.

Horgan, J.M. and Wareing, P.F. (1980) Cytokinins and the growth responses of seedlings of *Betula pendula* Roth and *Acer pseudoplatanus* L. to nitrogen and phosphorus deficiency. *J. exp. Bot.* **31**, 525–532.

Horsman, D.C., Nicholls, A.O. and Calder, D.M. (1980) Growth response of *Dactylis glomerata*, *Lolium perenne* and *Phalaris aquatica* to chronic ozone exposure. *Aust. J. Plant Physiol.* **7**, 511–517.

Hubick, K. and Farquhar, G.D. (1989) Carbon isotope discrimination and the ratio of carbon gained to water lost in barley cultivars. *Plant, Cell Environ.* **12**, 795–804.

Hughes, A.P. and Cockshull, K.E. (1971) The variation in response to light intensity and carbon dioxide concentration shown by two cultivars of *Chrysanthemum morifolium* grown in controlled environments at two times of year. *Ann. Bot.* **35**, 933–945.

Hunt, R., Nicholls, A.O. and Fathy, S.A. (1987) Growth and root–shoot partitioning in eighteen British grasses. *Oikos* **50**, 53–59.

Ingestad, T. (1981) Nutrition and growth of birch and grey alder seedlings in low conductivity solutions and at varied relative rates of nutrient addition. *Physiol. Plant.* **52**, 454–466.

Jackson, R.B. and Caldwell, M.M. (1989) The timing and degree of root proliferation in fertile-soil microsites for three cold-desert perennials. *Oecologia* **81**, 149–153.

Jackson, R.B., Manwaring, J.H. and Caldwell, M.M. (1990) Rapid physiological adjustment of roots to localized soil enrichment. *Nature* **344**, 58–60.

Jackson, W.A., Kwik, K.D. and Volk, R.J. (1976) Nitrate uptake during recovery from nitrogen deficiency. *Plant Physiol.* **36**, 174–181.

de Jager, A. (1982) Effects of a localized supply of H_2PO_4, NO_3, SO_4, Ca and K on the production and distribution of dry matter in young maize plants. *Neth. J. agric. Sci.* **30**, 193–203.

de Jager, A. and Posno, M. (1979) A comparison of the reaction to a localized supply of phosphate in *Plantago major*, *Plantago lanceolata* and *Plantago media*. *Acta Bot. Neerlandica* **28**, 479–489.

Janzen, H.H. and Radder, G.D. (1989) Nitrogen mineralization in a green manure amended soil as influenced by cropping history and subsequent crop. *Plant Soil* **120**, 125–131.

Johnson, N.D., Liu, B. and Bentley, B.L. (1987) The effects of nitrogen fixation, soil nitrate, and defoliation on the growth, alkaloids, and nitrogen levels of *Lupinus succulentus* (Fabaceae). *Oecologia* **74**, 425–431.

Jones, R.L. and Robinson, D.G. (1989) Protein secretion in plants. *New Phytol.* **111**, 567–597.

Jupe, S.C., Causton, D.R. and Scott, I.M. (1988) Cellular basis of the effects of gibberellin and the pro gene on stem growth in tomato. *Planta* **174**, 106–111.

Kakes, P. (1987) On the polymorphism for cyanogenesis in natural populations of *Trifolium repens* L. in The Netherlands. I. Distribution of the genes *Ac* and *Li*. *Acta Bot. Neerlandica* **36**, 59–69.

Karlsson, P.S. (1991) Intraspecific variation in photosynthetic nitrogen utilization in the mountain birch, *Betula pubescens* ssp. *tortuosa*. *Oikos* **60**, 49–54.

Karlsson, P.S. and Nordell, K.O. (1987) Growth of *Betula pubescens* and *Pinus sylvestris* seedlings in a subarctic environment. *Funct. Ecol.* **1**, 37–44.

Khan, M.A. and Tsunoda, S. (1970a) Growth analysis of cultivated wheat species and their wild relatives with special reference to dry matter distribution among different plant organs and to leaf area expansion. *Tohoku J. Agric. Res.* **21**, 47–59.

Khan, M.A. and Tsunoda, S. (1970b) Growth analysis of six commercially cultivated wheats of West Pakistan with special reference to a semi-dwarf modern wheat variety, mexi pak. *Tohoku J. Agric. Res.* **21**, 60–72.

Kimmerer, T.W. and Potter, D.A. (1987) Nutritional quality of specific leaf tissues and selective feeding by a specialist leafminer. *Oecologia* **71**, 548–551.

Kleis-San Francisco, S.M. and Tierney, M.L. (1990) Isolation and characterization of a proline-rich cell wall protein from soybean seedlings. *Plant Physiol.* **94**, 1897–1902.

Konings, H. (1989) Physiological and morphological differences between plants with a high NAR or a high LAR as related to environmental conditions. In: *Causes and Consequences of Variation in Growth Rate and Productivity of Higher Plants* (Ed. by H. Lambers, M.L. Cambridge, H. Konings and T.L. Pons), pp. 101–123. SPB Academic Publishing, The Hague.

Konings, H., Koot, E. and Tijman-de Wolf, A. (1989) Growth characteristics, nutrient allocation and photosynthesis of *Carex* species from floating fens. *Oecologia* **80**, 111–121.

Koornneef, M., Bosma, T.D.G., Hanhart, C.J., van der Veen, J.H. and Zeevaart, J.A.D. (1990) Isolation and characterization of gibberellin-deficient mutants in tomato. *Theoret. App. Genetics* **80**, 852–857.

Körner, C. and Diemer, M. (1987) *In situ* photosynthetic responses to light, temperature and carbon dioxide in herbaceous plants from low and high altitude. *Funct. Ecol.* **1**, 179–194.

Körner, C. and Pelaez Menendez-Riedl, S. (1989) The significance of developmental aspects in plant growth analysis. In: *Causes and Consequences of Variation in Growth Rate and Productivity of Higher Plants* (Ed. by H. Lambers, M.L. Cambridge, H. Konings and T.L. Pons), pp. 141–157. SPB Academic Publishing, The Hague.

Körner, C., Farquhar, G.D. and Roksandic, Z. (1988) A global survey of carbon isotope discrimination in plants from high altitude. *Oecologia* **74**, 623–632.

Kuehn, G.D., Rodriguez-Garay, B., Bagga, S. and Phillips, G.C. (1990) Novel occurrence of uncommon polyamines in higher plants. *Plant Physiol.* **94**, 855–857.

Kuiper, D. and Smid, A. (1985) Genetic differentiation and phenotypic plasticity in *Plantago major* ssp. *major*: 1. The effect of differences in level of irradiance on growth, photosynthesis, respiration and chlorophyll content. *Physiol. Plant.* **65**, 520–528.

Kuiper, D., Kuiper, P.J.C., Lambers, H., Schuit, J.T. and Staal, M. (1988) Cytokinin contents in relation to mineral nutrition and benzyladenine addition in *Plantago major* ssp. *pleiosperma*. *Physiol. Plant* **75**, 511–517.

Kuiters, A.T. (1990) Role of phenolic substances from decomposing forest litter in plant–soil interactions. *Acta Bot. Neerlandica* **39**, 329–348.

Lambers, H. (1983) The "functional equilibrium": Nibbling on the edges of a paradigm. *Neth. J. agric. Sci.* **31**, 305–311.

Lambers, H. (1985) Respiration in intact plants and tissues: Its regulation and dependence on environmental factors, metabolism and invaded organisms.

In: *Encyclopedia of Plant Physiology* (Ed. by R. Douce and D.A. Day), Vol. 18, pp. 418–473. Springer-Verlag, Berlin.

Lambers, H. (1987) Growth, respiration, exudation and symbiotic associations: The fate of carbon translocated to the roots. In: *Root Development and Function–Effects of the Physical Environment* (Ed. by P.J. Gregory, J.V. Lake and D.A. Rose), pp. 125–145. Cambridge University Press, Cambridge.

Lambers, H. and Dijkstra, P. (1987) A physiological analysis of genetic variation in relative growth rate: Can growth rate confer ecological advantage? In: *Disturbance in Grasslands. Causes, Effects, and Processes* (Ed. by J. van Andel, J.P. Bakker and R.W. Snaydon), pp. 237–252. Dr W. Junk Publishers, Dordrecht.

Lambers, H. and Posthumus, F. (1980) The effects of light intensity and relative humidity on growth rate and root respiration of *Plantago lanceolata* and *Zea mays*. *J. exp. Bot.* **31**, 1621–1630.

Lambers, H. and Rychter, A. (1989) The biochemical background of variation in respiration rate: respiratory pathways and chemical composition. In: *Causes and Consequences of Variation in Growth Rate and Productivity of Higher Plants* (Ed. by H. Lambers, M.L. Cambridge, H. Konings and T.L. Pons), pp. 199–225. SPB Academic Publishing, The Hague.

Lambers, H., Posthumus, F., Stulen, I., Lanting, L., van de Dijk, S.J. and Hofstra, R. (1981a) Energy metabolism of *Plantago major major* as dependent on the supply of nutrients. *Physiol. Plant.* **51**, 245–252.

Lambers, H., Blacquière, T. and Stuiver, C.E.E. (1981b) Interactions between osmoregulation and the alternative respiratory pathway in *Plantago coronopus* as affected by salinity. *Physiol. Plant.* **51**, 63–68.

Lambers, H., Simpson, R.J., Beilharz, V.C. and Dalling, M.J. (1982) Translocation and utilization of carbon in wheat (*Triticum aestivum*). *Physiol. Plant.* **56**, 18–22.

Lambers, H., Freijsen, N., Poorter, H., Hirose, T. and van der Werf, A. (1989) Analyses of growth based on net assimilation rate and nitrogen productivity. In: *Causes and Consequences of Variation in Growth Rate and Productivity of Higher Plants* (Ed. by H. Lambers, M.L. Cambridge, H. Konings and T.L. Pons), pp. 1–17. SPB Academic Publishing, The Hague.

Lamont, B. (1982) Mechanisms for enhancing nutrient uptake in plants, with particular reference to mediterranean South Africa and Western Australia. *Bot. Rev.* **48**, 597–689.

Lamport, D.T. (1980) Structure and function of plant glycoproteins. In: *The Biochemistry of Plants, Vol. 3. Carbohydrates* (Ed. by J. Preis), pp. 501–541. Academic Press, New York.

Lamport, D.T. and Catt, J.W. (1981) Glycoproteins and enzymes of the cell wall. In: *Encyclopedia of Plant Physiology* (Ed. by W. Tanner and F.A. Loewus), Vol. 13B, pp. 133–165. Springer-Verlag, Berlin.

Langenheim, J.H., Osmond, C.B., Brooks, A. and Ferrar, P.J. (1984) Photosynthetic responses to light in seedlings of selected Amazonian and Australian rainforest tree species. *Oecologia* **63**, 215–224.

Lee, R.B. (1982) Selectivity and kinetics of ion uptake by barley plants following nutrient deficiency. *Ann. Bot.* **50**, 429–449.

Liljeroth, E., Baath, E., Mathiasson, I. and Lundborg, T. (1990) Root exudation and rhizoplane bacterial abundance of barley (*Hordeum vulgare* L.) in relation to nitrogen fertilization and root growth. *Plant Soil* **127**, 81–89.

Lipton, D.S., Blanchar, R.W. and Blevins, D.G. (1987) Citrate, malate and succinate concentrations in exudates from P-sufficient and P-stressed *Medicago sativa* L. seedlings. *Plant Physiol.* **85**, 315–317.

Lolkema, P.C., Donker, M.H., Schouten, A.J. and Ernst, W.H.O. (1984) The possible role of metallothioneins in copper tolerance of *Silene cucubales*. *Planta* **162**, 174–179.

Lynch, J.M. and Whipps, J.M. (1990) Substrate flow in the rhizosphere. *Plant Soil* **129**, 1–10.

McClure, P.R., Kochian, L.V., Spanswick, R.M. and Shaff, J.E. (1990) Evidence for cotransport of nitrate and protons in maize roots. II. Measurements of NO_3^- and H^+ fluxes with ion-selective microelectrodes. *Plant Physiol.* **93**, 290–294.

McCully, M.E. and Canny, M.J. (1985) Localisation of translocated ^{14}C roots of field-grown maize. *Physiol. Plant.* **65**, 380–392.

McDonald, A.J.S., Lohammer, T. and Ericsson, A. (1986) Growth response to step decrease in nutrient availability in small birch (*Betula pendula* Roth). *Plant, Cell Environ.* **9**, 427–432.

McDougall, B.M. (1970) Movement of ^{14}C-photoassimilate into the roots of wheat seedlings and exudation from the roots. *New Phytol.* **69**, 37–46.

McNeilly, T., Ashraf, M. and Veltkamp, C. (1987) Leaf micromorphology of sea cliff and inland plants of *Agrostis stolonifera* L., *Dactylis glomerata* L. and *Holcus lanatus* L. *New Phytol.* **106**, 261–269.

Mahmoud, A. and Grime, J.P. (1976) An analysis of competitive ability in three perennial grasses. *New Phytol.* **77**, 431–435.

Makino, A., Mae, T. and Ohira, K. (1988) Differences between wheat and rice in the enzymic properties of ribulose-1, 5-bisphosphate carboxylase/oxygenase and the relationship to photosynthetic gas exchange. *Planta* **174**, 30–38.

Margna, U., Margna, E. and Vainjärv, T. (1989) Influence of nitrogen nutrition on the utilization of L-phenylalanine for building flavonoids in buckwheat seedlings. *J. Plant Physiol.* **134**, 697–702.

Marschner, H., Treeby, M. and Römheld, V. (1989) Role of root-induced changes in the rhizosphere for iron acquisition in higher plants. *Z. Pflanzenernähr. Bodenk.* **152**, 197–204.

Merino, J., Field, C. and Mooney, H.A. (1984) Construction and maintenance costs of Mediterranean-climate evergreen and deciduous leaves. *Acta Oecol./Oecol. Plant.* **5**, 211–229.

Métraux, J.-P. (1987) Gibberellins and plant cell elongation. In: *Plant Hormones and Their Role in Plant Growth and Development* (Ed. by P.J. Davies), pp. 296–317. Martinus Nijhoff, Dordrecht.

Metzger, J.D. and Hassebrock, A.T. (1990) Selection and characterization of a gibberellin-deficient mutant of *Thlaspi arvense* L. *Plant Physiol.* **94**, 1655–1662.

Mole, S. and Waterman, P.G. (1988) Light-induced variation in phenolic levels in foliage of rain-forest plants. II. Potential significance to herbivores. *J. chem. Ecol.* **14**, 23–34.

Monk, C.D. (1966) An ecological significance of evergreenness. *Ecology* **47**, 504–505.

Monson, R.K. and Fall, R. (1989) Isoprene emission from aspen leaves. Influence of environment and relation to photosynthesis and photorespiration. *Plant Physiol.* **90**, 267–274.

Mooney, H.A. and Chu, C. (1983) Stomatal responses to humidity of coastal and interior populations of a Californian shrub. *Oecologia* **57**, 148–150.

Mooney, H.A., Ferrar, P.J. and Slatyer, R.O. (1978) Photosynthetic capacity and carbon allocation patterns in diverse growth forms of *Eucalyptus*. *Oecologia* **36**, 103–111.

Mooney, H.A., Field, C., Gulmon, S.L., Rundel, P. and Kruger, F.J. (1983) Photosynthetic characteristics of South African sclerophylls. *Oecologia* **58**, 398–401.

Morgan, J.A. (1986) The effects of N-nutrition on the water relations and gas exchange characteristics of wheat (*Triticurn aestivum* L.). *Plant Physiol.* **80**, 52–58.

Morrow, P.A. and Fox, L.R. (1980) Effects of variation in *Eucalyptus* essential oil yield on insect growth and grazing damage. *Oecologia* **45**, 209–219.

Muller, B. and Garnier, E. (1990) Components of relative growth rate and sensitivity to nitrogen availability in annual and perennial species of *Bromus*. *Oecologia* **84**, 513–518.

Mulroy, T.W. (1979) Spectral properties of heavily glaucous and non-glaucous leaves of a succulent rosette plant. *Oecologia* **38**, 349–357.

Nassery, H. (1970) Phosphate absorption by plants from habitats of different phosphate status. *New Phytol.* **69**, 197–203.

Newman, E.I. (1985) The rhizosphere: carbon sources and microbial populations. In: *Ecological Interactions in Soil.* Special Publication No. 4 of the British Ecological Society (Ed. by A.H. Fitter), pp. 107–121. Blackwell Scientific Publications, Oxford.

Nicolai, V. (1988) Phenolic and mineral content of leaves influences decomposition in European forest ecosystems. *Oecologia* **75**, 575–579.

Niemann, G.J., Pureveen, J.B.M., Eijkel, G.B., Poorter, H. and Boon, J.J. (1992) Differences in relative growth rate in 11 grasses correlate with differences in chemical composition as determined by mass spectrometry. *Oecologia* **89**, 567–573.

Oberbauer, S.F., Strain, B.R. and Fetcher, N. (1985) Effect of CO_2-enrichment on seedling physiology and growth of two tropical tree species. *Physiol. Plant.* **65**, 352–356.

Oscarson, P., Ingemarsson, B., af Ugglas, M. and Larsson, C.-M. (1987) Short-term studies of NO_3^- uptake in *Pisum* using $^{13}NO_3^-$. *Planta* **170**, 550–555.

Oscarson, P., Ingemarsson, B. and Larsson, C.-M. (1989) Growth and nitrate uptake properties of plants grown at different relative addition rates of nitrogen supply. II. Activity and affinity of the nitrate uptake system in *Pisum* and *Lemna* in relation to nitrogen availability and nitrogen demand. *Plant, Cell Environ.* **12**, 787–794.

Ovington, J.D. and Olson, J.S. (1970) Biomass and chemical content of El Verde lower montane rain forest plants. In: *A Tropical Rainforest* (Ed. by H.T. Odum and R.F. Pigeon), pp. h53–h77. USEAC, Oak Ridge.

Pammenter, N.W., Drennan, P.M. and Smith, V.R. (1986) Physiological and anatomical aspects of photosynthesis of two *Agrostis* species at a sub-antarctic island. *New Phytol.* **102**, 143–160.

Parkhurst, D.F. and Mott, K.A. (1990) Intercellular diffusion limits to CO_2 uptake in leaves. Studies in air and helox. *Plant Physiol.* **94**, 1024–1032.

Parkhurst, D.F., Wong, S.C., Farquhar, G.D. and Cowan, I.R. (1988) Gradients of intercellular CO_2 levels across the leaf mesophyll. *Plant Physiol.* **86**, 1032–1037.

Pavlik, B.M. (1983) Nutrient and productivity relations of the dune grasses *Ammophila arenaria* and *Elymus mollis*. I. Blade photosynthesis and nitrogen use efficiency in the laboratory and field. *Oecologia* **57**, 227–232.

Pearson, C.J., Volk, R.J. and Jackson, W.A. (1981) Daily changes in nitrate influx, efflux and metabolism in maize and pearl millet. *Planta* **152**, 319–324.

Penning de Vries, F.W.T. (1975) The cost of maintenance processes in plant cells. *Ann. Bot.* **39**, 77–92.

Penning de Vries, F.W.T., Brunsting, A.H.M. and van Laar, H.H. (1974) Products, requirements and efficiency of biosynthetic processes: a quantitative approach. *J. theoret. Biol.* **45**, 339–377.

Penning de Vries, F.W.T., van Laar, H.H. and Chardon, M.C.M. (1983) Bioenergetics of growth of seeds, fruits, and storage organs. In: *Proc. Symp. Potential Productivity of Field Crops under Different Environments*, pp. 37–59. International Rice Research Institute, Manila.

Philipson, J.J. and Coutts, M.P. (1977) The influence of mineral nutrition on the root development of trees. II. The effect of specific nutrient elements on the growth of individual roots of sitka spruce. *J. exp. Bot.* **105**, 864–871.

Plaut, Z., Mayoral, M.L. and Reinhold, L. (1987) Effect of altered sink:source ratio on photosynthetic metabolism of source leaves. *Plant Physiol.* **85**, 786–791.

Pons, T.L. (1977) An ecophysiological study in the field layer of ash coppice. II. Experiments with *Geum urbanum* and *Cirsium palustre* in different light intensities. *Acta Bot. Neerlandica* **26**, 29–42.

Pons, T.L., Schieving, F., Hirose, T. and Werger, M.J.A. (1989) Optimization of leaf nitrogen allocation for canopy photosynthesis in *Lysimachia vulgaris*. In: *Causes and Consequences of Variation in Growth Rate and Productivity of Higher Plants* (Ed. by H. Lambers, M.L. Cambridge, H. Konings and T.L. Pons), pp. 175–186. SPB Academic Publishing, The Hague.

Poorter, H. (1989) Interspecific variation in relative growth rate: On ecological causes and physiological consequences. In: *Causes and Consequences of Variation in Growth Rate and Productivity of Higher Plants* (Ed. by H. Lambers, M.L. Cambridge, H. Konings and T.L. Pons), pp. 45–68. SPB Academic Publishing, The Hague.

Poorter, H. (1991) *Interspecific Variation in the Relative Growth date of Plants: The Underlying Mechanisms*. PhD thesis, University of Utrecht.

Poorter, H. and Bergkotte, M. (1992) Chemical composition of 24 wild species differing in relative growth rate. *Plant. Cell Environ.* **15**, 221–229.

Poorter, H. and Lambers, H. (1986) Growth and competitive. ability of a highly plastic and a marginally plastic genotype of *Plantago major* in a fluctuating environment. *Physiol. Plant.* **67**, 217–222.

Poorter, H. and Lambers, H. (1991) Is Interspecific variation in relative growth rate positively correlated with biomass allocation. *Am. Nat.* **138**, 1264–1268.

Poorter, H. and Remkes, C. (1990) Leaf area ratio and net assimilation rate of 24 wild species differing in relative growth rate. *Oecologia* **83**, 553–559.

Poorter, H., Remkes, C. and Lambers, H. (1990) Carbon and nitrogen economy of 24 wild species differing in relative growth rate. *Plant Physiol.* **94**, 621–627.

Poorter, H., van der Werf, A., Atkin, O.K. and Lambers, H. (1991) Respiratory energy requirements of roots vary with the potential growth rate of a plant species. *Physiol. Plant.* **83**, 469–475.

Popma, J. and Bongers, F. (1988) The effects of canopy gaps on growth and morphology of seedlings of rain forest species. *Oecologia* **75**, 625–632.

Potter, D.A. and Kimmerer, T.W. (1989) Inhibition of herbivory on young holly leaves: evidence for the defensive role of saponins. *Oecologia* **78**, 322–329.

Potter, J.R. and Jones, J.W. (1977) Leaf area partitioning as an important factor in plant growth. *Plant Physiol.* **59**, 10–14.

Poulton, J.E. (1990) Cyanogenesis in plants. *Plant Physiol.* **94**, 401–405.

Powell, C.L. (1974) Effect of P-fertilizer on root morphology and P-uptake of *Carex coriacea*. *Plant Soil* **41**, 661–667.

Prudhomme, T.I. (1983) Carbon allocation to antiherbivore compounds in a deciduous and an evergreen subarctic shrub species. *Oikos* **40**, 344–356.

Putnam, A. and Tang, C.-S. (Eds.) (1986) *The Science of Allelopathy*. John Wiley, New York.

Rawson, H.M., Gardner, P.A. and Long, M.J. (1987) Sources of variation in specific leaf area in wheat grown at high temperature. *Aust. J. Plant Physiol.* **14**, 287–298.

Ratnayaka, M., Leonard, R.T. and Menge, J.A. (1978) Root exudation in relation to supply of phosphorus and its possible relevance to mycorrhizal formation. *New Phytol.* **81**, 543–552.

Reimann-Philipp, U., Schrader, G., Martinoia, E., Barkholt, V. and Apel, K. (1989) Intracellular thionins of barley. A second group of leaf thionins closely related to but distinct from cell wall-bound thionins. *J. biol. Chem.* **264**, 8978–8984.

Reinink, K., Groenwold, R. and Bootsma, A. (1987) Genotypical differences in nitrate content in *Lactuca sativa* L. and related species and correlation with dry matter content. *Euphytica* **36**, 11–18.

Rhoades, D.F. (1985) Offensive–defensive interactions between herbivores and plants: their relevance in herbivore population dynamics and ecological theory. *Amer. Natural.* **125**, 205–238.

Richards, R.A., Rawson, H.M. and Johnson, D.A. (1986) Glaucousness in wheat: Its development and effect on water-use efficiency, gas exchange and photosynthetic tissue temperatures. *Aust. J. Plant Physiol.* **13**, 465–473.

Richardson, A.E., Djordjevic, M.A., Rolfe, B.G. and Simpson, R.J. (1988) Effects of pH, Ca and Al on the exudation from clover seedlings of compounds that induce the expression of nodulation genes in *Rhizobium trifolii*. *Plant Soil* **109**, 37–47.

Robinson, D. and Rorison, I.H. (1985) A quantitative analysis of the relationships between root distribution and nitrogen uptake from soil by two grass species. *J. Soil Sci.* **36**, 71–85.

Robinson, D. and Rorison, I.H. (1987) Root hairs and plant growth at low nitrogen availabilities. *New Phytol.* **107**, 681–693.

Robinson, D. and Rorison, I.H. (1988) Plasticity in grass species in relation to nitrogen supply. *Funct. Ecol.* **2**, 249–257.

Robinson, N.J. and Jackson, P.J. (1986) Metallothinein-like" metal complexes in angiosperms; their structure and function. *Physiol. Plant.* **67**, 499–506.

Rodgers, C.O. and Barneix, A.J. (1988) Cultivar differences in the rate of nitrate uptake by intact wheat plants as related to growth rate. *Physiol. Plant.* **72**, 121–126.

Römheld, V. (1987) Different strategies for iron acquisition in higher plants. *Physiol. Plant.* **70**, 231–234.

Römheld, V. and Marschner, H. (1986) Evidence for a specific uptake system for iron phytosiderophores in roots of grasses. *Plant Physiol.* **80**, 175–180.

Römheld, V. and Marschner, H. (1990) Genotypical differences among gramineceous species in release of phytosidereophores and uptake of iron phytosiderophores. *Plant Soil* **123**, 147–153.

Rood, S.B. and Pharis, R.P. (1987) Hormones and heterosis in plants. In: *Plant Hormones and Their Role in Plant Growth and Development* (Ed. by P.J. Davies), pp. 463–473. Martinus Nijhoff, Dordrecht.

Rood, S.B., Pharis, R.P., Koshioka, M. and Major, D.J. (1983) Gibberellin and heterosis in maize. I. Endogenous gibberellin-like substances. *Plant Physiol.* **71**, 639–644.

Rood, S.B., Buzzell, R.I., Major, D.J. and Pharis, R.P. (1990a) Gibberellins and heterosis in maize: Quantitative relationships. *Crop Sci.* **30**, 281–286.

Rood, S.B., Zanewich, K.P. and Bray, D.F. (1990b) Growth and development of *Brassica* genotypes differing in endogenous gibberellin content. II. Gibberellin content, growth analyses and cell size. *Physiol. Plant.* **79**, 679–685.

Rood, S.B., Pearce, D.P., Williams, P.H. and Pharis, R.P. (1990c) A. Gibberellin-deficient *Brassica* mutant—rosette. *Plant Physiol.* **89**, 482–1187.

Rorison, I.H. (1968) The response of phosphorus of some ecologically distinct plant species. I. Growth rates and phosphorus absorption. *New Phytol.* **67**, 913–923.

Rozijn, N.A.M.G. and van der Werf, D.C. (1986) Effect of drought during different stages in the life cycle on the growth and biomass of two *Aira* species. *J. Ecol.* **74**, 507–523.

Rufty, T.W., Huber, S.C. and Volk, R.J. (1988) Alterations in leaf carbohydrate metabolism in response to nitrogen stress. *Plant Physiol.* **88**, 725–730.

Saab, I.N., Sharp, R.E., Pritchard, J. and Voetberg, G.S. (1990) Increased endogenous abscisic acid maintains primary root growth and inhibits shoot growth of maize seedlings at low water potentials. *Plant Physiol.* **93**, 1329–1336.

Sage, R.F. and Pearcy, R.W. (1987) The nitrogen use efficiency of C_3 and C_4 plants. II. Leaf nitrogen effects on the gas exchange characteristics of *Chenopodium album* and *Amaranthus retroflexus* (L.). *Plant Physiol.* **84**, 959–963.

Sage, R.F. and Sharkey, T.D. (1987) The effect of temperature on the occurrence of O_2 and CO_2 insensitive photosynthesis in field grown plant. *Plant Physiol.* **84**, 658–664.

Salvucci, M.E. (1989) Regulation of Rubisco activity *in vivo*. *Physiol. Plant.* **77**, 164–171.

Sanadze, G.A. (1969) Light-dependent excretion of molecular isoprene. *Progress in Photosynthesis Research* **2**, 701–706.

Schmidt, J.E. and Kaiser, W.M. (1987) Response of the succulent leaves of *Peperomia magnoliaefolia* to dehydration. Water relations and solute movement in chlorenchyma and hydrenchyma. *Plant Physiol.* **83**, 190–194.

Schulze, E.-D. (1982) Plant life forms and their carbon, water and nutrient relations. In: *Encyclopedia of Plant Physiology* (Ed. by O.L. Lange, P.S. Nobel, C.B. Osmond and H. Ziegler), Vol. 12B, pp. 615–676. Springer-Verlag, Berlin.

Schulze, E.-D. and Chapin, F.S. III (1987) Plant specialization to environments of different resource availability. In: *Potentials and Limitations of Ecosystem Analysis* (Ed. by E. D. Schulze and Zwölfer), pp. 120–148. Springer-Verlag, Berlin.

Schulze, E.-D., Turner, N.C., Gollan, T. and Shackel, K.A. (1987) Stomatal responses, to air humidity and to soil drought. In: *Stomatal Function* (Ed. by E. Zeiger, G.D. Farquhar and I.R Cowan), pp. 311–321. Stanford University Press, Stanford.

Seemann, J.R. (1989) Light adaptation/acclimation of photosynthesis and the regulation of ribulose-1,5-bisphosphate carboxylase activity in sun and shade plants. *Plant Physiol.* **91**, 379–386.

Seemann, J.R., Sharkey, T.D., Wang, J.L. and Osmond, C.B. (1987) Environmental effects on photosynthesis, nitrogen-use efficiency, and metabolite pools in leaves of sun and shade plants. *Plant Physiol.* **84**, 796–802.

Servaites, J.C. (1990) Inhibition of ribulose 1,5-bisphosphate carboxylase/oxygenase by 2-carboxyarabinitol-1-phosphate. *Plant Physiol.* **92**, 867–870.

Sharkey, T.D., Stitt, M., Heineke, D., Gerhardt, R., Raschke, K. and Heldt, H.W. (1986) Limitation of photosynthesis by carbon metabolism. II. CO_2-insensitive CO_2 uptake results from limitation of triose phosphate utilization. *Plant Physiol.* **81**, 1123–1129.

Shipley, B. and Peters, R.H. (1990) A test of the Tilman model of plant strategies: Relative growth rate and biomass partitioning. *Am. Nat.* **136**, 139–153.

Sibly, RM. and Grime, J.P. (1986) Strategies of resource capture by plants—Evidence for adversity selection. *J. theoret. Biol.* **118**, 247–250.

Siddiqi, M.Y., Glass, A.D.M., Ruth, T.J. and Fernando, M. (1990) Studies on the regulation of nitrate influx by barley seedlings using $^{13}NO_3^-$. *Plant Physiol.* **56**, 806–813.

Simpson, R.J., Lambers, H. and Dalling, M.J. (1982) Kinetin application to roots and its effects on uptake, translocation and distribution of nitrogen in wheat (*Triticum aestivum*) grown with a split root system. *Physiol. Plant.* **56**, 430–435.

Smedley, M.P., Dawson, T.E., Comstock, J.P., Donovan, L.A., Sherrill, D.E., Cook, C.S. and Ehleringer, J.R. (1991) Seasonal carbon isotope discrimination in a grassland community. *Oecologia* **85**, 314–320.

Smirnoff, N. and Stewart, G.R. (1985) Nitrate assimilation and translocation by higher plants: Comparative physiology and ecological consequences. *Physial. Plant.* **64**, 133–140.

Smith, R.I.L. and Walton, D.W.H. (1975) A growth analysis technique for assessing habitat severity in tundra regions. *Ann. Bot.* **39**, 831–843.

Sparling, G.P., Cheshire, M.V. and Mundie, C.M. (1982) Effect of barley plants on the decomposition of ^{14}C-labelled soil organic matter. *J. Soil Sci.* **33**, 89–100.

Spitters, C.J.T. and Kramer, T. (1986) Differences between spring wheat cuitivars in early growth. *Euphytica* **35**, 273–292.

Stanton, M.L. (1984) Seed variation in wild radish: Effect of seed size on components of seedling and adult fitness. *Ecology* **65**, 1105–1112.

Staswick, P.E. (1988) Soybean vegetative storage protein structure and gene expression. *Plant Physiol.* **87**, 250–254.

Stienstra, A.W. (1986) Does nitrate play a role in osmoregulation? In: *Fundamental, Ecological and Agricultural Aspects of Nitrogen Metabolism in Higher Plants* (Ed. by H. Lambers, J.J. Neeteson and I. Stulen), pp. 481–484. Martinus Nijhof/ Dr W. Junk, The Hague.

Stulen, I., Lanting, L., Lambers, H., Posthumus, F., van de Dijk, S.J. and Hofstra, R. (1981) Nitrogen metabolism of *Plantago major* L. as dependent on the supply of mineral nutrients. *Physiol. Plant.* **52**, 108–114.

Taper, M.L. and Case, T.J. (1987) Interactions between oak tannins and parasite community structure: Unexpected benefits of tannins to cynipid gall-wasps. *Oecologia* **71**, 254–261.

Tester, M., Smith, S.E. and Smith, F.A. (1987) The phenomenon of "nonmycorrhizal" plants. *Can. J. Bot.* **65**, 419–431.

Thompson, K. (1987) Seeds and seed banks. *New Phytol.* **106** (suppl), 23–34.

Tilman, D. and Cowan, M.L. (1989) Growth of old field herbs on a nitrogen gradient. *Funct. Ecol.* **3**, 425–438.

Tingey, D.T., Evans, R.C. and Gumpertz, M.L. (1981) Effects of environmental conditions on isoprene emissions from live oak. *Planta* **152**, 565–570.

Tingey, D.T., Evans, R.C., Bates, E.H. and Gumpertz, M.L. (1987) Isoprene emissions and photosynthesis in three ferns—The influence of light and temperature. *Physiol. Plant.* **69**, 609–616.

Tognoni, F., Halevy, A.H. and Wittwer, S.H. (1967) Growth of bean and tomato plants as affected by root absorbed growth substances and atmospheric carbon dioxide. *Planta* **72**, 43–52.

Tsunoda, S. (1959) A developmental analysis of yielding ability in varieties of field crops. I. Leaf area per plant and leaf area ratio. *Jap. J. Breeding* **9**, 161–168.

Tukey, H.B. (1970) The leaching of substances from plants. *Ann. Rev. Plant Physiol.* **21**, 305–324.

Veen, B.W. and Kleinendorst, A. (1986) The role of nitrate in osmoregulation of Italian ryegrass. In: *Fundamental, Ecological and Agricultural Aspects of Nitrogen*

Metabolism in Higher Plants (Ed. by H. Lambers, J.J. Neeteson and I. Stulen), pp. 477–480. Martinus Nijhof/Dr W. Junk, The Hague.

van Veen, J.A., Merckx, R. and van de Geijn, S.C. (1989) Plant- and soil related controls of the flow of carbon from roots through the soil microbial biomass. *Plant Soil* **115**, 179–188.

Verkleij, J.A.C. and Prast, J.E. (1989) Cadmium tolerance and co-tolerance in *Silene vulgaris* (Moench.) Garcke (= *S. cucubalus* (L.) Wib.). *New Phytol.* **111**, 637–645.

Verkleij, J.A.C. and Schat, H. (1990) Mechanisms of metal tolerance in higher plants. In: *Heavy Metal Tolerance in Plants* (Ed. by A.J. Shaw), pp. 179–193. CRC Press Inc, Boca Raton.

Verkleij, J.A.C., Koevoets, P., van't Riet, J., Bank, R., Nijdam, Y. and Ernst, W.H.O. (1990) Poly(γ-glutamylcysteinyl)glycines or phytochelatines and their role in cadmium tolerance of *Silene vulgaris*. *Plant, Cell Environ.* **13**, 913–921.

Vertregt, N. and Penning de Vries, F.W.T. (1987) A rapid method for determining the efficiency of biosynthesis of plant biomass. *J. theoret. Biol.* **128**, 109–119.

Vitousek, P.M., Matson, P.A. and van Cleve, K. (1989) itrogen availability and nitrification during succession: Primary, secondary, and old field seres. *Plant Soil.* **115**, 229–239.

Waller, G.R. and Nowacki, E.K. (1978) *Alkaloids. Biology and Metabolism.* Plenum Press, New York.

Waring, R.H., McDonald, A.J.S., Larsson, S., Ericsson, T., Wiren, A., Arwidsson, E. *et al.* (1985) Differences in chemical composition of plants grown at constant relative growth rates with stable mineral nutrition. *Oecologia* **66**, 157–160.

Warren Wilson, J. (1966) An analysis of plant growth and its control in arctic environments. *Ann. Bot.* **30**, 384–402.

Waterman, P.G. and McKey, D. (1989) Herbivory and secondary compounds in rain-forest plants. In: *Tropical Rain Forest Ecosystems* (Ed. by H. Lieth and M.J.A. Werner), pp. 513–536. Elsevier, Amsterdam.

Welch, M.B. (1920) *Eucalyptus* oil glands. *J. Proc. R. Soc. N.S.W.* **54**, 208–217.

van der Werf, A., Kooijman, A., Welschen, R. and Lambers, H. (1988) Respiratory costs for the maintenance of biomass, for growth and for ion uptake in roots of *Carex diandra* and *Carex acutiformis*. *Physiol. Plant.* **72**, 483–491.

van der Werf, A., Hirose, T. and Lambers, H. (1989) Variation in root respiration; Causes and consequences for growth. In: *Causes and Consequences of Variation in Growth Rate and Productivity of Higher Plants* (Ed. by H. Lambers, M.L. Cambridge, H. Konings and T.L. Pons), pp. 227–240. SPB Academic Publishing, The Hague.

van der Werf, A., Visser, A.J., Schieving, F. and Lambers, H. (1992a) Evidence for optimal partitioning of biomass and nitrogen at a range of nitrogen availabilities for a fast- and slow-growing species. *Funct. Ecol.* (in press).

van der Werf, A., Welschen, R. and Lambers, H. (1992b) Respiratory losses increase with decreasing inherent growth rate of a species and with decreasing nitrate supply: A search for explanations for these observations. In: *Molecular, Biochemical and Physiological Aspects of Plant Respiration* (Ed. by H. Lambers, H. and L.H.W. Van der Plas). SPB Academic Publishing, The Hague (in press).

van der Werf, A., van den Berg, G., Ravenstein, H.J.L., Lambers, H. and Eising, R. (1992c) Protein turnover: A significant component of maintenance respiration in roots? In: *Molecular, Biochemical and Physiological Aspects of Plant Respiration* (Ed. by H. Lambers, H. and L.H.W. Van der Plas). SPB Academic Publishing, The Hague (in press).

Werner, P., Rebele, F. and Bornkamm, R. (1982) Effects of light intensity and light quality on the growth of the shadow plant *Lamium galeobdolon* (L.) Crantz and the half-shadow plant *Stellaria holostea* L. *Flora* **172**, 235–249.

Whitehouse, D.G., Fricaud, A.-C. and Moore, A.L. (1989) Role of nonohmicity in the regulation of electron transport in plant mitochondria. *Plant Physiol.* **91**, 487–492.

von Willert, D.J., Eller, B.M., Werger, M.J.A., Brinckmann, E. and Ihlenfeldt, H.-D. (1992) *Life Strategies of Succulents in Deserts with Special Reference to the Namib Desert.* Cambridge University Press, Cambridge.

Williamson, P. (1976) Above-ground primary production of chalk grassland allowing for leaf death. *J. Ecol.* **64**, 1059–1075.

Wilson, J.B. (1988) The cost of heavy-metal tolerance: An example. *Evolution* **42**, 408–413.

Woldendorp, J.W. (1963) The influence of living plants on denitrification. *Mededelingen Landbouwhogeschool. Wageningen* **63**, 1–100.

Woodman, R.L. and Fernandes, G.W. (1991) Differential mechanical defense: herbivory, evapotranspiration, and leaf-hairs. *Oikos* **60**, 11–19.

Woodward, F.I. (1979) The differential temperature responses of the growth of certain plant species from different altitudes. I. Growth analysis of *Phleum alpinum* L., *P. bertolonii* D.C., *Sesleria albicans* Kit. and *Dactylis glomerata* L. *New Phytol.* **82**, 385–395.

Woodward, F.I. (1983) The significance of interspecific differences in specific leaf area to the growth of selected herbaceous species from different altitudes. *New Phytol.* **95**, 313–323.

Wyn Jones, R.G. and Gorham, J. (1983) Osmoregulation. In: *Encyclopedia of Plant Physiology* (Ed. by O.L. Lange, P.S. Nobel, C.B. Osmond and H. Ziegler), Vol. 12C, pp. 35–58. Springer-Verlag, Berlin.

Young, D.R. and Smith, W.K. (1980) Influence of sunlight on photosynthesis, water relations and leaf structure in the understory species *Arnica cordifolia*. *Ecology* **61**, 1380–1390.

Zanewich, K.P., Rood, S.B. and Williams, P.H. (1990) Growth and development of *Brassica* genotypes differing in endogenous gibberellin content. II. Leaf and reproductive development. *Physiol. Plant.* **79**, 673–678.

Zhang, F., Römheld, V. and Marschner, H. (1989) Effects of zinc deficiency in wheat on the release of zinc and iron mobilizing root exudates. *Z. Pflanzener. Bodenk.* **152**, 205–210.

Advances in Ecological Research Volume 1–34

Cumulative List of Titles